Frontiers in Deep-Sea Equipment and Technology II

Frontiers in Deep-Sea Equipment and Technology II

Editors

Weicheng Cui
Lian Lian
Dahai Zhang

MDPI

Basel • Beijing • Wuhan • Barcelona • Belgrade • Novi Sad • Cluj • Manchester

Editors

Weicheng Cui
School of Engineering
Westlake University
Hangzhou
China

Lian Lian
School of Oceanography
Shanghai Jiao Tong
University
Shanghai
China

Dahai Zhang
Ocean College
Zhejiang University
Hangzhou
China

Editorial Office
MDPI AG
Grosspeteranlage 5
4052 Basel, Switzerland

This is a reprint of articles from the Special Issue published online in the open access journal *Journal of Marine Science and Engineering* (ISSN 2077-1312) (available at: https://www.mdpi.com/journal/jmse/special_issues/DMUL9SL9MW).

For citation purposes, cite each article independently as indicated on the article page online and as indicated below:

Lastname, A.A.; Lastname, B.B. Article Title. *Journal Name* **Year**, *Volume Number*, Page Range.

ISBN 978-3-7258-1603-3 (Hbk)
ISBN 978-3-7258-1604-0 (PDF)
doi.org/10.3390/books978-3-7258-1604-0

Contents

About the Editors

Weicheng Cui

Weicheng Cui is a chair professor at the School of Engineering, Westlake University, China. Currently, his lab carries out research on and the development of bionic intelligent robot fish submersibles and frontier basic research on the unified theory of complex systems. He has more than 30 years of experience in the research and development of marine structures and deep-sea submersibles. In 1999, he was appointed by Shanghai Jiaotong University as the special professor of the Changjiang Scholarship Scheme and served as the Vice Dean of the School of Naval Architecture and Ocean Engineering. In June 2002, he was appointed as the Director of the China Ship Scientific Research Center and also the project leader and First Deputy Chief Designer of the Jiaolong deep manned submersible. In 2013, he joined Shanghai Ocean University and founded the Hadal Science and Technology Research Center. He initiated the project "Rainbowfish Challenging the Challenger Deep". He has published more than 400 research papers with collaborators and published six books.

Lian Lian

Lian Lian is a professor at the School of Oceanography, Shanghai Jiao Tong University, China. Her research interests include underwater vehicles and marine observation. She has been leading her team in concentrating their efforts on underwater vehicles, such as ROVs, deep-sea towing systems, underwater gliders, multimodal vehicles, underwater sampling, and tooling systems, and has built the first series of Chinese brand working-class ROVs—HAIMA ROVs. By using these ROV systems, scientific progress and discoveries have been made, such as the discovery of the "HAIMA Cold Seep".

Dahai Zhang

Dahai Zhang is a professor at Ocean College, Zhejiang University, China. His research primarily focuses on ocean renewable energy equipment technology, such as offshore wind turbines, wave energy converters, and tidal current turbines. He is the first author of 42 research papers, including 35 SCI papers published in *Renewable Energy*, *Renewable & Sustainable Energy Reviews*, *Physics of Fluids*, *Ocean Engineering*, *Coastal Engineering*, etc.

Preface

The conflict between populations, resources, and the environment in the 21st century has made the ocean a strategic space and resource treasure for human society to realize sustainable development. In order to study the ocean environment and exploit ocean resources, a fundamental understanding of complex and interwoven ocean processes across a broad range of spatial and temporal observational scales is required. This understanding is heavily reliant on various research fleets and equipment to support increasingly complex, multidisciplinary, multi-investigator research projects, including those in support of autonomous technologies, ocean observation systems, process studies, remote sensing, and modeling. Various underwater submersibles are the main working force for research fleets.

In response to the promotion of deep-sea technological development and to better meet the requirements for a sustainable society, we launched a Special Issue on this topic, and 30 papers were ultimately published within the time period of the Special Issue. Following the success of our previous edition, this Special Issue continues to highlight the most recent advancements in deep-sea equipment and technology. Over the course of this Special Issue, we have published 12 high-quality papers, each contributing important insights and findings to the field.

The topics discussed encompass four critical areas: the wireless power transfer systems of autonomous underwater vehicles (AUVs), the water entry dynamics of freely falling unmanned aerial–underwater vehicles and the descent dynamics of benthic landers, the connectors utilized in subsea production systems, and cleaning equipment designed to combat fouling on marine steel piles. Each one of these subjects represents a key technological challenge to be addressed in the ongoing development of deep-sea equipment.

While the number of papers in this Special Issue is slightly fewer than anticipated, it is nonetheless a substantial contribution for a Special Issue. Alongside the papers published in the previous Special Issue, these two editions provide a comprehensive and up-to-date overview of the state of deep-sea equipment and technology.

As Guest Editors, we are deeply grateful for the invaluable contributions of our authors and reviewers, as well as the unwavering support of the Editorial Office. It is our hope that these two Special Issues will serve as a catalyst for further advancements in this field, inspiring new research and technological innovation in deep-sea equipment.

Weicheng Cui, Lian Lian, and Dahai Zhang
Editors

Journal of
Marine Science and Engineering

MDPI

Editorial

Frontiers in Deep-Sea Equipment and Technology II

Weicheng Cui [1,2,*], Lian Lian [3,4] and Dahai Zhang [5]

1 Research Center for Industries of the Future, Westlake University, Hangzhou 310030, China
2 Key Laboratory of Coastal Environment and Resources of Zhejiang Province (KLaCER), School of Engineering, Westlake University, Hangzhou 310024, China
3 School of Oceanography, Shanghai Jiao Tong University, Shanghai 200240, China; llian@sjtu.edu.cn
4 China State Key Laboratory of Ocean Engineering, Shanghai Jiao Tong University, Shanghai 200240, China
5 Ocean College, Zhejiang University, Zhoushan 316021, China; zhangdahai@zju.edu.cn
* Correspondence: cuiweicheng@westlake.edu.cn

Citation: Cui, W.; Lian, L.; Zhang, D. Frontiers in Deep-Sea Equipment and Technology II. *J. Mar. Sci. Eng.* **2024**, *12*, 635. https://doi.org/10.3390/jmse12040635

Received: 29 October 2023
Accepted: 8 April 2024
Published: 9 April 2024

1. Introduction

The vast oceans could play a pivotal role in resolving the conflicts between population, resources, and the environment in the twenty-first century. This positions the ocean as a strategic space and a treasure trove of resources for human society to achieve sustainable development. To study the ocean environment and exploit its resources, a fundamental understanding of complex and intertwined oceanic processes across a wide range of spatial and temporal observational scales is necessary [1]. This understanding heavily relies on various research fleets and equipment to support increasingly complex, multidisciplinary, multi-investigator research projects. These projects include those in support of autonomous technologies, ocean observing systems, process studies, remote sensing, and modeling [2].

In response to this need, the *Journal of Marine Science and Technology* (*JMSE*) launched a Special Issue on the topic of "Frontiers in Deep-Sea Equipment and Technology II". The feedback from the authors was very positive, resulting in numerous submissions for this Special Issue. Ultimately, 30 papers were published, falling into the following six categories: ROV and its variants, AUVs for multi-agent cooperation, gliders and other specific equipment, pressure-resistant hulls, component technology, and robotic fish [3].

Given that many authors were unable to submit their papers within the deadline and considering the rapid technological advancements in this field, we decided to launch another Special Issue to continue this topic, albeit with a slightly different focus as announced. While the previous Special Issue had many papers focused on the design and development of various submersibles, in this issue, we emphasized that "we particularly welcome papers on the design, analysis, and testing of various new methods, new theories, new sensors, and new equipment used in deep seas".

Following the launch, we received numerous submissions and, based on the same standards established by the *JMSE Journal*, ultimately selected 12 papers for publication within the time frame of this Special Issue. We are deeply grateful to all the authors who contributed to the success of this Special Issue, and we are pleased to know that *JMSE* intends to compile all these papers into a book for further promotion.

The purpose of this Editorial is to provide a brief introduction to all the papers included in this issue. Based on a preliminary analysis, these papers are grouped into four categories: AUV's wireless power transfer system, water entry and descent process, subsea production system, and marine cleaning equipment. In the following sections, each topic is discussed in the context of the papers that have been published in this Special Issue.

2. AUV's Wireless Power Transfer System

Autonomous underwater vehicles (AUVs) are increasingly critical in marine resource utilization, scientific exploration, and military applications. However, power supply remains a key issue, limiting their long-term operation. Recently, wireless power transfer

(WPT) technology has seen extensive research and application in various fields. In this paper, Wen et al. (contribution 1) proposed a compact WPT system for AUVs, featuring a planar magnetic coupler. They integrated a passive induction (PI) coil into the circular transmitter (Tx) coil to create a uniform magnetic field (UMF), enhancing the stability of the WPT system under radial and axial misalignments. They also provided a method for constructing a UMF with a PI coil, aiming to stabilize the magnetic field intensity. Their analysis showed the integrated coil improves radial misalignment tolerance compared to a conventional coil. Adjusting the series capacitor connected with the Tx coil can achieve a zero-phase angle input condition. Experiments indicated the proposed magnetic coupler with an integrated coil enhances output power stability and power transfer efficiency under possible misalignments compared to a conventional coupler.

In underwater charging, AUV docking devices are used for homing, energy supply, and data exchange, enhancing AUV endurance and enabling continuous, large-scale operation. Du et al. (contribution 2) designed a funnel-shaped, multi-degree freedom underwater docking device based on a deep-sea platform to improve the AUV docking success rate. The device adjusts its heading, pitch, and roll angles according to the current flow direction and AUV position. An underwater hydraulic system was developed to drive these adjustments, powering hydraulic cylinders, motors, and other components. They established a model for the heading angle adjustment circuit and derived its open-loop transfer function. After simulating the hydraulic circuit's dynamic response performance, they used an optimized PID algorithm for improvement. The model's accuracy and the control algorithm's effectiveness were validated through pool experiments with the docking device.

Liu et al. (contribution 3) explored the issue of fault-tolerant control (FTC) for autonomous underwater vehicles (AUVs) equipped with multiple thrusters. They specifically focused on scenarios involving current disturbances, thruster faults, and modeling uncertainty. The main objectives of their research were to minimize energy consumption, which can be exacerbated by control signal chattering, and to enhance the tracking accuracy of AUVs operating in deep-sea environments. To address these challenges, they proposed a fault-tolerant control method for AUVs with multiple thrusters, grounded in a finite-time extended state observer (FTESO). In more detail, they designed a FTESO based on an integral sliding mode surface to estimate the generalized uncertainty, which is a composite of current disturbances, thruster faults, and modeling uncertainty. They thoroughly analyzed the fast, finite-time, uniform, and ultimately bounded stability of the proposed FTESO. Subsequently, using the estimated value derived from FTESO, they developed an FTC method for AUVs that is based on non-singular fast terminal sliding mode surfaces.

3. Water Entry and Descent Process

Dong et al. (contribution 4) conducted a comprehensive study on the hydrodynamic and motion characteristics of an unmanned aerial-underwater vehicle (UAUV), a novel vehicle capable of both flight and underwater cruising, expected to perform continuous, uninterrupted observations and sampling by crossing the free water surface multiple times. They employed the Reynolds-averaged Navier–Stokes method to investigate the multi-degree-of-freedom UAUV's complete water-entry process at various velocities and pitch angles. To balance computational accuracy and cost, they employed adaptive mesh refinement and an adaptive time-stepping strategy, capturing the slamming pressure accurately with reasonable computational effort. Their research involved simulations of the vehicle at different initial velocities and pitch angles, followed by an analysis of the variable physical properties. They found that both initial velocity and pitch angle significantly influence the hydrodynamic behavior, including time-varying force, while the thickness ratio greatly impacts added mass and pressure. Their results indicated that a higher entry velocity yields a larger peak vertical force, and the transverse hydrodynamic characteristics vary significantly for oblique water entries at different pitch angles.

Benthic landers, compared to remotely operated vehicles (ROVs) and autonomous underwater vehicles (AUVs), offer cost-effective operations and increased endurance. Establishing an appropriate descent velocity is crucial for their design, as it can prevent retrieval failures and enhance sea trial efficiencies. In their research, Zhang et al. (contribution 5) proposed an effective strategy for the configuration and optimization of a self-developed benthic lander. Sea trial data was used to validate their scheme, confirming its accuracy and efficiency. Overall, their findings offer valuable insights for the logical configuration and practical deployment of benthic landers.

Guo et al. (contribution 6) developed an improved constitutive model for shape memory alloys (SMAs), crucial for deep-sea intelligent actuators in marine engineering. They extended the one-dimensional thermodynamic equation for SMAs using an improved logistic nonlinear function to construct macroscopic variable-speed phase transition relations. The model was tested against traditional models and experimental data under constant load and temperature. The results indicated that the improved model accurately describes slow phase transitions with fewer variable parameters and adjustable phase transition rates, providing valuable theoretical support for designing SMA deep-sea actuators.

4. Subsea Production System

Hao et al. (contribution 7) conducted an in-depth analysis of the mechanical behavior and sealing performance of core-sealing components in subsea production systems (SPS), which are interconnected by subsea connectors. They examined the loading conditions of the subsea clamp connector to understand the load transfer relationship between components under different modes. Mathematical models were developed for load transfer in both preloading and operation modes, introducing the concept of mechanical transfer efficiency. A three-dimensional finite element model of the subsea clamp connector was established to analyze the impact of complex loads. The simulation results showed that internal pressure loading enhances the sealing of the subsea connector, and the stress distribution in core-sealing components under bending moment loading is significantly asymmetric. Axial tensile loads reduce the effect of the bending moment on the strength of the core seal member but further weaken the seal. These simulations were validated through an experimental system.

To address the sensitivity of radial seals to axial overload failure, which can lead to leaks, Li et al. (contribution 8) proposed two semi-analytical methods. They developed an analytical model based on membrane theory for the joint strength of the connection and seal under axial tension and compression and a finite element model for the radial metal seal's joint strength. The overload sealing performance method was derived using a finite element model and the Reynolds equation for laminar flow. They analyzed the effects of critical parameters on joint strength and overload sealing performance and conducted experiments to evaluate these aspects. Results showed that an internally turned sealing surface and pipe deflection can improve joint strength and that the compression-type connector can remain sealed under maximum axial overload. The proposed methods offer insights into the radial metal seal's behavior under axial overload.

A similar method is extended to predict the sealing performance of the subsea pipeline compression connector (contribution 9). This method integrates a macroscopic analytical model for the interference process, a mesoscopic finite element analysis of the internally turned sealing surface, and a calculation of the leakage ratio using the Abbott–Firestone curve. They analyzed the impact of geometry parameters and conducted experiments to validate their method. Results indicated that compression connectors designed using this method could reliably seal a rough internally turned surface within five times the pipeline's thin-wall thickness threshold. This approach saves 57% of design time and reduces machining time and costs.

Subsea clamp connectors are prone to sealing and locking failures due to harsh marine environments and complex loading conditions. This poses risks to subsea oil and gas field safety. Therefore, accurately predicting their reliability under realistic operating conditions is crucial. An et al. (contribution 10) used finite element numerical simulation and multiple response surface methodology (MRSM) to analyze the reliability of subsea clamp connectors, considering key structural parameters and based on seal failure and yield failure criteria. The approach was validated in the Bohai Sea, showing that the system's failure probability was primarily influenced by seal radius, flange contact angle, and internal pressure. The connector's reliability was calculated to be 98.73%, as confirmed by a sealing performance test. This study offers a practical method for analyzing the reliability of subsea clamp connectors, considering multiple factors, and contributes to ensuring the safety of subsea oil and gas fields.

Underwater wet-mateable connectors are essential for cost- and time-efficient installation, maintenance, and reconfiguration in industries like oil and gas, offshore renewable energy, and undersea observatories. Song et al. (contribution 11) proposed a methodology for designing and testing such connectors. They introduced an innovative wet-mateable electrical connector with dual-bladder pressure-balanced oil-filled technology, derived generalized equations of differential pressure, and proposed a thermal-electric-structure coupling simulation procedure. They conducted finite element analysis involving coupled multi-field problems. A prototype connector was developed, and its electrical performance was tested successfully under 3000 m of ocean depth pressure, achieving a leading level in China. This comprehensive study on wet-mateable connectors' design, theory, simulation, and testing provides valuable insights for ocean scientists, especially in developing countries.

5. Marine Cleaning Equipment

Addressing the challenge of marine fouling on marine steel piles, Li et al. (contribution 12) proposed an innovative configuration for marine steel pile cleaning equipment. This novel design employs a scraping method and a telescopic mechanism, applying a multi-cylinder synchronous control strategy to the cleaning equipment. They successfully produced a test prototype of the cleaning equipment, designed to overcome issues of eccentricity and tilt often encountered in ocean engineering.

To evaluate the multi-cylinder synchronous control performance under various working conditions, they established a simulation model of the operation process of the equipment's telescopic mechanism using the MATLAB Simulink module. They conducted tests to preliminarily verify the synchronous working performance of the telescopic mechanism of the cleaning equipment under no-load conditions.

The test results were promising. Under no-load conditions, the relative errors between the three cylinders and the target displacement were as low as 0.8%, 0.4%, and 0.2%, respectively. The cleaning equipment could reach the specified working position at the given working speed, and the displacement synchronization error between each cylinder was controlled within 1 mm.

These results demonstrate that the telescopic mechanism exhibits excellent synchronization, ensuring the stability of the cleaning equipment during operation and effectively preventing eccentricity and tilt. As such, this study provides a valuable reference for the manufacturing of similar cleaning equipment, potentially enhancing the efficiency and effectiveness of marine fouling clean-up efforts.

6. Conclusions

Following the success of our previous edition [3], this Special Issue continues to highlight the most recent advancements in deep-sea equipment and technology. Over the course of this issue, we have published twelve high-quality papers, each contributing important insights and findings to the field.

The topics discussed encompass four critical areas: the wireless power transfer systems of autonomous underwater vehicles (AUVs), the water entry dynamics of freely falling unmanned aerial–underwater vehicles and the descent dynamics of Benthic Landers, the connectors utilized in subsea production systems, and the cleaning equipment designed to combat fouling on marine steel piles. Each of these subjects represents a key technological challenge to be addressed in the ongoing development of deep-sea equipment.

While the number of papers in this issue is slightly fewer than anticipated, it is nonetheless a substantial contribution for a Special Issue. Alongside the papers published in the previous Special Issue, these two editions provide a comprehensive and up-to-date overview of the state of deep-sea equipment and technology.

We, the editors, are deeply grateful for the invaluable contributions of our authors and reviewers, as well as the unwavering support of our Editorial Office. It is our hope that these two Special Issues will serve as a catalyst for further advancements in this field, inspiring new research and technological innovation in deep-sea equipment.

Funding: This work was supported by the National Key Research and Development Program (2022YFC2805200), start-up funding from Westlake University under grant number 041030150118, and the Scientific Research Funding Project of Westlake University under Grant No. 2021WUFP017.

Acknowledgments: All the authors are greatly acknowledged for their support, and the arrangement of publishing this Special Issue as a book by MDPI is appreciated.

Conflicts of Interest: The authors declare no conflicts of interest.

List of Contributions:

1. Wen, H.; Li, J.; Zhang, K.; Ye, J.; Yan, Z.; Song, B.; Tong, X. Enhancing Power Transmission Stability of AUV's Wireless Power Transfer System with Compact Planar Magnetic Coupler. *J. Mar. Sci. Eng.* **2023**, *11*, 566. https://doi.org/10.3390/jmse11030566.
2. Du, X.; Zong, C.; Zhang, B.; Shi, M. Design, Simulation, and Experimental Study on the Hydraulic Drive System of an AUV Docking Device with Multi-Degree Freedom. *J. Mar. Sci. Eng.* **2022**, *10*, 1790. https://doi.org/10.3390/jmse10111790.
3. Liu, X.; Zhang, M.; Liu, X.; Zhao, W. Finite-Time Extended State Observe Based Fault Tolerant Control for Autonomous Underwater Vehicle with Unknown Thruster Fault. *J. Mar. Sci. Eng.* **2022**, *10*, 1624. https://doi.org/10.3390/jmse10111624.
4. Dong, L.; Wei, Z.; Zhou, H.; Yao, B.; Lian, L. Numerical Study on the Water Entry of a Freely Falling Unmanned Aerial-Underwater Vehicle. *J. Mar. Sci. Eng.* **2023**, *11*, 552. https://doi.org/10.3390/jmse11030552.
5. Zhang, Q.; Dong, C.; Shao, Z.; Zhou, D. Analysis of the Descent Process and Multi-Objective Optimization Design of a Benthic Lander. *J. Mar. Sci. Eng.* **2023**, *11*, 224. https://doi.org/10.3390/jmse11010224.
6. Guo, J.; Pan, B.; Cui, W.; Hu, S. A Nonlinear Phase Transition Dynamic Model for Shape Memory Alloys Based Deep Sea Actuators. *J. Mar. Sci. Eng.* **2022**, *10*, 1951. https://doi.org/10.3390/jmse10121951.
7. Hao, X.; Yun, F.; Jiao, K.; Chen, X.; Jia, P.; Wang, X.; Wang, L. Mechanical Behavior and Sealing Performance Study of Subsea Connector Core-Sealing Components under the Combined Action of Internal Pressure, Bending Moment, and Axial Load. *J. Mar. Sci. Eng.* **2023**, *11*, 1691. https://doi.org/10.3390/jmse11091691.
8. Li, Z.; Wang, G.; Wang, X.; Yao, S.; Yun, F.; Jia, P.; Wang, L. Semi-Analytical Methods for the Joint Strength and Sealing Performance in the Failure Process of the Subsea Pipeline Compression Connector. *J. Mar. Sci. Eng.* **2023**, *11*, 1417. https://doi.org/10.3390/jmse11071417.
9. Li, Z.; Wang, G.; Yao, S.; Yun, F.; Jia, P.; Li, C.; Wang, L. A Semi-Analytical Method for the Sealing Performance Prediction of Subsea Pipeline Compression Connector. *J. Mar. Sci. Eng.* **2023**, *11*, 854. https://doi.org/10.3390/jmse11040854.
10. An, W.; Wang, Y.; Chen, B.; Duan, M.; Zhang, X. Analysis of the Reliability of Subsea Clamp Connector Based on Multiple Response Surface Methodology. *J. Mar. Sci. Eng.* **2023**, *11*, 1378. https://doi.org/10.3390/jmse11071378.
11. Song, W.; Yang, C.; Cui, W.; Song, C.; Yang, P.; Hong, J.; Lei, Y.; Liu, Q.; Wang, Z. An Underwater Wet-Mateable Electrical Connector with Dual-Bladder Pressure-Balanced Oil-Filled (PBOF) Technology. *J. Mar. Sci. Eng.* **2023**, *11*, 156. https://doi.org/10.3390/jmse11010156.
12. Li, C.; Pang, N.; Xu, K.; Geng, Q.; Wang, X.; Yun, F.; Gao, L. Application of Multi-Cylinder Synchronous Control for Telescopic Mechanism of Marine Steel Pile Cleaning Equipment. *J. Mar. Sci. Eng.* **2023**, *11*, 1010. https://doi.org/10.3390/jmse11051010.

References

1. Committee on Evolution of the National Oceanographic Research Fleet (CENORF); National Research Council. *Science at Sea: Meeting Future Oceanographic Goals with a Robust Academic Research Fleet*; The National Academies Press: Washington, DC, USA, 2009.
2. Lin, M.; Yang, C. Ocean Observation Technologies: A Review. *Chin. J. Mech. Eng.* **2020**, *33*, 32. [CrossRef]
3. Cui, W.C.; Lian, L.; Pan, G. Frontiers in Deep-Sea Equipment and Technology. *J. Mar. Sci. Eng.* **2023**, *11*, 715. [CrossRef]

Journal of
Marine Science and Engineering

MDPI

Article

Design, Simulation, and Experimental Study on the Hydraulic Drive System of an AUV Docking Device with Multi-Degree Freedom

Xiaofei Du [1,2], Chaoyong Zong [3], Bo Zhang [4] and Maolin Shi [5,*]

1 School of Mechanical Engineering, Nanjing Institute of Technology, Nanjing 211167, China
2 School of Mechanical Engineering, Southeast University, Nanjing 211189, China
3 School of Mechanical Engineering, Dalian University of Technology, Dalian 116024, China
4 College of Mechanical and Electrical Engineering, Harbin Engineering University, Harbin 150001, China
5 School of Agricultural Engineering, Jiangsu University, Zhenjiang 212013, China
* Correspondence: maolin@ujs.edu.cn

Abstract: AUV docking devices have the ability to achieve homing of the AUV, supply the AUV with energy, and exchange data with it, thus improving the endurance and ensuring the continuous long-term and large-scale operation of the AUV. To improve the successful docking rate of the AUV, a funnel-shaped underwater docking device with multi-degree freedom based on a deep-sea platform was designed in this paper. The heading angle, pitch angle, and roller angle of the docking device could be adjusted in a timely manner according to the current flow direction and the position of the AUV. In order to realize the timely adjustment of the heading angle, pitch angle, and roll angle of the docking device, a set of underwater hydraulic systems was developed as the power source to drive the corresponding hydraulic cylinders, hydraulic motors, and other executive components. The model of the heading angle adjustment circuit of the hydraulic system was established and the open-loop transfer function of the heading angle adjustment circuit was derived. The dynamic response performance of the hydraulic circuit was simulated and the optimized PID algorithm was used to improve the dynamic response performance. Finally, the accuracy of the heading angle adjustment circuit model and the effectiveness of the control algorithm were validated by experiment of the docking device in a water pool.

Keywords: AUV docking device; hydraulic drive system; modelling and simulation; PID control algorithm; experimental validation

Citation: Du, X.; Zong, C.; Zhang, B.; Shi, M. Design, Simulation, and Experimental Study on the Hydraulic Drive System of an AUV Docking Device with Multi-Degree Freedom. *J. Mar. Sci. Eng.* **2022**, *10*, 1790. https://doi.org/10.3390/jmse10111790

Academic Editor: Mohamed Benbouzid

Received: 28 October 2022
Accepted: 18 November 2022
Published: 21 November 2022

1. Introduction

Autonomous underwater vehicles (AUVs) are important underwater operating equipment, which play an irreplaceable role in the fields of underwater observation, underwater rescue, and deep-sea sampling. Owing to the restrictions of the volume of AUVs, the energy of its own self-carrying cannot support its long-term and large-scale underwater operations, and it is usually necessary to recover the AUV to the mothership for energy supply and data exchange to allow it continue working, which not only reduces the operating efficiency, but also increases the use cost of the AUV. The AUV underwater docking device can be connected with the AUV underwater, so as to supply the AUV with energy and exchange data with it, thus improving the endurance of the AUV and ensuring the continuous long-term and large-scale operation of the AUV.

Many researchers have designed different docking devices and proposed different docking strategies for AUVs. Stokey et al. developed a docking device composed of a conical guide cover and a cylindrical butt joint cylinder for the Remus AUV [1,2]. The Remus AUV relies on the USBL to enter the guide cover and the docking cylinder to complete the docking process, and then the docking device supplements the energy and exchanges data with the AUV through the underwater cable. Feezor et al. constructed an electromagnetic homing (EM) system for the docking device to complete the docking with Odyssey IIB AUV and employed the decoupled PID control loop to supply the control response of the AUV [3]. Cowen et al. designed an optical terminal guidance docking station and the Odyssey IIB and the NRaD Flying Plug were used to validate the accuracy of terminal guidance and docking [4]. Singh et al. designed a bar-type docking device and installed a latching mechanism at the head of the AUV to capture the positioning bar [5]. The AUV and the docking device can perform 360° all-round docking in a horizontal plane. McEwen et al. designed a docking device with a bellmouth guide cover for Bluefin AUV with a diameter of 540 mm and tested the homing and docking control system at sea [6]. Jantapremjit et al. proposed an optimal high-order sliding mode control technique based on the state-dependent Riccati equation [7] and a line-of-sight method to optimize the trajectory of the AUV [8], and carried out a simulation of the docking process. Park et al. developed a set of underwater docking devices for the ISiMI AUV, which uses the guide cover and charge coupled device (CCD) camera to guide the AUV to enter into the docking device to complete the docking, and they also conducted the docking experiment in an ocean engineering basin [9,10]. Raspante designed a mobile funnel-shaped docking device for the REMUS-100 AUV and proposed a four-step docking approach to enable the recovery of the AUV in moderate sea conditions [11]. Rigaud et al. developed a mobile funnel-shaped docking device for the Ifremer Asterx AUV, which incorporated USBL and vision-based docking strategies as well as induction charging and Wi-Fi data connection technologies [12]. Batista et al. proposed a two-step control approach for an intervention type AUV docking and carried out docking simulation, which shows high accuracy in the presence of sensor noise [13]. Ferreira et al. presented a control law for the MARES AUV based on the range-only measurement docking approach and the docking experiment was carried out in the Douro river to validate the robustness of the control strategy [14]. Li et al. performed a docking experiment based on vision positioning with two cameras, achieving an 83% successful docking rate with the micro-AUV WL-3 [15]. Yang et al. developed an active stationary funnel-shaped docking device with adjustable orientation for the hybrid underwater glider and studied docking guidance algorithms to ensure a successful docking rate [16]. Sans-Muntadas et al. [17] designed a fixed funnel-shaped docking device for the torpedo-shaped AUV, and Piskura et al. [18] and Sarda et al. [19] designed a line capture docking station for REMUS-600 and REMUS-100 AUV, respectively. With the use of a light beacon localization module and a single beacon range-only localization algorithm, Hurtós et al. developed a funnel-shaped docking device for the Sparus II AUV [20,21]. Zuo et al. proposed unified strategy to docking an over-actuated AUV that included task planning, guiding, and thrust distribution [22]. Lin et al. developed a funnel-shaped docking station for the AUV with a length of 2.15 m and a diameter of 0.18 m, and they carried out kinematic simulation, electromagnetic guidance system design, and docking experiments [23,24]. Anderlini et al. investigated two reinforcement learning strategies in a simulated environment to regulate the docking of an AUV onto a stationary platform [25]. Palomeras et al. designed a docking station for the Girona 500 I-AUV and a docking test was performed at sea using the USBL system to locate both the AUV and the panel [26]. Zhang et al. developed a cone-shaped docking device and established a dynamic model in ADAMS to carry out the docking simulation [27]. Li et al. designed a prototype AUV docking system for an AUV with a diameter of 200 mm and a length of 2.7 m, and they tested entrance adjustment, clamping, non-penetrating power, and data transfer in a lake environment [28]. Roy et al. proposed a robust control strategy for the AUV-150 and conducted a comparative study on the positioning control of the AUV [29,30].

In order to improve the success rate of AUV underwater docking, an AUV underwater docking device with multi-degree freedom based on the submarine platform is designed in this paper. The multi-freedom AUV docking device designed in this paper can adjust the heading angle, pitch angle, and roller angle of the docking device in a timely manner according to the direction of the current flow and the position of the AUV, so that the opening direction of the docking device horn cover can better adapt to the attitude position of the AUV and avoid the adverse impact of the lateral current on the AUV docking. In order to realize the timely adjustment of the heading angle, pitch angle, and roll angle of the docking device, a set of underwater hydraulic system was developed as the power source to drive the corresponding hydraulic cylinders, hydraulic motors, and other executive components. The dynamic response performance of the hydraulic drive system of the designed docking device was simulated and the optimized PID algorithm was used to improve the dynamic response performance of the hydraulic system. Finally, the effectiveness of the designed hydraulic system and control algorithm was verified by the experiment of the docking device in the water pool.

This paper is organized as follows. Section 2 is the design of the docking device and hydraulic drive system. In Section 3, we constructed the simulation model of the heading angle adjustment circuit of the hydraulic system. Section 4 is devoted to the dynamic response simulation of the designed heading angle adjustment circuit and the application of the optimized PID control algorithm. Section 5 aims through experimental validation to verify the accuracy of the model and effectiveness of the PID control algorithm. The conclusions end the paper in Section 6.

2. Design of Docking Device and Hydraulic Drive System

The designed multi-freedom AUV docking device is shown in Figure 1. The docking device has three bottom corners to ensure it can be located on the seafloor. The hydraulic system cabin and control system cabin are fixed on the disc at the bottom of the device. The docking device mainly consists of the guide port, docking barrel, underwater camera, communication antenna, electronic compass, charging and data exchange module, locking module, adjustment module of roll angle and pitch angle, and adjustment module of heading angle.

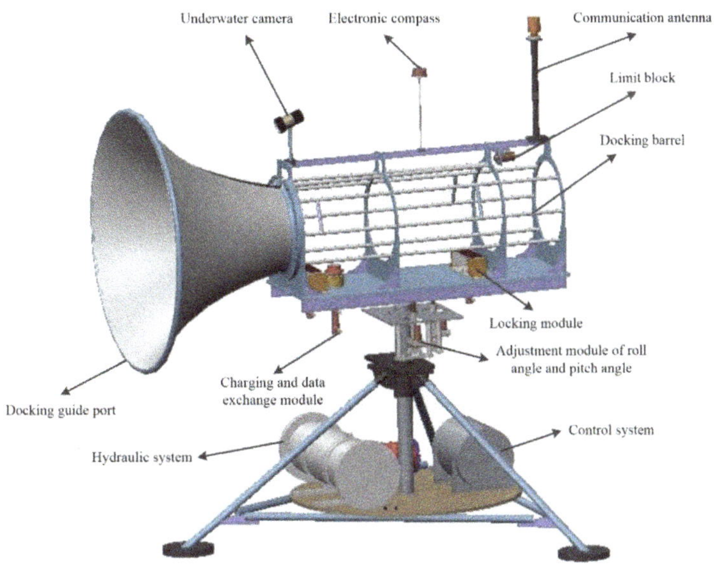

Figure 1. Designed AUV docking device with multi-degree freedom.

The adjustment module of heading angle is also located on the disc of the device, as shown in Figure 2, and it can lock the shaft of the docking device by one hydraulic motor driving the locking rod, and then drive the shaft to rotate by another hydraulic motor driving the worm gear mechanism, so as to achieve the heading angle adjustment of the guide port and the docking barrel. Owing to the self-locking characteristics of the worm gear mechanism, the module can achieve stable self-locking when the hydraulic motor does not rotate. The hydraulic cabin is connected with the executive components of each module through hydraulic pipelines and the control system cabin is connected with the hydraulic cabin, sensors, underwater camera, and other corresponding components through watertight cables. The height of the docking device is 3400 mm and the guide port is conical with a thickness of 8 mm. The maximum diameter of the guide port is 2000 mm and the diameter of the interface between the guide port and the docking barrel is 640 mm.

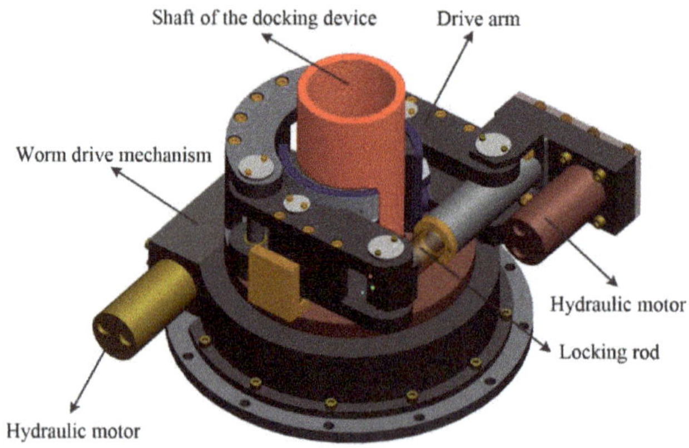

Figure 2. Adjustment module of the heading angle.

The three views of the target AUV to be docked are shown in Figure 3. The diameter of the AUV is 560 mm and the total length of the AUV is 5100 mm. The shell of the AUV is mainly made of fiber-reinforced plastic. The bottom of the AUV is flat and equipped with a charging interface, the upper of the AUV is equipped with a round lifting lug, and the tail is equipped with a communication antenna and a main propeller.

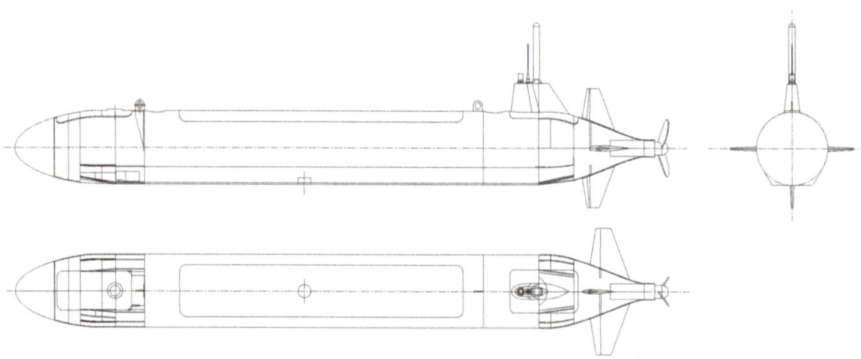

Figure 3. Three views of the target AUV to be docked.

The docking flow chart of the AUV and the docking device is shown in Figure 4. After the docking device is placed on the seafloor, the current direction meter on the docking device will measure the current direction at this time and the electronic compass will measure the heading angle, roll angle, and pitch angle of the guide port and docking barrel. The heading angle of the guide port and docking barrel will be changed through the heading angle adjustment module to make it the same as the current flow direction angle. Then, the roll angle and pitch angle of the guide port and docking barrel will be adjusted through the adjustment module of roll angle and pitch angle so that the guide port and docking barrel are in the horizontal state. When the AUV drives into the docking barrel and reaches the designated position, the locking module first completes the locking of the AUV, and then the charging and data exchange module will charge the AUV and exchange data with it. Finally, the locking module, charging, and data exchange module move in the opposite direction, making the AUV in a non-locking state and the AUV exits the docking device under the force generated by the reverse movement of the propeller. The actions of the heading angle adjustment module, the adjustment module of roll angle and pitch angle, locking module, and charging and data exchange module are all realized by the movement of executive components such as hydraulic cylinder or hydraulic motor powered by the hydraulic system in the hydraulic cabin.

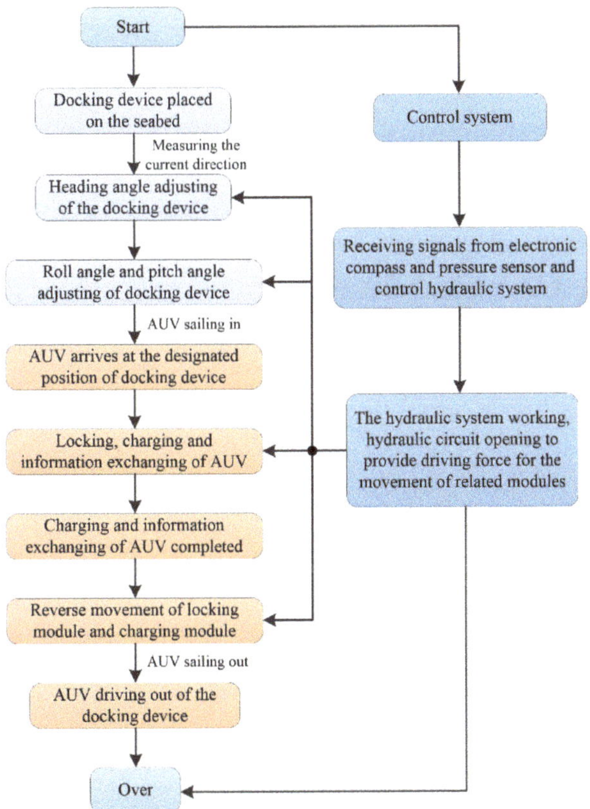

Figure 4. Docking flow chart of the AUV and the designed docking device.

Figure 5 displays the schematic design of the hydraulic system of the docking device. The hydraulic system has the characteristics of small volume and large output power. Compared with using the motor drive in which we must solve the sealing problem of the motor in the deep-sea environment, the hydraulic cylinder and hydraulic motor of the hydraulic system designed in this paper are made of 316 stainless steel, which can withstand seawater corrosion for a long time and have high stability and low cost. The seawater pressure compensator of the hydraulic system can apply seawater pressure outside the hydraulic cabin to the pipelines of the hydraulic system, thus making the hydraulic system suitable for the deep-sea environment. The six independent hydraulic circuits can achieve shaft locking, heading angle adjustment, roll angle adjustment, pitch angle adjustment, AUV locking, and AUV charging, respectively. The six hydraulic circuits share a common hydraulic power source and each hydraulic circuit is controlled by its corresponding valve for opening, closing, direction changing, or flow speed control. The hydraulic pump, electric machinery, electromagnetic directional control valve, proportional servo valve, flow speed control valve, and some other components are sealed in the hydraulic system cabin and the hydraulic motor, hydraulic cylinder, and other actuating components can be resistant to seawater corrosion and directly exposed to seawater. The control system sends the control signal to the hydraulic cabin through a watertight cable, which can realize the start and stop of the electric machinery, the control of the proportional servo valve, and the electromagnetic directional control valve. Through the work of each circuit of the hydraulic system, a series of docking actions between the docking device and the AUV in Figure 4 can be achieved, thus completing the whole docking process.

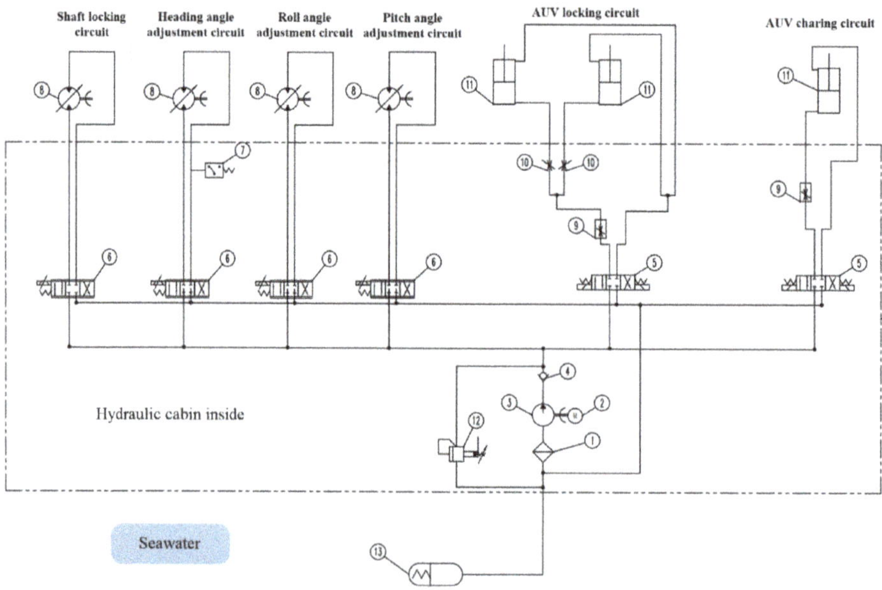

Figure 5. Designed underwater hydraulic system of the AUV docking device: 1. filter; 2. electric machinery; 3. hydraulic pump; 4. unilateral valve; 5. electromagnetic directional control valve; 6. proportional servo valve; 7. pressure sensor; 8. hydraulic motor; 9. flow speed control valve; 10. throttle valve; 11. hydraulic cylinder; 12. relief valve; 13. seawater pressure compensator; 14. hydraulic system cabin.

3. Modeling of Heading Angle Adjustment Circuit of the Hydraulic System

The heading angle adjustment, roll angle adjustment, and pitch angle adjustment of the AUV docking device can be achieved by the positive and negative rotation of the hydraulic motors controlled by the proportional servo valves. The heading angle adjustment circuit is composed of a proportional servo valve, a hydraulic motor, a worm gear mechanism, and an angle sensor. The control schematic diagram of the heading angle adjustment circuit is shown in Figure 6.

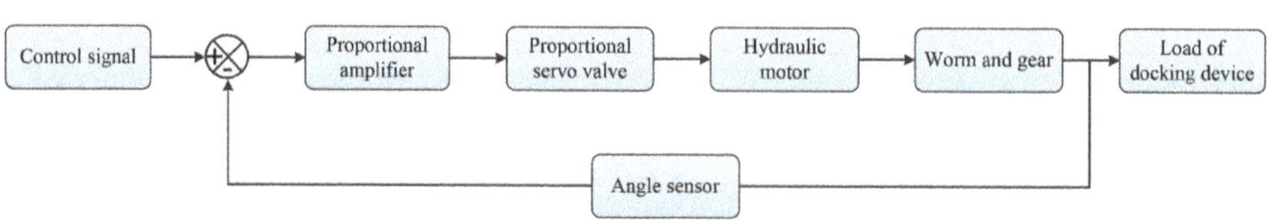

Figure 6. Control schematic diagram of heading angle adjustment circuit.

The schematic diagram of the power mechanism composed of a four-way proportional servo valve and a two-way hydraulic motor is shown in Figure 7. The linearized flow equation of the proportional servo valve can be expressed as follows:

$$Q_L = K_q x_v - K_c p_L \tag{1}$$

where Q_L is the load flow, K_q is the flow coefficient of the proportional servo valve, K_c is the flow pressure coefficient of the proportional servo valve, and p_L is the load pressure drop. Because of the compression of hydraulic oil, the flow Q_1 flowing into the motor is not equal to the flow Q_2 flowing out of the motor, that is $Q_1 \neq Q_2$. However, the compression amount of hydraulic oil is very small compared with Q_1 and Q_2, so the load flow can be defined as $Q_L = (Q_1 + Q_2)/2$.

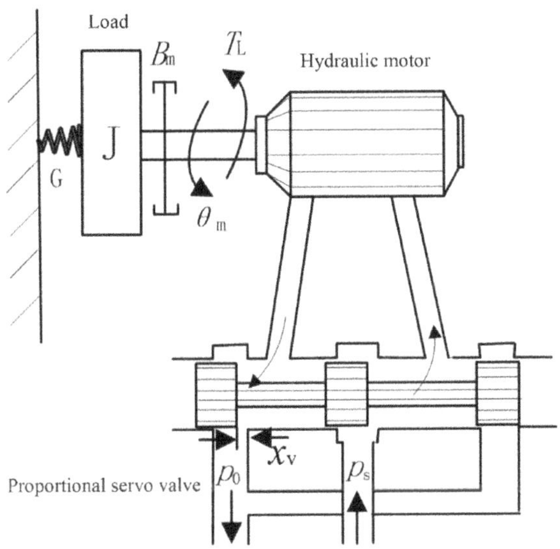

Figure 7. Schematic diagram of the valve-controlled hydraulic motor.

The hydraulic pipeline of the docking device is short, and the friction, fluid mass loss and dynamic pressure loss in the pipeline can be ignored. Moreover, the hydraulic motor has a short continuous working time and a small oil temperature rise, so the change of oil temperature and bulk modulus can also be ignored. The linearized flow equation of the motor can be expressed as follows:

$$Q_L = D_{tM}\frac{d\theta_M}{dt} + C_{tM}p_L + \frac{V_t}{4\beta_e}\frac{dp_L}{dt} \tag{2}$$

where D_{tM} is the theoretical displacement of the hydraulic motor; θ_M is the rotation angle of the hydraulic motor shaft; β_e is the elastic modulus of the effective volume, $\beta_e = 700$ Mpa; V_t is the volume of oil inlet chamber of the hydraulic motor (including valve, pipeline and oil inlet chamber); C_{tM} is the total leakage coefficient of the hydraulic motor $C_{tM} = C_{iM} + C_{eM}/2$; C_{iM} is the internal leakage coefficient of the hydraulic motor; and C_{eM} is the external leakage coefficient of the hydraulic motor.

The torque balance equation of the hydraulic motor is as follows:

$$T_g = D_{tM}p_L = J\frac{d^2\theta_M}{dt^2} + B_M\frac{d\theta_M}{dt} + G\theta_M + T_L \tag{3}$$

where T_g is the theoretical torque of the hydraulic motor, J is the total inertia of the hydraulic motor and the load, θ_M is the rotation angle of the hydraulic motor shaft, B_M is the total viscous damping coefficient of the hydraulic motor and the load, G is the torsional spring stiffness of the load, and T_L is the external load force acting on the motor.

The following equation be obtained after the Laplacian transformation and combination of Equations (1)–(3):

$$\theta_M = \frac{\frac{K_q}{D_M}X_v - \frac{K_{ce}}{D_M^2}(1+\frac{V_t}{4\beta_e K_{ce}}s)T_L}{\frac{V_t J}{4\beta_e D_M^2}s^3 + (\frac{K_{ce}J}{D_M^2} + \frac{B_M V_t}{4\beta_e D_M^2})s^2 + (1 + \frac{B_M K_{ce}}{D_M^2} + \frac{KV_t}{4\beta_e D_M^2})s + \frac{K_{ce}G}{D_M^2}} \quad (4)$$

where K_{ce} is the total flow-pressure coefficient, $K_{ce} = K_c + C_{tM}$.

Equation (4) shows the response characteristics of the hydraulic motor rotation angle to the valve input displacement and load force and takes into account the influence of the system's physical characteristics such as inertial load, elastic load, viscous friction load, hydraulic motor leakage, and oil compressibility on the motor speed. The load of the docking device can be considered as an inelastic load, that is $G = 0$. In addition, the damping coefficient D_M^2/K_{ce} is much larger than B_M, so $B_M K_{ce}/D_M^2 \ll 1$. In order to study the control characteristics of the system, the load is considered as the interference term in this paper, so the transfer function of motor shaft rotation angle to valve displacement without load can be obtained as follows:

$$G_M(s) = \frac{\theta_M}{X_v} = \frac{K_q/D_M}{s(\frac{s^2}{\omega_h^2} + \frac{2\xi_h}{\omega_h}s + 1)} \quad (5)$$

where ω_h is the natural frequency of the system, $\omega_h = \sqrt{\frac{4\beta_e D_M^2}{V_t J}}$; ξ_h is the damping of the system, $\xi_h = \frac{K_{ce}}{D_M}\sqrt{\frac{\beta_e J}{V_t}} + \frac{B_M}{4D_M}\sqrt{\frac{V_t}{\beta_e J}}$; and, in general, B_M is very small, so ξ_h can be simplified as $\xi_h = \frac{K_{ce}}{D_M}\sqrt{\frac{\beta_e J}{V_t}}$.

The hydraulic system parameters in Equation (5) are listed in Table 1.

Table 1. Parameters of the hydraulic system.

D_M	V_t	J	β_e	K_{ce}
16 mL/r	1.5×10^{-4} m^3	0.083 kg·m/s^2	700 MPa	5.1×10^{-11} m^5/N·s

In practical engineering, generally $\xi_h \approx 0.1 \sim 0.2$, and in this paper we set $\xi_h = 0.1$. Finally, the transfer function of the hydraulic motor can be obtained as follows:

$$G_M(s) = \frac{K_q/D_M}{s(\frac{s^2}{38.2^2} + \frac{2\times 0.1}{38.2}s + 1)} \quad (6)$$

Equation (6) is composed of a proportional component, an integral component, and a shock component, and it can be concluded that it is a type I system.

In addition to the proportional servo valve and the hydraulic motor, mathematical models of other components in Figure 6 also need to be established. The proportional amplifier is used together with the proportional servo valve to convert the input voltage signal into a current signal and output it to the proportional servo valve.

In this paper, the rated input voltage of the proportional amplifier is U_0 = 10 V and the rated output current is I_0 = 3 A, then the proportional amplification gain is as follows:

$$K_a = U_0/I_0 = \frac{3}{10} = 0.3\ A/V \quad (7)$$

The transfer function of the proportional servo valve is a typical second-order model, which could be expressed as follows:

$$K_{xv}G_{sv}(s) = Q(s)/I(s) = \frac{K_{xv}}{\frac{s^2}{\omega_{sv}^2} + \frac{2\zeta_{sv}}{\omega_{sv}}s + 1} \tag{8}$$

where K_{xv} is the flow gain of the proportional servo valve, $K_{xv} = \Delta q_n\sqrt{\Delta p/\Delta p_n}/I$; Δq_n is the flow gain under the specified valve port pressure drop of the proportional servo valve, Δq_n = 25 L/min; Δp is the actual valve port pressure drop of the proportional servo valve, Δp = 3.7 MPa; Δp_n is the specified valve port pressure drop for the proportional servo valve, Δp_n = 10 MPa; $G_{sv}(s)$ is the transfer function of the proportional servo valve; Q is the flow of proportional servo valve; I is the input current of the proportional servo valve; ω_{sv} is the natural frequency of the proportional servo valve, ω_{sv} = 502.4Hz; and ζ_{sv} is the damping ratio of the proportional servo valve, with a value range of 0.5~0.7.

Therefore, Equation (8) can be rewritten as follows:

$$K_{xv}G_{sv}(s) = Q(s)/I(s) = \frac{5 \times 10^{-3}}{\frac{s^2}{502.4^2} + \frac{1.2}{502.4}s + 1} \tag{9}$$

The deviation signal input by the proportional amplifier is as follows:

$$\theta_e = \theta_r - \theta_f \tag{10}$$

where θ_r is the control voltage signal and θ_f is the feedback voltage signal.

The feedback signal of the angle sensor is as follows:

$$\theta_f = K_\theta \theta \tag{11}$$

where K_θ is the gain of the angle sensor, K_θ = 1.44 V/rad.

The input current equation of the proportional servo valve is as follows:

$$\Delta I = K_a \theta_e \tag{12}$$

The speed reduction ratio of the worm gear mechanism is i = 80.

The angle control block diagram of the heading angle adjustment circuit can be obtained by combining Equations (5), (7), (8) and (10)–(12), as shown in Figure 8.

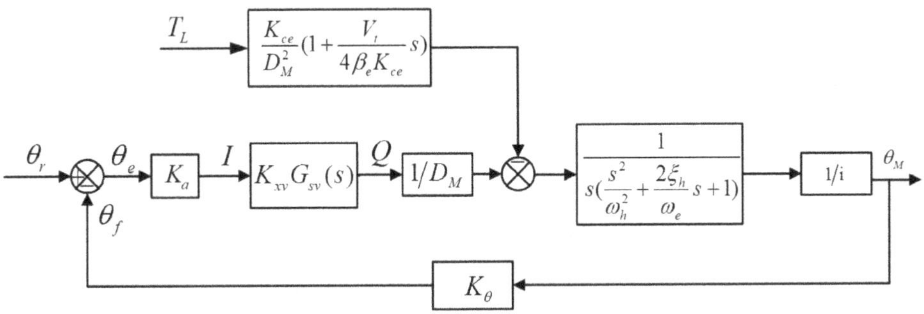

Figure 8. The angle control block diagram of the heading angle adjustment circuit.

Taking the load as interference, when the external load torque T_L is zero, the block diagram of the system can be simplified as Figure 9.

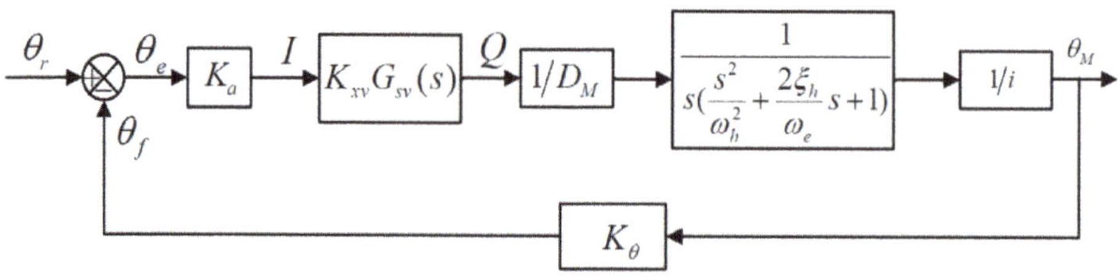

Figure 9. Simplified angle control block diagram of the heading angle adjustment circuit.

Therefore, the open-loop transfer function of the output angle θ_M to the input signal θ_r of the system is as follows:

$$G_K(s) = \frac{K_a K_{xv} G_{sv}(s) \frac{1}{D_{Mi}} K_\theta}{s(\frac{s^2}{\omega_0^2} + \frac{2\zeta_0}{\omega_0}s + 1)} = \frac{1.69}{s(\frac{s^2}{38.2^2} + \frac{0.2}{38.2}s + 1)(\frac{s^2}{502.4^2} + \frac{2\times 0.6}{502.4}s + 1)} \quad (13)$$

4. Simulation of Dynamic Characteristics of the Hydraulic Control System

The open-loop Bode diagram of Equation (13) was calculated using Matlab, as shown in Figure 10. It can be seen from Figure 10 that the phase margin and amplitude margin of the system are 33.9 db and 89.9 db, respectively, so the system meets the stability requirements. The open-loop crossing frequency of the system is 0.476 rad/s, which is much lower than the natural frequency of the hydraulic actuator (38.2 rad/s) and the natural frequency of the proportional servo valve (502.4 rad/s). Therefore, the system bandwidth should be increased to speed up the response of the system. The phase margin of the system is 89.9 rad/s, which is large and has good stability. However, the transient response to interference is slow and the acceleration error of the system will also be large.

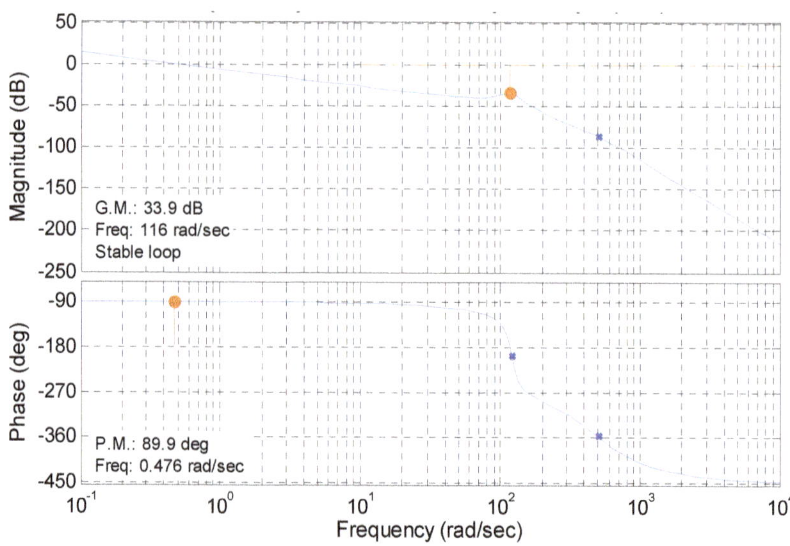

Figure 10. Open-loop Bode diagram of the established model.

In order to ensure the stability of the system, reduce the steady-state error, and improve the response speed, the PID controller was used to improve the dynamic response performance of the system. The module block diagram of the heading angle adjustment circuit was established through Simulink in Matlab, as shown in Figure 11. The PID parameters were optimized using the Simulink Design Optimization tool and the toolbox can select the best controller parameters that meet the requirements through continuous simulation. Finally, we obtained $K_p = 9.5$, $K_i = 0.03$, and $K_d = 0.01$.

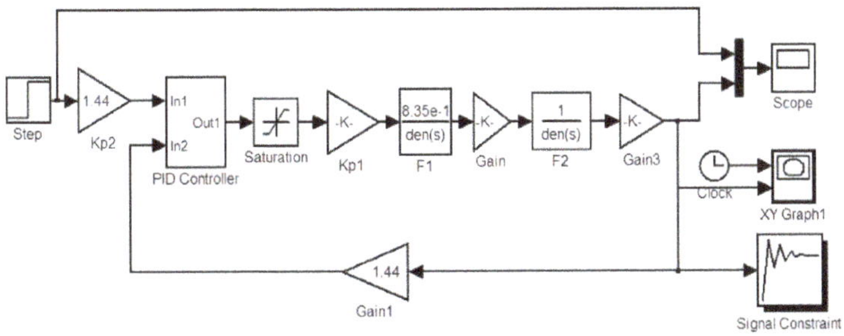

Figure 11. Module block diagram of the heading angle adjustment circuit.

Figure 12 shows the response of the system to the unit step signal with and without the optimized PID control algorithm and the response curve under the pulse interference. The amplitude of the applied pulse interference is 10,000 and the duration time is 0.1 s. It can be seen from Figure 12 that the system with optimized PID has no overshoot, and the time to reach stability is about 1 s. Compared with the system without PID, its response time to the unit step signal is reduced by 9 s. The maximum deviation of the system is about 0.1 at 12.5 s and it takes less than 2 s for the system to return to the stable value at nearly 14 s. Compared with the system without PID, the deviation amplitude of the system from the target value under pulse signal interference is reduced from 0.4 to 0.1, the deviation amplitude is reduced by 75%, the time to recover to the stable value is reduced from 8 s to 2 s, and the time is shortened by 75%. The anti-interference ability of the system is significantly improved after the optimized PID control algorithm is applied.

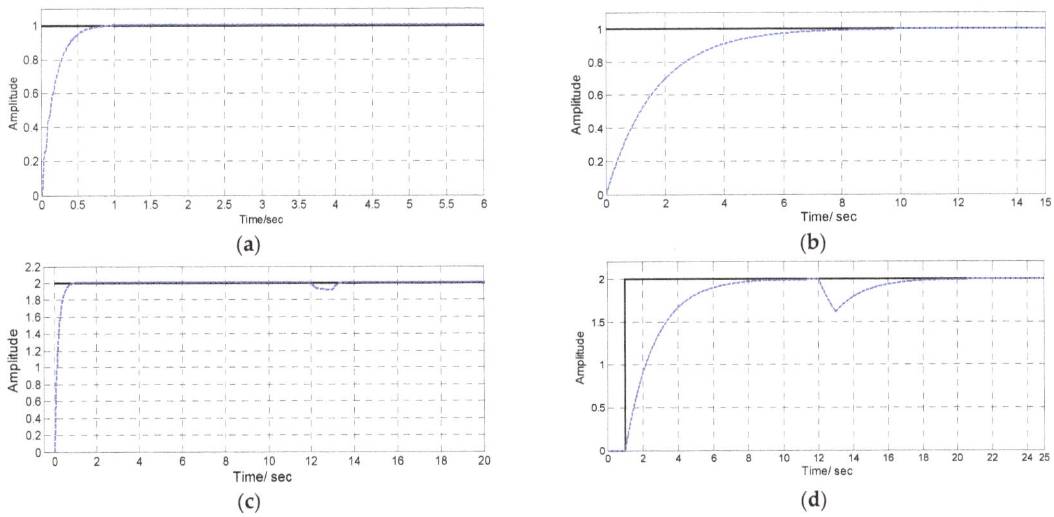

Figure 12. Response of the system to the unit step signal and response curve under the pulse interference: (**a**) unit step response of the system with optimized PID; (**b**) unit step response of the system without PID; (**c**) response curve of the system with optimized PID under pulse interference; (**d**) response curve of the system without PID under pulse interference.

5. Experimental Validation

The experiment of the heading angle adjustment of the docking device was conducted in a water pool to validate the accuracy of the modeling and the effectiveness of the control algorithm. The physical experiment diagram of the docking device in the water pool is shown in Figure 13. The three feet of the docking device are located at the bottom of the pool and the docking device is completely submerged in the water.

Figure 13. Physical experiment diagram of the docking device in the water pool.

In order to test the tracking performance of the system, a sine signal was input to the system, and the equation of the signal is $y = 60\sin(0.05t)$. The response curve and control voltage curve of the system are shown in Figures 14 and 15, respectively. It can be seen from Figures 14 and 15 that, in the first cycle, the maximum positive error is 5° (at 10 s), the corresponding control voltage is 0.25 V, the maximum negative error is 5° (at 61 s), and the corresponding control voltage is −0.3 V. In the second cycle, the tracking error of the system increases. The maximum positive error is 10° (at 130 s), the corresponding control voltage is 0.45 V, the maximum positive error is 10° (at 175 s), and the corresponding control voltage is −0.45 V. The maximum error of the angle tracking of the system is 8.3% under the excitation of the above sinusoidal signal. As a whole, the angle tracking error of the system is always small, which shows that the model of the hydraulic circuit and the control system established in this paper has achieved high accuracy, and the accuracy of modeling and the effectiveness of the control algorithm were validated. The performance of the docking device and the effectiveness of the control algorithm in the deep-sea environment need to be further verified.

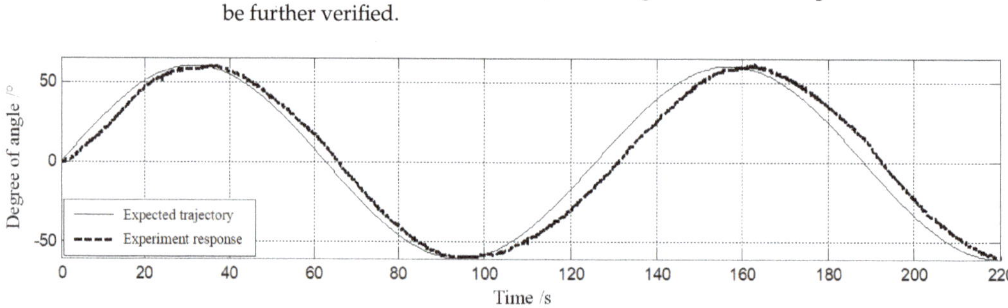

Figure 14. The expected trajectory and experiment response of the system.

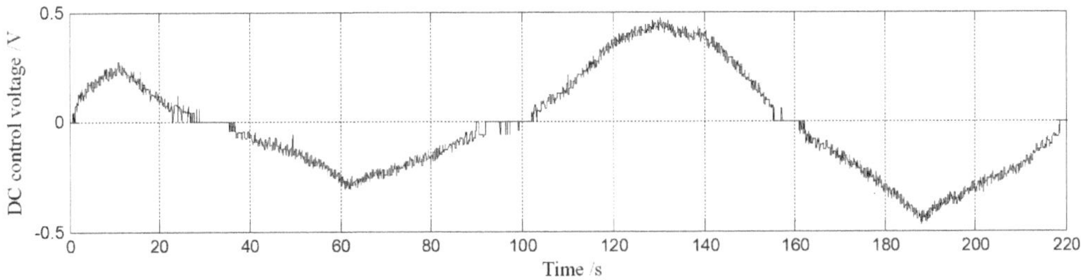

Figure 15. Control voltage of the proportional servo valve.

6. Conclusions

In this paper, a funnel-shaped underwater docking device with multi-degree freedom based on a submarine platform was designed. In order to realize the timely adjustment of the heading angle, pitch angle, and roll angle of the docking device, a set of underwater hydraulic system was developed as the power source to drive the corresponding hydraulic cylinders, hydraulic motors, and other executive components. The model of the heading angle adjustment circuit of the hydraulic system was established and the open-loop transfer function of the circuit was derived. In order to ensure the stability of the system, reduce the steady-state error and improve the response speed, the optimized PID controller was employed to increase the system's dynamic response performance. The simulation results demonstrate that the anti-interference ability of the system has been greatly increased as the optimized PID control algorithm is applied. To validate the accuracy of the modeling and the effectiveness of the control algorithm, the experiment of the heading angle adjustment of the docking device was carried out in a water pool. The maximum error of the angle tracking of the system is 8.3% under the excitation of the sinusoidal signal. It was concluded from the experiment results that the angle tracking error of the system is small, showing that the model of the hydraulic circuit and the control system established has achieved high accuracy and the control algorithm is effective.

In future research, we will focus on the research of the collision between the AUV and the docking device during the docking process and the AUV docking control strategy, in order to achieve more accurate underwater docking.

Author Contributions: Conceptualization, X.D. and C.Z.; software, B.Z.; validation, X.D. and C.Z.; formal analysis, X.D.; investigation, X.D. and B.Z.; writing—original draft preparation, X.D. and B.Z.; writing—review and editing, M.S. and C.Z. All authors have read and agreed to the published version of the manuscript.

Funding: This research was funded by the National Key Research and Development Program of China (grant number 2019YFB2006402) and Scientific Research Foundation for High-level Talents of Nanjing Institute of Technology (grant number YKJ202102).

Institutional Review Board Statement: Not applicable.

Informed Consent Statement: Not applicable.

Data Availability Statement: Not applicable.

Conflicts of Interest: The authors declare no conflict of interest.

References

1. Stokey, R.; Allen, B.; Austin, T.; Goldsborough, R.; Forrester, N.; Purcell, M.; Von Alt, C. Enabling technologies for REMUS docking: An integral component of an autonomous ocean-sampling network. *IEEE J. Ocean. Eng.* **2001**, *26*, 487–497. [CrossRef]
2. Stokey, R.; Purcell, M.; Forrester, N.; Austin, T.; Goldsborough, R.; Allen, B.; von Alt, C. A docking system for REMUS, an autonomous underwater vehicle. In Proceedings of the Oceans' 97. MTS/IEEE Conference Proceedings, Halifax, NS, Canada, 6–9 October 2017; Volume 2, pp. 1132–1136.
3. Feezor, M.D.; Sorrell, F.Y.; Blankinship, P.R.; Bellingham, J.G. Autonomous underwater vehicle homing/docking via electromagnetic guidance. *IEEE J. Ocean. Eng.* **2001**, *26*, 515–521. [CrossRef]

4. Cowen, S.; Briest, S.; Dombrowski, J. Underwater docking of autonomous undersea vehicles using optical terminal guidance. In Proceedings of the Oceans' 97. MTS/IEEE Conference Proceedings, Halifax, NS, Canada, 6–9 October 2017; Volume 2, pp. 1143–1147.
5. Singh, H.; Bellingham, J.G.; Hover, F.; Lemer, S.; Moran, B.A.; Von der Heydt, K.; Yoerger, D. Docking for an autonomous ocean sampling network. *IEEE J. Ocean. Eng.* **2001**, *26*, 498–514. [CrossRef]
6. McEwen, R.S.; Hobson, B.W.; McBride, L.; Bellingham, J.G. Docking control system for a 54-cm-diameter (21-in) AUV. *IEEE J. Ocean. Eng.* **2008**, *33*, 550–562. [CrossRef]
7. Jantapremjit, P.; Wilson, P.A. Optimal control and guidance for homing and docking tasks using an autonomous underwater vehicle. In Proceedings of the 2007 International Conference on Mechatronics and Automation, Harbin, China, 5–9 August 2007; pp. 243–248.
8. Jantapremjit, P.; Wilson, P.A. Guidance-control based path following for homing and docking using an autonomous underwater vehicle. In Proceedings of the OCEANS 2008-MTS/IEEE Kobe Techno-Ocean, Kobe, Japan, 8–11 April 2008; pp. 1–6.
9. Park, J.Y.; Jun, B.H.; Lee, P.M.; Oh, J. Experiments on vision guided docking of an autonomous underwater vehicle using one camera. *Ocean. Eng.* **2009**, *36*, 48–61. [CrossRef]
10. Park, J.Y.; Jun, B.H.; Lee, P.M.; Lim, Y.K.; Oh, J.H. Docking problem and guidance laws considering drift for an underactuated AUV. In Proceedings of the OCEANS 2011 IEEE-Spain, Santander, Spain, 6–9 June 2011; pp. 1–7.
11. Raspante, F. Underwater mobile docking of autonomous underwater vehicles. In Proceedings of the 2012 Oceans, Hampton Roads, VA, USA, 14–19 October 2012; pp. 14–19.
12. Rigaud, V.; Nicolas-Meunier, B. From manned to autonomous and hybrid underwater systems. In Proceedings of the 2015 IEEE Underwater Technology (UT), Chennai, India, 23–25 February 2015; pp. 1–10.
13. Batista, P.; Silvestre, C.; Oliveira, P. A two-step control approach for docking of autonomous underwater vehicles. *Int. J. Robust Nonlinear Control* **2015**, *25*, 1528–1547. [CrossRef]
14. Ferreira, B.M.; Matos, A.C.; Cruz, N.A.; Moreira, A.P. Homing a robot with range-only measurements under unknown drifts. *Robot. Auton. Syst.* **2015**, *67*, 3–13. [CrossRef]
15. Li, Y.; Jiang, Y.; Cao, J.; Wang, B.; Li, Y. AUV docking experiments based on vision positioning using two cameras. *Ocean Eng.* **2015**, *110*, 163–173. [CrossRef]
16. Yang, C.; Peng, S.; Fan, S.; Zhang, S.; Wang, P.; Chen, Y. Study on docking guidance algorithm for hybrid underwater glider in currents. *Ocean Eng.* **2016**, *125*, 170–181. [CrossRef]
17. Sans-Muntadas, A.; Pettersen, K.Y.; Brekke, E.; Henriksen, V.F. A hybrid approach to underwater docking of AUVs with cross-current. In Proceedings of the OCEANS 2016 MTS/IEEE Monterey, Monterey, CA, USA, 19–23 September 2016; pp. 1–7.
18. Piskura, J.C.; Purcell, M.; Stokey, R.; Austin, T.; Tebo, D.; Christensen, R.; Jaffre, F. Development of a robust Line Capture, Line Recovery (LCLR) technology for autonomous docking of AUVs. In Proceedings of the OCEANS 2016 MTS/IEEE Monterey, Monterey, CA, USA, 19–23 September 2016; pp. 1–5.
19. Sarda, E.I.; Dhanak, M.R. Launch and recovery of an autonomous underwater vehicle from a station-keeping unmanned surface vehicle. *IEEE J. Ocean. Eng.* **2018**, *44*, 290–299. [CrossRef]
20. Hurtos, N.; Mallios, A.; Palomeras, N.; Bosch, J.; Vallicrosa, G.; Vidal, E.; Ribas, D.; Gracias, N.; Carreras, M.; Ridao, P. LOON-DOCK: AUV homing and docking for high-bandwidth data transmission. In Proceedings of the OCEANS 2017-Aberdeen, Aberdeen, UK, 19–22 June 2017; pp. 1–7.
21. Palomeras, N.; Vallicrosa, G.; Mallios, A.; Bosch, J.; Vidal, E.; Hurtos, N.; Carreras, M.; Ridao, P. AUV homing and docking for remote operations. *Ocean Eng.* **2018**, *154*, 106–120. [CrossRef]
22. Zuo, M.; Wang, G.; Xiao, Y.; Xiang, G. A unified approach for underwater homing and docking of over-actuated AUV. *J. Mar. Sci. Eng.* **2021**, *9*, 884. [CrossRef]
23. Lin, R.; Zhao, Y.; Li, D.; Lin, M.; Yang, C. Underwater Electromagnetic Guidance Based on the Magnetic Dipole Model Applied in AUV Terminal Docking. *J. Mar. Sci. Eng.* **2022**, *10*, 995. [CrossRef]
24. Lin, M.; Lin, R.; Yang, C.; Li, D.; Zhang, Z.; Zhao, Y.; Ding, W. Docking to an underwater suspended charging station: Systematic design and experimental tests. *Ocean Eng.* **2022**, *249*, 110766. [CrossRef]
25. Anderlini, E.; Parker, G.G.; Thomas, G. Docking control of an autonomous underwater vehicle using reinforcement learning. *Appl. Sci.* **2019**, *9*, 3456. [CrossRef]
26. Palomeras, N.; Peñalver, A.; Massot-Campos, M.; Negre, P.L.; Fernández, J.J.; Ridao, P.; Sanz, P.J.; Oliver-Codina, G. I-AUV docking and panel intervention at sea. *Sensors* **2016**, *16*, 1673. [CrossRef] [PubMed]
27. Zhang, T.; Li, D.; Yang, C. Study on impact process of AUV underwater docking with a cone-shaped dock. *Ocean Eng.* **2017**, *130*, 176–187. [CrossRef]
28. Li, D.J.; Chen, Y.H.; Shi, J.G.; Yang, C.J. Autonomous underwater vehicle docking system for cabled ocean observatory network. *Ocean Eng.* **2015**, *109*, 127–134. [CrossRef]
29. Roy, S.; Shome, S.N.; Nandy, S.; Ray, R.; Kumar, V. Trajectory following control of AUV: A robust approach. *J. Inst. Eng. (India) Ser. C* **2013**, *94*, 253–265. [CrossRef]
30. Roy, S.; Nandy, S.; Shome, S.N.; Ray, R. Robust position control of an autonomous underwater vehicle: A comparative study. In Proceedings of the 2013 IEEE International Conference on Automation Science and Engineering (CASE), Madison, WI, USA, 17–20 August 2013; pp. 1002–1007.

Journal of
Marine Science and Engineering

MDPI

Article

Finite-Time Extended State Observe Based Fault Tolerant Control for Autonomous Underwater Vehicle with Unknown Thruster Fault

Xiaofeng Liu, Mingjun Zhang, Xing Liu * and Wende Zhao

College of Mechanical and Electrical Engineering, Harbin 150001, China
* Correspondence: liuxing20080724@gmail.com

Abstract: This paper investigates the problem of fault tolerant control (FTC) for autonomous underwater vehicles (AUVs) with multiple thrusters in the presence of current disturbances, thruster faults, and modelling uncertainty. This paper focuses on the problems of reducing the energy consumption caused by the chattering of control signals and improving the tracking accuracy of an AUV operating in deep-sea environments. In view of the problem of large energy consumption in some other methods, a fault tolerant control method for multiple-thruster AUVs based on a finite-time extended state observer (FTESO) is proposed. More specifically, a FTESO based on an integral sliding mode surface is designed to estimate the generalized uncertainty compounded using current disturbances, thruster faults, and modelling uncertainty. The fast finite-time uniformly ultimately bounded stability of the proposed FTESO is analyzed. Then, based on the estimated value of FTESO, an FTC method based on non-singular fast terminal sliding mode surfaces is developed for AUVs. The finite-time convergence of the closed-loop control system is proved theoretically. In this design, two different sliding mode surfaces are used to design FTESO and FTC, in order to avoid the appearance of singularities. Moreover, a parameter adjustment method is designed to improve tracking accuracy. Finally, comparative numerical simulations show that the proposed control scheme is effective at reducing energy consumption and improving tracking accuracy.

Keywords: autonomous underwater vehicles; fault tolerant control; finite-time extended state observer; nonsingular fast terminal sliding mode; parameter adjustment

Citation: Liu, X.; Zhang, M.; Liu, X.; Zhao, W. Finite-Time Extended State Observe Based Fault Tolerant Control for Autonomous Underwater Vehicle with Unknown Thruster Fault. *J. Mar. Sci. Eng.* **2022**, *10*, 1624. https://doi.org/10.3390/jmse10111624

Academic Editor: Sergei Chernyi

Received: 19 September 2022
Accepted: 28 October 2022
Published: 2 November 2022

1. Introduction

Because of the great advantages in terms of operating automatically in the deep-sea environments, autonomous underwater vehicles (AUVs) equipped with multiple thrusters play an important role in many tasks, including the detection of marine resources, seafloor mapping, and crashed aircraft search [1–3]. As AUVs operate in the unknown and complicated deep-sea environment, safety is one of the important issues to be considered [4]. Fault tolerant control (FTC) is the key technology to ensure the safety of AUVs [5]. Thrusters are the main power consumers on an AUV, and they are also the components that experience the heaviest loading and are thus the most prone to faults [6]. The failure of a thruster directly affects the safety of an AUV operating in deep areas [7]. Therefore, fault tolerant control for AUV thrusters is one of the current research hotspots in this field [6]. This paper focuses on fully actuated AUVs. Fully actuated AUVs equipped with manipulators have unique capabilities to complete missions, such as grasping or cutting [8].

Nowadays, two main FTC methods exist, that is, passive fault tolerant control (PFTC) and active fault-tolerant control (AFTC) [9]. The former can deal with some classes of faults specified and considered during the controller design process, but the fault tolerant control performance is poor for faults that are not of the specified class. As for the AFTC, it has attracted much attention in the past decades because of its feasibility and effectiveness to deal with unknown faults [10].

In the field of AUVs, AFTC is mainly divided into two types: AFTC based on thrust reallocation, and adaptive FTC [11]. The former is based on the premise of known fault information. However, in a complex marine environment, the dynamic performance of an AUV varies with the environment and is disturbed by random currents, so it is difficult to obtain accurate fault diagnosis results sufficiently quickly to compensate for the fault. Thus, this method has limitations in AUV fault tolerant control [12]. The adaptive FTC approach treats faults as a generalized uncertainty and then uses adaptive strategies to estimate its upper bounds along with its disturbances [13]. It is also suitable for vehicles such as AUVs that move in complex environments [14]. Adaptive FTC has developed rapidly and achieved good research results in recent years [15,16]. For AUVs subject to ocean current disturbances and modelling uncertainty, researchers have investigated a virtual closed-loop adaptive FTC method, which avoids serious chattering phenomena in control outputs. Su et al. [17] proposed an event-triggered adaptive FTC scheme, which solves the trajectory tracking control problem for AUVs with actuator faults and model uncertainties.

In the fault-tolerant control of AUVs, there are problems with system dynamic model uncertainty and external disturbances [16]. The adaptive FTC method based on observer estimation is suitable for moving vehicles in complex environments. The observer has a good effect in dealing with nonlinear dynamics due to disturbances and faults [18,19], and the so-called extended state observer (ESO) has been considered as an effective way to actively compensate for uncertainty and disturbances [20,21]. The key idea of ESO is that system uncertainties and disturbances are considered as an added or extended state of the system, and then all the states including the extended one will be estimated accurately and quickly by the ESO.

In recent years, ESO has been effectively applied in many fields [22]. In [22], a nonsingular fast terminal sliding mode control (NFTSMC) based on a third-order fast finite-time extended state observer was proposed for the trajectory tracking of AUVs with various hydrodynamic uncertainties and external disturbances. Specifically, ref. [22] constructed the extended state variable according to the velocity information and constructed a third-order FFTESO to realize the finite-time estimation of the extended state variable. Another typical example is that Li proposed robust fault-tolerant control for a spacecraft system based on a finite-time ESO (FTESO) [23], where the FTESO and controller design shared a non-singular terminal sliding mode surface (NTSMS). The proposed control scheme was continuous with the property of restraining chattering. However, in the research on fault-tolerant control of fully actuated AUV thrusters based on the above methods [23], it was found that the control law derived theoretically from the AUV had singularities, and from the experimental results, the control output chattering was serious, which would lead to energy consumption increases.

Motivated by the mentioned-above discussion, this paper studies the FTC problem based on FTESO for a fully actuated AUV in the presence of current disturbances, thruster faults, and dynamic model uncertainty. In this paper, an improved FTC method based on FTESO is proposed, which aims to reduce the energy consumption caused by the chattering of the control signal under the premise of ensuring a high tracking accuracy. The main contributions of this paper are as follows.

(1) In a previous approach [23], the design of FTESO and the controller share an NTSMS, but this paper uses two sliding mode surfaces respectively for the FTESO and the controller. In the design of FTESO in this paper, the integral sliding mode surface (ISMS) was used to replace the NTSMS, so that the coefficient of the generalized uncertainty term after the derivation of the sliding mode surface was 1, that is, the generalized uncertainty term was directly used in the design of the control law. The estimated value of the uncertain item replaces the real value. This is different to the approach of [23], which requires the estimated value to remove the fractional power term of the velocity error to obtain the real value. In particular, when the velocity error term tends to zero, the method in this paper effectively suppresses the fluctuation of the generalized uncertainty term and avoids the appearance of singular points.

(2) To design an FTESO based on ISMS, this paper replaces NTSMS with a non-singular fast terminal sliding mode surface (NFTSMS) in the design of a fault-tolerant controller. That is, the generalized uncertainty term estimated by FTESO is applied to the newly designed NFTSMS to derive the control law. This avoids the pseudo-inverse term of the velocity error with the fractional power that appears in the control gain of the method in [23], thus, it can effectively suppress the chattering of the control variable, thereby reducing the energy consumption.

(3) The FTESO-based fault-tolerant control (FTESO − FTC) method proposed in this paper has a good effect in reducing energy consumption, but it has the problem of reducing the tracking accuracy. Therefore, this paper adds a parameter adjustment method (PAM).

The remainder of this paper is structured as follows. Section 2 presents a mathematical model, notations, and some useful lemmas. In Section 3, the ISMS-based FTESO is designed and analyzed. The NFTSMS-based FTC design and convergence analysis and the PAM for AUV thruster control are described in Section 4. In Section 5, comparative simulations through the ODIN AUV were carried out to validate the effectiveness of the proposed method. Finally, some concluding statements are given in Section 6.

2. Mathematical Models and Preliminaries

The fault-tolerant AUV controller is obtained based on the AUV dynamic model. In this paper, the dynamic model of the AUV disturbed by the ocean current is described, then the thruster fault and the uncertainty of the AUV model are considered in turn, and finally the disturbed AUV dynamic model is converted into the AUV state space equation required for this paper.

2.1. Model Description

According to existing AUV research results, the dynamic model of an AUV disturbed by ocean currents in the body-fixed coordinate system [24] can be described as follows:

$$\begin{aligned} &\dot{\eta} = J(\eta)v \\ &M(v)\dot{v} + C_{RB}(v)v + g(\eta) + C_A(v_r)v_r + D(v_r)v_r = \tau \end{aligned} \tag{1}$$

where $J(\eta)$ is the rotation transformation matrix between the earth coordinate frame and the body-fixed coordinate frame; M is the inertia matrix including the added mass; C_{RB} is the rigid body centripetal force and Coriolis force matrix; C_A is the hydrodynamic centripetal force and Coriolis force matrix; D is the hydrodynamic resistance matrix; g is the restoring force matrix; τ is the control force and moment acting on the center of gravity of the AUV; $v = [u\ v\ w\ p\ q\ r]^T$ is the velocity vector relative to the body-fixed coordinate system; $\eta = [x\ y\ z\ \varphi\ \theta\ \psi]^T$ is relative to the position and attitude vector of the boat in earth coordinate system; and $v_r = v - v_c$, v_c is the current velocity relative to the body-fixed coordinate system.

In this paper, the fault-tolerant controller is designed in terms of the state quantities in the earth-fixed coordinate system. Therefore, the above AUV dynamic model needs to be converted into the description in the earth coordinate system. The results are expressed as follows:

$$M_\eta(\eta)\ddot{\eta} + C_{RB\eta}(\eta,\dot{\eta})\dot{\eta} + C_{A\eta}(\eta_r,\dot{\eta}_r)\dot{\eta}_r + D_\eta(\eta_r,\dot{\eta}_r)\dot{\eta}_r + g_\eta(\eta) = J^{-T}\tau \tag{2}$$

where $M_\eta(\eta) = J^{-T}MJ^{-1}$; $C_{RB\eta}(\eta,\dot{\eta}) = J^{-T}[C_{RB}(q) - MJ^{-1}\dot{J}]J^{-1}$; $C_{A\eta}(\eta_r,\dot{\eta}_r) = J^{-T}C_A(q_r)J^{-1}$; $D_\eta(\eta_r,\dot{\eta}_r) = J^{-T}D(q_r)J^{-1}$; $g_\eta(\eta) = J^{-T}g$; J^{-T} means inverting matrix J and then transposing it; $\dot{\eta}_r = \dot{\eta} - V_c$; V_c is the current vector in the earth coordinate system.

Equation (2) can be simplified as follows:

$$\ddot{\eta} = M_\eta^{-1}(\eta)\left(J^{-T}\tau - C_{RB\eta}(\eta,\dot{\eta})\dot{\eta} - C_{A\eta}(\eta_r,\dot{\eta}_r)\dot{\eta}_r - D_\eta(\eta_r,\dot{\eta}_r)\dot{\eta}_r - g_\eta(\eta)\right) \tag{3}$$

2.2. AUV Dynamic Model with Thruster Fault

This paper studies a thruster fault-tolerant control method based on an AUV dynamic model that considers the thruster fault and the uncertainty of the dynamic model. To make the logic of this paper complete, the following will briefly describe the mathematical expressions for the thruster fault and the uncertainty of the dynamic model.

2.2.1. Thruster Fault

Typical forms of thruster fault include blade winding, blade damage, jamming, and other faults. They are approximately equivalent to the condition that the thrust provided by the faulty thruster is less than that of the no-fault thruster [25]. The actual output of the thruster under fault conditions can be expressed as follows:

$$u' = u - u_f \tag{4}$$

where u is the expected output of the thruster, and u_f is the influence of the thruster fault.

Simulations of thruster faults often adopt the method of soft simulation. They generally introduce a diagonal coefficient matrix K on the control output, the element $k_i \in [0,1]$, that is, the thruster fault can be expressed as: $u_f = Ku$. When $k_i = 0$, it means that the i-th thruster has no fault; when $k_i = 1$, it means that the i-th thruster fails completely; intermediate values $k_i \in (0,1)$ indicate partial failure of the thruster.

When the thruster fails, the actual output of the thruster acts on the AUV, and the control moments and forces acting on the center of gravity of the vehicle can be calculated through the thruster configuration matrix B, described as follows:

$$\tau' = Bu' = Bu - Bu_f = Bu - BKu = (B - BK)u \tag{5}$$

where τ' the vector of control moments and forces acting on the center of gravity of the vehicle, and B is the thruster configuration matrix, which is a constant matrix.

2.2.2. Model Uncertainty

Because of the strong nonlinearity and mutual coupling characteristics of each degree of freedom of the AUV, the motion model of the AUV derived from the dynamic analysis method has great uncertainty. This paper considers the existence of model uncertainty in the modeling process, shown as follows:

$$M_\eta = \hat{M}_\eta + \Delta M_\eta; \quad C_\eta = \hat{C}_\eta + \Delta C_\eta; \quad D_\eta = \hat{D}_\eta + \Delta D_\eta; \quad g_\eta = \hat{g}_\eta + \Delta g_\eta \tag{6}$$

where the symbol ^ represents the estimated value of the variable, the symbol Δ represents the modeling uncertainty of the variable, and $C_\eta = C_{RB\eta} + C_{A\eta}$, $D_\eta = D_{\eta q} + D_{\eta c}$.

Based on these definitions, the AUV dynamic model is converted into the AUV state space equation to prepare for the subsequent fault-tolerant controller design.

The AUV state space equation derived from Equations (3)–(6) is described as follows:

$$\ddot{\eta} = \hat{M}_\eta^{-1}\left[J^{-T}Bu - \hat{C}_\eta\dot{\eta} - \hat{D}_\eta\dot{\eta} - \hat{g}_\eta\right] + d \tag{7}$$

where $d = -\hat{M}_\eta^{-1}(\Delta M_\eta\ddot{\eta} + \Delta C_\eta\dot{\eta}_r + \Delta D_\eta\dot{\eta}_r + D_{\eta q}V_c + D_{\eta c}\dot{\eta}_r + \Delta g_\eta) + \hat{M}_\eta^{-1}J^{-T}BKu$, d represents the lumped uncertainty composed of current disturbances, thruster faults, and dynamic model uncertainty. Here, d is supposed to be unknown but bounded, that is, there exists a positive constant d_1, such that $\|d\| \leq d_1$.

The above AUV state space equation can be described in the form of a second-order system, by defining new auxiliary variables as $x_1 = \eta$ and $x_2 = \dot{\eta}$, then Equation (7) can be rewritten as follows:

$$\dot{x}_1 = x_2 \tag{8}$$

$$\dot{x}_2 = \hat{M}_\eta^{-1}\left[J^{-T}Bu - \hat{C}_\eta\dot{\eta} - \hat{D}_\eta\dot{\eta} - \hat{g}_\eta\right] + d \tag{9}$$

The tracking error in this paper is defined as follows:

$$e_3 = x_1 - x_d \quad (10)$$

where x_1 represents the true position and attitude vector, and x_d represents the desired position and attitude vector.

Then, the first derivative of the tracking error is $\dot{e}_3 = \dot{x}_1 - \dot{x}_d = x_2 - \dot{x}_d$, and the second derivative is $\ddot{e}_3 = \dot{x}_2 - \ddot{x}_d$.

2.3. Notation and Preliminaries

The following notations and related theorems are used in the subsequent theoretical derivation. To make this paper more organized, the definitions of relevant notations and theorems are given here first.

2.3.1. Notations

The following notations are utilized for simplicity:

$$\langle x \rangle^{\alpha} = sign(x)|x|^{\alpha}$$

$$|x|^{\alpha} = \left[|x_1|^{\alpha}, |x_2|^{\alpha}, \ldots, |x_n|^{\alpha}\right]^T$$

where $x = [x_1, x_2, \ldots, x_n]^T$, $\alpha \in R$, and $sign(\cdot)$ is a sign function with $sign(0) = 0$. Particularly, $\langle x \rangle^0 = sign(x)$, $\langle x \rangle^0 |x|^{\alpha} = \langle x \rangle^{\alpha}$.

2.3.2. Lemmas and New Propositions

Consider the following nonlinear system:

$$\dot{x} = F(x(t)), x(0) = x_0, F(0) = 0, x \in R^n \quad (11)$$

where $F : U \to R^n$ is continuous on an open neighborhood U of the origin. Suppose that the system in (11) has a unique solution in forwarding time for all of the initial conditions.

Lemma 1 ([26]). *Consider the above nonlinear system (11), and suppose there exists a Lyapunov function $V(x)$ defined on the neighborhood $U \subset R^n$ of the origin, and $\dot{V}(x) + \beta_1 V(x)^{\alpha_1} < 0$, where $x \in U/\{0\}, 0 < \alpha_1 < 1, \beta_1 > 0$. Then, the system is locally finite-time stable, and the time required to reach $V(x) = 0$ is $T \leq \frac{1}{\beta_1(1-\alpha_1)}|V(x_0)|^{1-\alpha_1}$.*

Lemma 2 ([27]). *Consider System (11) and suppose there exists a Lyapunov candidate function $V(x)$, where $V(x_0)$ represents its initial value. (1) If $\dot{V}(x) \leq -\beta_1 V(x)^{\alpha_1} + \beta_2 V(x)^{\alpha_2}$, for $\alpha_1 > \alpha_2$, $\beta_1 > 0$, $\beta_2 > 0$, $\theta_1 \in (0, \beta_1)$, then the trajectory of the above nonlinear system is finite-time uniform and eventually bounded stable within the range of $Q_1 = \left\{x \middle| V(x)^{\alpha_1 - \alpha_2} < \frac{\beta_2}{\theta_1}\right\}$, and the stable time T_1 to reach the state of the stable residual set satisfies: $T_1 \leq \frac{V(x_0)^{1-\alpha_1}}{(\beta_1-\theta_1)(1-\alpha_1)}$. (2) If $\dot{V}(x) \leq -\beta_1 V(x)^{\alpha_1} - \beta_2 V(x) + \beta_3 V(x)^{\alpha_2}$, for $\beta_3 > 0$, then the trajectory of the system is fast and finally bounded and stable within the finite time T, and the convergence time T_2 is bounded: $T_2 \leq \frac{\ln\left[1+(\beta_2-\theta_2)V(x_0)^{1-\alpha_1}/(\beta_1-\theta_1)\right]}{(\beta_2-\theta_2)(1-\alpha_1)}$, with $\theta_2 \in (0, \beta_2)$. Then, it can be calculated that the stable residual region of the above nonlinear system is $Q_2 = \left\{x \middle| \theta_1 V(x)^{\alpha_1-\alpha_2} + \theta_2 V(x)^{1-\alpha_2} < \beta_3\right\}$.*

Lemma 3 ([26]). *Assume $V_1(x)$ and $V_2(x)$ are continuous real-valued functions, which are homogeneous of degrees $l_1 > 0$ and $l_2 > 0$, respectively, with respect to weight $(\gamma_1, \gamma_2, \ldots, \gamma_n)$, and $V_1(x)$ is positive definite. Then, for any $x \subset R^n$, there exists $c_1 V_1(x)^{l_2/l_1} \leq V_2(x) \leq c_2 V_1(x)^{l_2/l_1}$, with $c_1 = \min_{\{z; V_1(z)=1\}} V_2(z), c_2 = \max_{\{z; V_1(z)=1\}} V_2(z)$.*

3. FTESO Design Based on ISMS

This section explains the design ideas and implementation process of the FTESO in this paper, and then analyzes the convergence of the FTESO proposed in this paper.

3.1. Design Ideas and Implementation Process of FTESO

To compensate for the generalized uncertainty term combining ocean current disturbances, AUV thruster faults, and dynamic model uncertainty, an FTESO method based on integral sliding mode surface (ISMS) is proposed in this paper. Firstly, ISMS is constructed based on the tracking error. Then, the control system of the AUV is obtained by derivation of the established sliding mode surface. Finally, an extended state observer is used to estimate the generalized uncertainty to obtain the FTESO in this paper.

First, this paper selects an ISMS as follows [28]:

$$S_1 = \dot{e}_3 + K\int e_3 dt \tag{12}$$

where S_1 is the designed integral sliding mode surface, e_3 is the tracking error, $\dot{e}_3$ is the first derivative of the tracking error, and K is a constant value parameter. The meanings of the relevant symbols in the formulas are shown in Section 2.

By differentiation of the ISMS in Equation (12), noting that $\ddot{e}_3 = \dot{x}_2 - \ddot{x}_d$, we find that the state Equation (7) becomes the following:

$$\dot{S}_1 = \hat{M}_\eta^{-1}\left(J^{-T}Bu - \hat{C}_\eta\dot{\eta} - \hat{D}_\eta\dot{\eta} - \hat{g}_\eta\right) + d - \ddot{x}_d + Ke_3 \tag{13}$$

To simplify Equation (13), we define $Q = \hat{M}_\eta^{-1}[-\hat{C}_\eta\dot{\eta} - \hat{D}_\eta\dot{\eta} - \hat{g}_\eta] - \ddot{x}_d + Ke_3$, $R = \hat{M}_\eta^{-1}J^{-T}Bu$, $W = d$.

Using the above-mentioned notation, Equation (14) is given by the following:

$$\dot{S}_1 = Q + Ru + W \tag{14}$$

To use the ESO technique to estimate and compensate for the lumped disturbances or uncertainties, a new state variable is defined as $z_1 = S_1$, and an extended state variable $z_2 = W$ is defined with $\dot{z}_2 = \dot{W} = j(t)$. We suppose $j(t)$ is unknown but bounded, that is, there is a positive constant $\bar{j}$, such that $\|j(t)\| < \bar{j}$.

Then, the mathematic model of the AUV control system governed by (14) will be extended as follows:

$$\dot{z}_1 = \dot{S}_1 = Q + Ru + z_2 \tag{15}$$

$$\dot{z}_2 = j(t) \tag{16}$$

Let $\hat{z}_1$ and $\hat{z}_2$ be the observation values of z_1 and z_2 in the above extended system, respectively, and denote $e_1 = \hat{z}_1 - z_1$ as the observation error of the integral sliding mode surface $z_1 = S_1$.

The fast finite-time extended state observer (FTESO) is designed as follows:

$$\dot{\hat{z}}_1 = Q + Ru + \hat{z}_2 - \beta_1(\langle e_1\rangle^{\alpha_1} + e_1) \tag{17}$$

$$\dot{\hat{z}}_2 = -\beta_2(\langle e_1\rangle^{\alpha_2} + 2\langle e_1\rangle^{\alpha_1} + e_1) \tag{18}$$

where the gain parameters satisfy $\beta_1 > 0$, $\beta_2 > 0$, $\frac{1}{2} < \alpha_1 < 1$ and $\alpha_2 = 2\alpha_1 - 1$.

As a part of the proposed FTESO, $\hat{z}_2$ is utilized to estimate the generalized uncertainty term W in (14). Although the actual value of z_2 is unavailable, its observation value $\hat{z}_2$ can be obtained by the above FTESO. The analysis and proof that $\hat{z}_2$ perfectly tracks the actual fault z_2 in finite time is described below.

3.2. FTESO Convergence Analysis

Based on the above-constructed observer, the following will analyze the convergence of the observer's estimation error according to the Lyapunov theory.

Define another auxiliary variable $e_2 = \hat{z}_2 - z_2$. Combined with Equations (15)–(18), the dynamic error of the proposed FTESO can be obtained as follows:

$$\dot{e}_1 = e_2 - \beta_1\left(\langle e_1\rangle^{\alpha_1} + e_1\right) \tag{19}$$

$$\dot{e}_2 = -\beta_2\left(\langle e_1\rangle^{\alpha_2} + 2\langle e_1\rangle^{\alpha_1} + e_1\right) - \dot{j}(t) \tag{20}$$

The stability and convergence of the proposed observer (17) and (18) are stated as the following theorem.

Theorem 1. *Considering the AUV state space equation in (7) under the assumptions, design FTESO as (17) and (18), and select an appropriate gain to satisfy the constraints, then the observation error* $e = \left[e_1^T, e_2^T\right]^T$ *converges to the following region* Z_1 *in a finite time* T_1:

Proof of Theorem 1. First introduce another auxiliary state variable:

$$\bar{\varepsilon} = \left[\left(\langle e_1\rangle^{\alpha_1} + e_1\right)^T, {e_2}^T\right]^T \tag{21}$$

Obviously, if the new auxiliary state variables $\bar{\varepsilon}$ converge to the origin in a finite time, the observation errors e_1 and e_2 will also converge to the origin in a finite time.

According to Lemma 2 and Proof of Theorem 1 in [23], the trajectory of the proposed FTESO (20) is fast finite-time uniformly ultimately bounded stable. This also implies that the observation errors of FTESO e_1 and e_2 will converge to a small region of the origin in the finite time T_1.

The observation error gradually decreases with time and finally enters the domain:

$$Z_1 = \left\{ e \middle| \gamma_1 V_1(e)^{1-\frac{1}{2\alpha_1}} + \gamma_2 V_1(e)^{\frac{1}{2}} < \lambda_3 \right\} \tag{22}$$

where $\lambda_1 = \lambda_{\min}(Q_1)\lambda_{\max}(N)^{\frac{1}{2\alpha_1}-\frac{3}{2}}$, $\lambda_2 = \lambda_{\min}(Q_2)\lambda_{\max}(N)^{-1}$, $\lambda_3 = 2\sqrt{3}\bar{g}\lambda_{\min}(N)^{-\frac{1}{2}}$ $\|N\|$, $\gamma_1 \in (0, \lambda_1)$ and $\gamma_2 \in (0, \lambda_2)$ are arbitrary constants. Then, the observation error e_2 of the generalized uncertainty is theoretically bounded, and this bound is assumed to be $\bar{\delta} = \left[\bar{\delta}_1, \bar{\delta}_2, \bar{\delta}_3\right]$. Note that $\dot{e}_2$ is bounded under the assumption that $\dot{j}(t)$ is bounded. Then, the observation error $\bar{\varepsilon}$ eventually converges to the domain Z_1 within a finite time T_1. □

Remark 1. *Compared with [23], the integral sliding mode surface is used to replace the non-singular terminal sliding mode surface in the design of the extended state observer (17) and (18) in this paper. In this way, the coefficient of the generalized uncertainty term in the derivative of the sliding mode surface does not include the fractional power term of the velocity error and is 1. Therefore, after the generalized uncertainty term of AUV is estimated by the extended state observer, its estimated value is directly used to replace the true value in the design of the control law, and it is not necessary to remove the fractional power term of the degree error from the estimated value of the observer to obtain the true value. The method in this paper can effectively suppress the amplification and fluctuation of the generalized uncertainty term and reduce chattering.*

4. Fault-Tolerant Control Design Based on NFTSMS

This section briefly introduces the specific implementation of the fault-tolerant control (FTC) based on the non-singular fast terminal sliding mode surface (NFTSMS), then analyzes the convergence of the controller designed in this paper.

4.1. FTC Design

Based on the generalized AUV uncertainty term estimated by FTESO, an FTC design method based on NFTSMS is proposed in this paper. First, a new NFTSMS is designed; then, the sliding mode surface is derived; finally, using the second derivative of the tracking error, the control law of the AUV controller can be constructed.

Design a new NFTSMS [29]:

$$S = e_3 + \sigma_1 e_3{}^{\psi} + \sigma_2 \langle \dot{e}_3 \rangle^{\frac{L}{p}} \tag{23}$$

where σ_1 and σ_2 are positive constants, and L and p are positive odd integers that satisfy the conditions $1 < \frac{L}{p} < 2$, $\psi > \frac{L}{p}$.

Thus, the time derivative of (23) is given as:

$$\dot{S} = \dot{e}_3 + \sigma_1 \psi |e_3|^{\psi-1} \times \dot{e}_3 + \sigma_2 \tfrac{L}{p} \langle \dot{e}_3 \rangle^{\frac{L}{p}-1} \times \ddot{e}_3 \tag{24}$$

Using the second derivative of the tracking error, expression (24) can be written as follows:

$$\dot{S} = \dot{e}_3 + \sigma_1 \psi |e_3|^{\psi-1} \times \dot{e}_3 + \sigma_2 \frac{L}{p} \langle \dot{e}_3 \rangle^{\frac{L}{p}-1} \times \left(\hat{M}_\eta^{-1} \left(J^{-T} B u - \hat{C}_\eta \dot{\eta} - \hat{D}_\eta \dot{\eta} - \hat{g}_\eta \right) + d - \ddot{x}_d \right) \tag{25}$$

For the newly introduced NFTSM, the control law of AUV can be designed as follows:

$$\begin{aligned} u = {} & B^+ J^T \left\{ (\hat{C}_\eta \dot{\eta} + \hat{D}_\eta \dot{\eta} + \hat{g}_\eta) + \hat{M}_\eta \ddot{x}_d - \hat{M}_\eta \tfrac{\sigma_1}{\sigma_2} \psi \tfrac{p}{L} |\dot{e}_3|^{2-\frac{L}{p}} sign(\dot{e}_3) |\dot{e}_3|^{\psi-1} \right\} + \\ & B^+ J^T \left\{ -\hat{M}_\eta \hat{z}_2 + \left(\tfrac{1}{\sigma_2} \tfrac{p}{L} |\dot{e}_3|^{1-\frac{L}{p}} \right) \times \left(-\hat{M}_\eta k_1 f^{\frac{2}{3}} |S|^{\frac{1}{3}} sign(S) + \hat{M}_\eta z_3 \right) \right\} - \\ & B^+ J^T \left\{ \hat{M}_\eta \tfrac{1}{\sigma_2} \tfrac{p}{L} |\dot{e}_3|^{2-\frac{L}{p}} sign(\dot{e}_3) \right\} \end{aligned} \tag{26}$$

$$\dot{z}_3 = -k_2 f \langle S \rangle^0 \tag{27}$$

where k_1, k_2, and f are positive gains to be designed, $()^+$ represents the pseudo-inverse operation of the specified matrix, i.e., $(B)^+ = B^T (BB^T)^{-1}$. Matrix B represents the thruster configuration matrix, which is a constant matrix, so there is no singularity problem for B.

This paper defines $e_{22} = \sigma_2 \frac{L}{p} \langle \dot{e}_3 \rangle^{\frac{L}{p}-1} e_2$ and $Y = z_3 - e_{22}$. Incorporating (26) in (25) leads to the (28), taking the time derivative of $Y = z_3 - e_{22}$ and using (27) yields (29).

$$\dot{S} = -k_1 f^{\frac{2}{3}} \langle S \rangle^{\frac{1}{3}} + S_3 \tag{28}$$

$$\dot{Y} = -k_2 f \langle S \rangle^0 - \ddot{e}_{22} \tag{29}$$

As the right-hand side of the above system is discontinuous, its solutions will be understood as the meaning of Filippov [30]. Furthermore, the convergence analysis of the controller is as follows.

Remark 2. *In [23], the FTESO and the controller share an NTSMS, so in the obtained control law, the control gain contains a pseudo-inverse term of the fractional power term of the velocity tracking error. When the velocity tracking error is smaller, the gain becomes larger, which plays a role in amplifying the control jitter. In addition, when the velocity tracking error is zero, this gain tends to infinity and, theoretically, singularities will appear. However, compared with [23], the FTESO and controller in this paper are designed with two sliding mode surfaces, namely the ISMS and the NFTSMS, which can effectively suppress the chattering of the control quantity, thereby reducing the energy consumption and theoretically avoiding the emergence of singular points.*

Remark 3. *The NFTSMS in reference [29] has many advantages, including the fast convergence property inherited from terminal attractors' rapid response, finite time converge, and non-singularity. Therefore, this paper also selects the sliding mode surface to design the control law.*

4.2. FTC Stability Analysis

Theorem 2. *Consider the AUV control system in (14) and assume* $\|j(t)\| < \bar{j}, \bar{j} > 0$. *If NFTSMS is selected as Equation (31), and the control law is selected as (34) and (35), the control system of the AUV will converge to the origin in a finite time globally and uniformly.*

Proof of Theorem 2. Consider introducing the method of continuously differentiable and homogeneous Lyapunov functions to prove Theorem 2.

Consider the following Lyapunov candidate function:

$$V_2 = \frac{2}{5}\alpha|S|^T|S|^{\frac{3}{2}} - \frac{1}{k_1^3}S^TY^3 + \kappa|Y|^T|Y|^4 \tag{30}$$

where $\kappa > 0$.

According to the Proof of Theorem 2 in [23], it is easy to calculate that the above candidate Lyapunov function V_2 is bounded and positive definite.

Thus, the time derivative of the Lyapunov candidate function V_2 is given as follows:

$$\begin{aligned}\dot{V}_2 &= \alpha\left(\langle S\rangle^{\frac{3}{2}}\right)^T\dot{S} - \tfrac{1}{k_1^3}\left(\left(Y^3\right)^T\dot{S} + 3S^TYY^T\dot{Y}\right) + 5\kappa\left(\langle Y\rangle^4\right)^T\dot{Y} \\ &= -\left(\alpha\left(\langle S\rangle^{\frac{3}{2}}\right)^T - \tfrac{1}{k_1^3}\left(Y^3\right)^T\right)\left(k_1\langle S\rangle^{\frac{1}{3}} - Y\right) - Y^TY\left(5\kappa\left(\langle Y\rangle^2\right)^T - \tfrac{3}{k_1^3}S^T\right)\left(k_2\langle S\rangle^0 + \tfrac{\ddot{e}_{22}}{f}\right)\end{aligned} \tag{31}$$

According to the Proof of Theorem 2 in [23] and Lemma 3 in this paper, one can obtain the following inequality:

$$\dot{V}_2 \le -\eta_2 V_2^{\frac{4}{5}} \tag{32}$$

According to Lemma 1, combined with the basic Lyapunov theories for differential inclusions, the AUV control system in (14) is globally uniformly finite-time stable. □

4.3. Parameter Adjustment Method for AUV Thrusters

This section introduces the reasons for the proposed parameter adjustment method (PAM) for AUV thrusters and its specific implementation steps.

The simulation results show that, compared with the method in [23], FTESO − FTC is effective at reducing energy consumption, but there is a problem of a low tracking accuracy. Moreover, the controller parameters of the referenced methods are mostly constant values, which have the problem of a low tracking accuracy. Specifically, a large value of k_1 will lead to an increase in tracking accuracy, but also increased chattering and energy consumption, while a small value of k_1 will lead to a decrease in tracking accuracy, but correspondingly decreased chattering and energy consumption, so it is not ideal to adopt constant values. Through the analysis, this problem can be effectively solved by analyzing k_1 as a piecewise function in the shape of a ramp, as shown in (34) designed in this paper.

The controller parameters are adaptive parameters instead of constant values, and the design of PAM can be expressed as follows:

$$k_1 = \begin{cases} a, (0 \le t < t_1) \\ a - \frac{(t-t_1)(a-b)}{(t_2-t_1)}, (t_1 \le t < t_2) \\ b, (t_2 \le t < t_3) \end{cases} \tag{33}$$

where a, b are constant values; t_1, t_2 and t_3 are time values.

Remark 4. *Therefore, on the basis of the FTESO − FTC proposed in this paper, we add the PAM for AUV thrusters, which aim to reduce energy consumption while maintaining a good tracking accuracy.*

5. Simulation and Analysis

To verify the effectiveness of the method proposed in this paper, this paper uses ODIN AUV [31] to conduct trajectory tracking control simulation experiments in the simulated ocean current environment. This section first introduces the preparation of the simulation experiment. Then, the comparison and simulation verifications between FTESO − FTC − APAM and FTESO − FTC and previous method in [23] are carried out.

5.1. Preparations for Simulation Experiments

5.1.1. Introduction of ODIN AUV

In this paper, ODIN AUV was selected as the simulated vehicle. At present, this AUV has been used as a research object in many literature works for simulation experiments, such as [31,32]. ODIN AUV is an open-frame underwater vehicle with four lateral thrusters and four vertical thrusters, and its model parameters are presented in [33]. In the simulation, the initial state of the ODIN AUV is set as $\eta\ (0) = [\,1\,;\,1\,;-1\,;\,\pi/18\,;\,\pi/18\,;\,2\pi/9\,]$ and $\dot{\eta}\ (0) = [\,0.2\,;\,0.2\,;\,0.2\,;\,0.2\,;\,0.2\,;\,0.2\,]$. Furthermore, it is assumed that the thruster amplitude limits are ± 200 N.

5.1.2. Desired Trajectory

In this simulation, ODIN AUV is controlled to track a spiral trajectory to test the performance of the proposed control scheme, in comparison with the controller in [23]. The desired spiral trajectory is described as: $\eta_d = [x_d, y_d, z_d, 0, 0, 0]^T$, $x_d = 3(1 - \cos(0.15t))$, $y_d = 3\sin(0.15t)$, $z_d = -0.2t$.

5.1.3. Simulate Current Disturbance

Based on the current simulation method given in the literature [24], in this paper, a first-order Gauss–Markov process was used to simulate the ocean current, and the specific expression is shown as follows:

$$\dot{V}_c + u_c V_c = \omega_c \tag{34}$$

where V_c is the simulated current value; ω_c is Gaussian white noise (mean 1.5 and variance 1); $u_c = 3$; two angles involved in ocean currents are β_c and α_c; β_c is generated by the sum of Gaussian noise with mean 0 and variance 50, and $\alpha_c = \beta_c/2$.

5.1.4. Model Uncertainty and Fault Simulation

This paper considers inevitable model uncertainty. In the simulations, 30% modelling uncertainty was considered, i.e., the parameters of ODIN AUV in the controller were only 70% of the nominal system dynamics. The thruster faults generally include the incipient thruster fault, abrupt thruster fault, etc. [34]. In this paper, two different thruster fault simulation methods were used to conduct the simulation experiments. The specific abrupt thruster fault was described, as Equation (36), which indicates that the thruster is healthy from the 0th s to 20th s, and then the magnitude of the thruster fault persists until the end of the simulation. However, the controller does not know this fault a priori.

The mathematical description of different fault degrees under an abrupt fault is given as follows:

$$\beta = \begin{cases} 1 & 0 \le t < 20 \\ 1 - \sigma & 20 \le t < 120 \end{cases} \tag{35}$$

where a 30% fault degree is expressed as $\sigma = 0.3$; when the fault degree is 50%, $\sigma = 0.5$.

5.1.5. Control Parameters

This paper involves multiple parameters, and the result of the parameter selection affects the control effect. The following is a brief introduction to each parameter selection method.

1. Parameter selection in sliding surface

When designing the FTESO, the parameter of the ISMS is selected, $K = 3$. When designing the controller, the parameters of the NFTSMS are selected as follows: $\sigma_1 = 1$, $\sigma_2 = 1$, $\psi = 7$, $L = 7$, $p = 6$.

2. FTESO parameter selection

According to reference [23], the specific parameters in the observer are $\alpha_1 = 0.66$, $\beta_1 = 0.2$, $\beta_2 = 0.07$.

3. FTC parameter selection

The control parameters are determined by trial and error, given as follows:$f = 2$ and $k_2 = 0.001$ are constants, and k_1 is determined by PAM instead of a constant value. The specific design of the adaptive parameter is shown as (33), where $a = 25 \times \text{diag}([5,5,5,1,1,5])$, $b = 0.3 \times \hat{M}_\eta$, $t_1 = 5$, $t_2 = 60$, $t_3 = 120$.

5.2. Comparative Simulation Verification

In order to verify the effectiveness of the FTESO − FTC − PAM proposed in this paper in reducing the energy consumption of fully actuated AUVs, this paper adopts the fault-tolerant control method proposed in [23], in which the FTESO and the controller share a terminal sliding surface to carry out comparative simulation experiments.

There are two types of fault: abrupt fault and incipient fault, and the thruster fault degree is divided into three kinds: 0, 30%, and 50%. In the simulation experiments, it was found that the simulation results of the abrupt and incipient fault types are similar under the same fault degree. Therefore, the fault type of the thruster in this paper is chosen as the abrupt fault.

To validate the performance of the proposed FTESO − FTC − PAM for fully actuated AUV, three cases were considered in this subsection.

Case 1: 30% fault degree at thruster T1.

Case 2: 50% fault degree at thruster T2.

Case 3: no fault at any thruster.

The simulation results of FTESO − FTC − APAM and FTESO − FTC and the compared method for Case 1 are presented in Figures 1–3, respectively. In addition, the simulation results of these three methods for Case 2 are presented in Figures 4–6, respectively. Finally, to fully illustrate the effectiveness of the method in this paper, the tracking results of these three methods for Case 3 are presented in Table 1.

Table 1. Compared results of the three methods under Cases 1–3.

Methods	**Tracking Error**	**Distance Errors (m)**			**SSCS ($10^7 \times N^2$)**		
		Case 1	Case 2	Case 3	Case 1	Case 2	Case 3
FTESO − FTC − PAM	Mean	0.0224	0.0227	0.0212	2.77	3.49	2.43
	Mean square error	0.1502	0.1522	0.1482			
FTESO − FTC	Mean	0.0356	0.0367	0.0328	2.74	3.48	2.34
	Mean square error	0.1991	0.2008	0.1863			
Compared method [23]	Mean	0.0225	0.0236	0.0215	3.81	4.71	3.07
	Mean square error	0.1516	0.1527	0.1496			

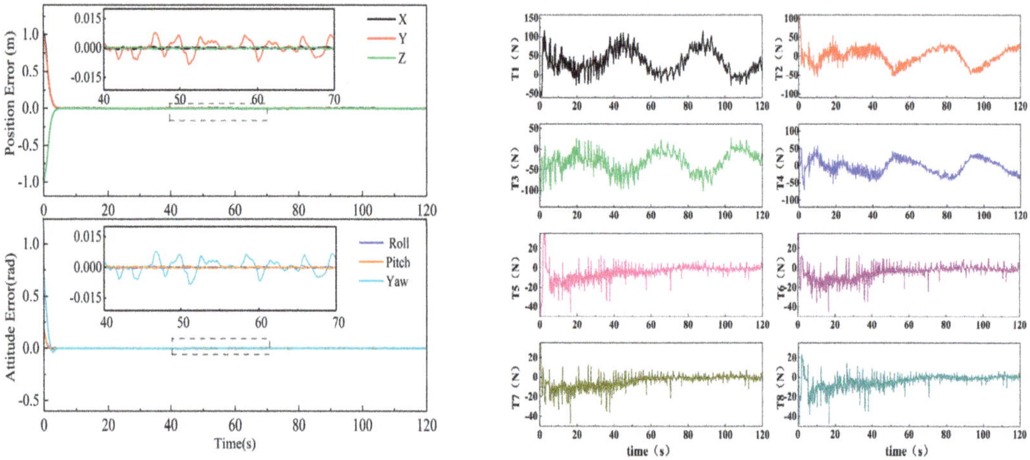

(**a**) Tracking error (**b**) Control output

Figure 1. The control effect of FTESO − FTC − APAM under Case 1.

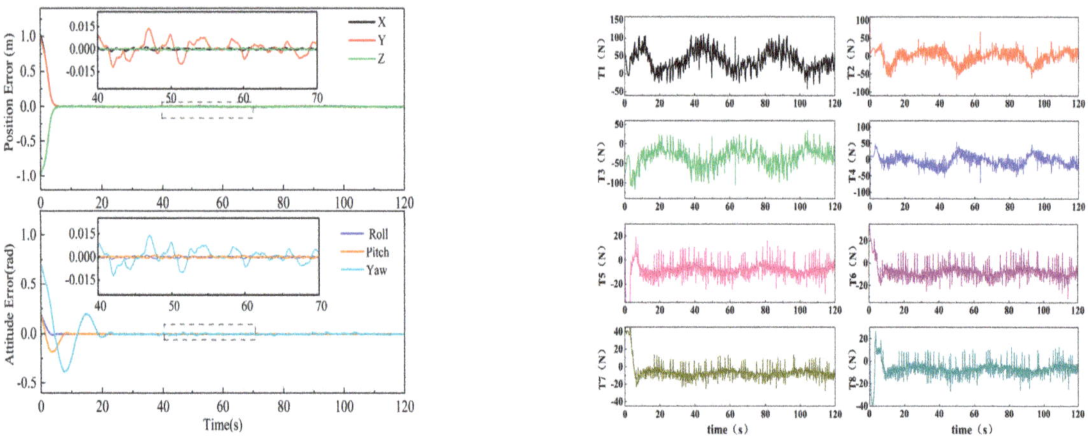

(**a**) Tracking error (**b**) Control output

Figure 2. The control effect of FTESO − FTC under Case 1.

Referring to [35], in this paper, the mean and mean square error of the absolute value of the tracking error were used to measure the tracking effect. If the chattering phenomenon in the control signals became serious, the energy consumption caused by the chattering of the control signal would be increased if the environmental parameters had no change, according to [36]. The energy consumption value was calculated by the sum of the squares of the control signal (SSCS), shown as follows:

$$SSCS = \sum \sum_{i=1}^{8} |u_i|^{2} \tag{36}$$

For the convenience of analysis, this paper organizes and summarizes the mean, mean square error of the absolute value of the tracking error, and energy consumption (SSCS) in Case 1–Case 3 into Table 1.

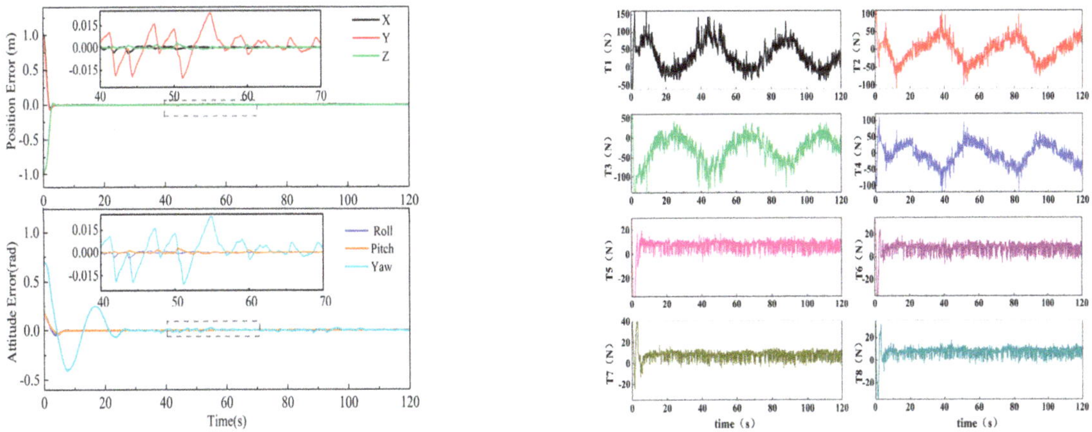

(**a**) Tracking error (**b**) Control output

Figure 3. The control effect of the compared method under Case 1.

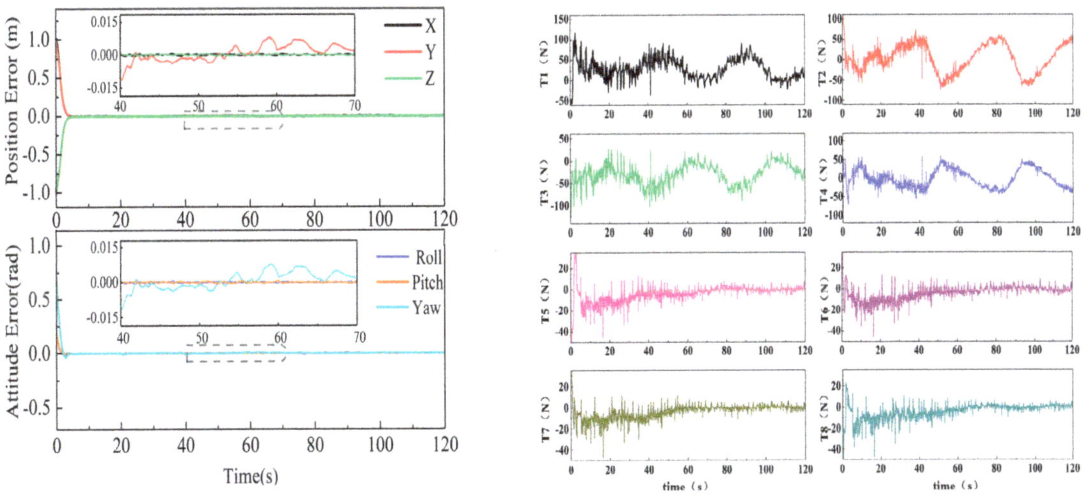

(**a**) Tracking error (**b**) Control output

Figure 4. The control effect of FTESO − FTC − PAM under Case 2.

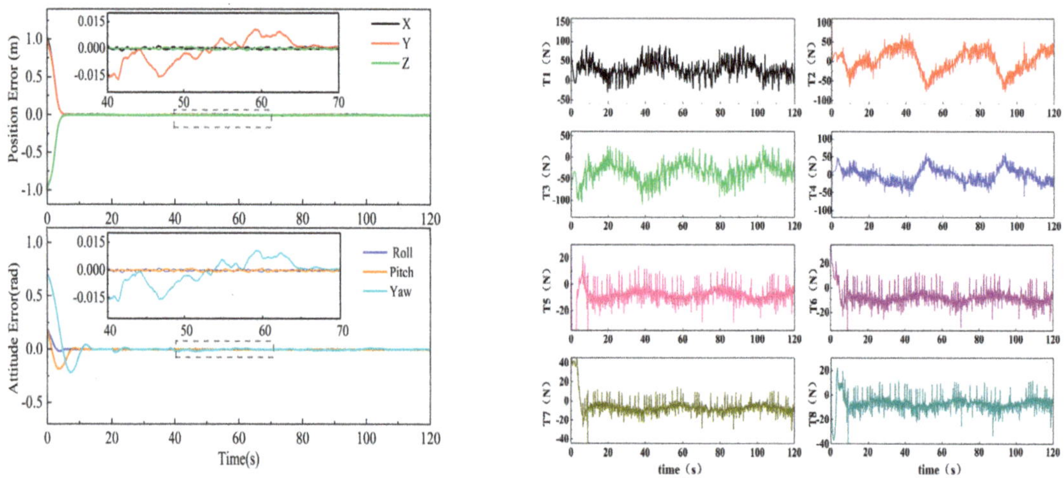

(**a**) Tracking error (**b**) Control output

Figure 5. The control effect of FTESO − FTC under Case 2.

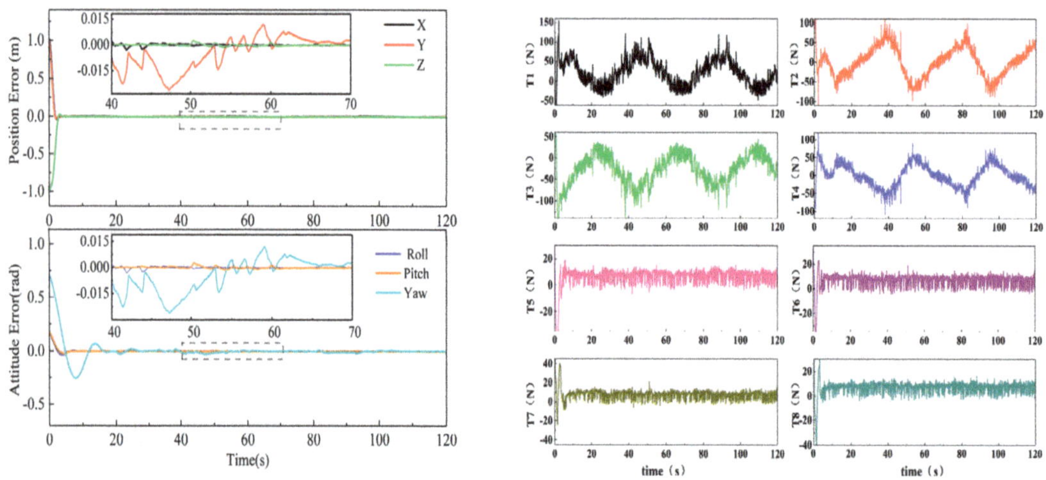

(**a**) Tracking error (**b**) Control output

Figure 6. The control effect of the compared method under Case 2.

In addition, in this paper, the position errors of the three degrees of freedom, X, Y, and Z, are synthesized into distance errors.

In terms of the tracking error and energy consumption, the simulation results of FTESO − FTC − APAM and FTESO − FTC and the compared method are analyzed.

Firstly, compared with the method in [23], the analysis of Cases 1–3 in Table 1 shows that FTESO − FTC has an obvious effect on reducing the energy consumption; however, the tracking accuracy was reduced. Taking Case 1 as an example, compared with the method in [23], the mean and mean square error of the distance error of FTESO − FTC increased by 58.22% and 31.33%, respectively, so the tracking accuracy decreased. However, it can be found that compared with the method in [23], FTESO + FTC reduced the energy consumption by 28.08%.

Compared with [23], it can be concluded that FTESO − FTC was effective at reducing the energy consumption, but there was the problem of reducing the tracking accuracy.

Therefore, this paper proposed PAM based on FTESO − FTC to reduce the energy consumption of the control output while maintaining a good tracking accuracy.

Then, compared with FTESO − FTC, it can be seen from Cases 1–3 in Table 1 that FTESO − FTC − PAM significantly improved the tracking accuracy and slightly increased the energy consumption. Similarly, taking Case 1 as an example, compared with FTESO − FTC, FTESO − FTC − PAM reduced the mean value and mean square error of the distance error by 37.07% and 24.56%, respectively, that is, the tracking accuracy was significantly improved. FTESO − FTC − APAM was close to FTESO − FTC in terms of the energy consumption (slightly increased).

Furthermore, compared with the method in [23], it can be obtained from Cases 1–3 in Table 1 that FTESO − FTC − APAM can significantly reduce energy consumption on the premise of a slightly improved tracking accuracy. Then, taking Case 1 as an example, compared with [23], the mean value and mean square error of the absolute value of the distance error in the FTESO − FTC − APAM were slightly smaller, and the tracking accuracy was slightly improved. In terms of energy consumption, compared with the method in [23], FTESO − FTC − APAM was reduced by 29.92%, and the effect of reducing the energy consumption was obvious.

It can be concluded that FTESO − FTC − PAM was a further improved method based on the FTESO − FTC of this paper. Overall, based on the simulation results of the above three methods, it is concluded that compared with FTESO − FTC, the FTESO − FTC − PAM could greatly improve the tracking accuracy under the condition of a slight increase in energy consumption. Compared with the method in [23], the FTESO − FTC − PAM of this paper made up for the shortcomings of FTESO − FTC in reducing the tracking accuracy when reducing the energy consumption and finally could achieve the control goal of reducing the energy consumption while maintaining a better tracking accuracy.

6. Conclusions

This paper studies the problem of fault tolerant control for fully actuated AUV thrusters in the presence of current disturbances, thruster faults, and dynamic model uncertainty. In view of the problems of high energy consumption caused by the chattering of the control signal when the method in [23] is used in AUV, this paper proposes FTESO − FTC − PAM, i.e., a fault-tolerant control method based on a finite-time extended state observer for AUV with unknown thruster faults. Through theoretical analysis and simulation experiments, the following conclusions can be drawn.

(1) Aiming at the problem of large energy consumption when a previous method is applied to AUVs, this paper proposes the FTC method based on FTESO (FTESO − FTC). In this paper, the design of the FTESO and the FTC adopts ISMS and NFTSMS, respectively. The stability of the finite-time extended state observer and the finite-time convergence of the closed-loop control system formed by the fault-tolerant control method are proved by the Lyapunov stability theory. The simulation comparison experiment results using an ODIN AUV show that under the simulated thruster fault and ocean current interference, FTESO − FTC is more effective at reducing the energy consumption than the method in [23], but it has the problem of reducing the accuracy of tracking.

(2) To this end, the above-mentioned FTESO − FTC is further improved, and the parameter adjustment method (PAM) for AUV thrusters is added. The simulation and compared experimental results for the ODIN AUV show that after adding PAM, the tracking accuracy can be effectively restored, and the energy consumption is basically not increased. Moreover, the simulation results show that the FTESO − FTC − PAM is superior to the method in [23] in terms of reducing the energy consumption and improving the tracking accuracy.

Author Contributions: Conceptualization, M.Z.; methodology, X.L. (Xiaofeng Liu); software,X.L. (Xing Liu); validation, M.Z. and X.L. (Xing Liu); formal analysis, W.Z.; investigation, X.L. (Xiaofeng Liu); resources, M.Z. and X.L. (Xing Liu); data curation, X.L. (Xiaofeng Liu); writing—original draft preparation, X.L. (Xiaofeng Liu); writing—review and editing, X.L. (Xiaofeng Liu) and M.Z.; visualization, W.Z.; supervision, X.L. (Xing Liu); project administration, X.L. (Xing Liu); funding acquisition, M.Z. All authors have read and agreed to the published version of the manuscript.

Funding: This research was funded by the National Natural Science Foundation of China (52201357, 51839004), Fundamental Research Funds for the Central Universities under Grant 3072022TS0701.

Institutional Review Board Statement: Not applicable.

Informed Consent Statement: Informed consent was obtained from all subjects involved in the study.

Data Availability Statement: The data and any statistical analysis can be made available from the corresponding author upon request.

Conflicts of Interest: The authors declare no conflict of interest.

References

1. Zeng, Z.; Lian, L.; Sammut, K.; He, F.P.; Tang, Y.H.; Lammas, A. A survey on path planning for persistent autonomy of autonomous underwater vehicles. *Ocean Eng.* **2015**, *110*, 303–313. [CrossRef]
2. Hong, L.; Song, C.H.; Yang, P.; Cui, W.C. Two-Layer Path Planner for AUVs Based on the Improved AAF-RRT Algorithm. *J. Mar. Sci. Appl.* **2022**, *21*, 102–115. [CrossRef]
3. Zhang, H.W.; Zeng, Z.; Yu, C.Y.; Jiang, Z.N.; Han, B.; Lian, L. Predictive and sliding mode cascade control for cross-domain locomotion of a coaxial aerial underwater vehicle with disturbances. *Appl. Ocean Res.* **2020**, *100*, 102183. [CrossRef]
4. Chu, Z.; Li, Z.; Gu, Z.; Chen, Y.; Zhang, M. A fault diagnosis method for underwater thruster based on RFR-SVM. *Proc. Inst. Mech. Eng. Part M J. Eng. Marit. Environ.* **2022**. [CrossRef]
5. Zhu, C.; Huang, B.; Zhou, B.; Su, Y.; Zhang, E. Adaptive model-parameter-free fault-tolerant trajectory tracking control for autonomous underwater vehicles. *ISA Trans.* **2021**, *114*, 57–71. [CrossRef]
6. Zhang, M.J.; Yin, B.J.; Liu, X.; Guo, J. Thruster fault identification method for autonomous underwater vehicle using peak region energy and least square grey relational grade. *Adv. Mech. Eng.* **2015**, *7*, 1–11. [CrossRef]
7. Chu, Z.; Wang, F.; Lei, T.; Luo, C. Path Planning based on Deep Reinforcement Learning for Autonomous Underwater Vehicles under Ocean Current Disturbance. *IEEE Trans. Intell. Veh.* **2022**. [CrossRef]
8. Xiang, X.B.; Lapierre, L.; Jouvencel, B. Smooth transition of AUV motion control: From fully-actuated to under-actuated configuration. *Robot. Auton. Syst.* **2015**, *67*, 14–22. [CrossRef]
9. Xiao, B.; Yin, S.; Gao, H.J. Reconfigurable Tolerant Control of Uncertain Mechanical Systems with Actuator Faults: A Sliding Mode Observer-Based Approach. *IEEE Trans. Control Syst. Technol.* **2018**, *26*, 1249–1258. [CrossRef]
10. Ghaf-Ghanbari, P.; Mazare, M.; Taghizadeh, M. Active fault-tolerant control of a Schonflies parallel manipulator based on time delay estimation. *Robotica* **2021**, *39*, 1518–1535. [CrossRef]
11. Li, J.Q.; Zhang, G.Q.; Li, B. Robust Adaptive Neural Cooperative Control for the USV-UAV Based on the LVS-LVA Guidance Principle. *J. Mar. Sci. Eng.* **2022**, *10*, 51. [CrossRef]
12. Benjemaa, R.; Elhsoumi, A.; Naoui, S. Active Fault Tolerant Control for uncertain neutral time delay system. In Proceedings of the 17th International Multi-Conference on Systems, Signals and Devices (SSD), Sfax, Tunisia, 20–23 July 2020; pp. 285–289.
13. Yu, Z.Q.; Zhang, Y.M.; Jiang, B.; Fu, J.; Jin, Y.; Chai, T.Y. Composite Adaptive Disturbance Observer-Based Decentralized Fractional-Order Fault-Tolerant Control of Networked UAVs. *IEEE Trans. Syst. Man Cybern. Syst.* **2022**, *52*, 799–813. [CrossRef]
14. Liu, Z.Q.; Cai, W.Y.; Zhang, M.Y.; Lv, S.S. Improved Integral Sliding Mode Control-Based Attitude Control Design and Experiment for High Maneuverable AUV. *J. Mar. Sci. Eng.* **2022**, *10*, 795. [CrossRef]
15. Teng, J.; Li, C.; Feng, Y.; Yang, T.; Zhou, R.; Sheng, Q.Z. Adaptive Observer Based Fault Tolerant Control for Sensor and Actuator Faults in Wind Turbines. *Sensors* **2021**, *21*, 8170. [CrossRef]
16. Al-Mahturi, A.; Santoso, F.; Garratt, M.A.; Anavatti, S.G. Self-Learning in Aerial Robotics Using Type-2 Fuzzy Systems: Case Study in Hovering Quadrotor Flight Control. *IEEE Access* **2021**, *9*, 119520–119532. [CrossRef]
17. Su, Y.B.; Liang, H.J.; Pan, Y.N.; Chen, D.X. Event-triggered adaptive fuzzy fault-tolerant control for autonomous underwater vehicles with prescribed tracking performance. *Int. J. Syst. Sci.* **2021**, *52*, 1145–1155. [CrossRef]
18. Zhirabok, A.N.; Zuev, A.V.; Filaretov, V.F.; Shumsky, A.E. Fault Identification in Nonlinear Systems Based on Sliding Mode Observers with Weakened Existence Conditions. *J. Comput. Syst. Sci. Inter.* **2022**, *61*, 313–321. [CrossRef]
19. Zuev, A.; Zhirabok, A.N.; Filaretov, V.; Protsenko, A. Fault Identification in Electric Servo Actuators of Robot Manipulators Described by Nonstationary Nonlinear Dynamic Models Using Sliding Mode Observers. *Sensors* **2022**, *22*, 317. [CrossRef]
20. Cheng, Y.; Ren, X.M.; Zheng, D.D.; Li, L.W. Non-Linear Bandwidth Extended-State-Observer Based Non-Smooth Funnel Control for Motor-Drive Servo Systems. *IEEE Trans. Ind. Electron.* **2022**, *69*, 6215–6224. [CrossRef]

21. Wang, L.; Kellett, C.M. Adaptive Semiglobal Nonlinear Output Regulation: An Extended-State Observer Approach. *IEEE Trans. Autom. Control* **2020**, *65*, 2670–2677. [CrossRef]
22. Ali, N.; Tawiah, I.; Zhang, W.D. Finite-time extended state observer based nonsingular fast terminal sliding mode control of autonomous underwater vehicles. *Ocean Eng.* **2020**, *218*, 10. [CrossRef]
23. Li, B.; Hu, Q.L.; Yang, Y.S. Continuous finite-time extended state observer based fault tolerant control for attitude stabilization. *Aerosp. Sci. Technol.* **2019**, *84*, 204–213. [CrossRef]
24. Fossen, T.I. *Handbook of Marine Craft Hydrodynamics and Motion Control*; John Wiley & Sons, Ltd.: Chichester, UK, 2011.
25. Yin, B.; Yao, F.; Wang, Y.; Zhang, M.; Zhu, C. Fault degree identification method for thruster of autonomous underwater vehicle using homomorphic membership function and low frequency trend prediction. *Proc. Inst. Mech. Eng. Part C J. Mech. Eng. Sci.* **2018**, *233*, 1426–1440. [CrossRef]
26. Bhat, S.P.; Bernstein, D.S. Geometric homogeneity with applications to finite-time stability. *Math. Control Signals Syst.* **2005**, *17*, 101–127. [CrossRef]
27. Hu, Q.L.; Jiang, B.Y. Continuous Finite-Time Attitude Control for Rigid Spacecraft Based on Angular Velocity Observer. *IEEE Trans. Aerosp. Electron. Syst.* **2018**, *54*, 1082–1092. [CrossRef]
28. Cui, R.X.; Chen, L.P.; Yang, C.G.; Chen, M. Extended State Observer-Based Integral Sliding Mode Control for an Underwater Robot with Unknown Disturbances and Uncertain Nonlinearities. *IEEE Trans. Ind. Electron.* **2017**, *64*, 6785–6795. [CrossRef]
29. Nguyen, V.T.; Su, S.F.; Nguyen, A.T.; Nguyen, V.T. Adaptive Nonsingular Fast Terminal Sliding Mode Tracking Control for Parallel Manipulators with Uncertainties. In Proceedings of the International Conference on System Science and Engineering (ICSSE), Dong Hoi City, Vietnam, 20–21 July 2019; pp. 522–525.
30. Kamal, S.; Moreno, J.A.; Chalanga, A.; Bandyopadhyay, B.; Fridman, L.M. Continuous terminal sliding-mode controller. *Automatica* **2016**, *69*, 308–314. [CrossRef]
31. Qiao, L.; Zhang, W.D. Adaptive non-singular integral terminal sliding mode tracking control for autonomous underwater vehicles. *IET Control Theory Appl.* **2017**, *11*, 1293–1306. [CrossRef]
32. Liu, X.; Zhang, M.J.; Wang, S.M. Adaptive region tracking control with prescribed transient performance for autonomous underwater vehicle with thruster fault. *Ocean Eng.* **2020**, *196*, 106804. [CrossRef]
33. Sarkar, N.; Podder, T.K.; Antonelli, G. Fault-accommodating thruster force allocation of an AUV considering thruster redundancy and saturation. *IEEE Trans. Robot. Autom.* **2002**, *18*, 223–233. [CrossRef]
34. Liu, X.; Zhang, M.J.; Yao, F. Adaptive fault tolerant control and thruster fault reconstruction for autonomous underwater vehicle. *Ocean Eng.* **2018**, *155*, 10–23. [CrossRef]
35. Wang, Y.J.; Zhang, M.J.; Wilson, P.A.; Liu, X. Adaptive neural network-based backstepping fault tolerant control for underwater vehicles with thruster fault. *Ocean Eng.* **2015**, *110*, 15–24. [CrossRef]
36. Liu, X.; Zhang, M.J.; Yao, F.; Yin, B.J.; Chen, J.W. Barrier Lyapunov function based adaptive region tracking control for underwater vehicles with thruster saturation and dead zone. *J. Frankl. Inst. Eng. Appl. Math.* **2021**, *358*, 5820–5844. [CrossRef]

Journal of
Marine Science and Engineering

MDPI

Article

A Nonlinear Phase Transition Dynamic Model for Shape Memory Alloys Based Deep Sea Actuators

Jian Guo [1], Binbin Pan [1,*], Weicheng Cui [1,2] and Shengbing Hu [3]

1 College of Marine Sciences, Shanghai Ocean University, Shanghai 201306, China
2 School of Engineering, Westlake University, Hangzhou 310024, China
3 Shanghai OceanHome Technology Co., Ltd., Shanghai 201306, China
* Correspondence: bbpan@shou.edu.cn

Abstract: A constitutive relation for shape memory alloys (SMAs) that is simple, accurate, and effective is the basis for deep-sea intelligent actuators used in marine engineering applications. The existing kinetic models of phase transition all have common drawbacks, such as sharp change at the turning point of the phase transition, constant phase transition rate, and many variable parameters. In this study, the one-dimensional thermodynamic constitutive equation for SMAs is extended based on the thermodynamic framework of the Boyd–Lagoudas constitutive model. In addition, the traditional phase transition function is replaced by an improved logistic nonlinear function in order to construct the relation for the macroscopic variable-speed phase transition that constitutes deep-sea actuator driving wires. The logistic model is compared to other models and verified by the numerical fitting results of the traditional constitutive model and the experimental data for two scenarios: (1) constant load and (2) constant temperature. The results show that the improved constitutive model has more advantages and better adaptability than the traditional models. Consequently, it can accurately describe the slow and gradual phase transitions in the initial and final regions of the phase transition with fewer variable parameters and has the ability to flexibly adjust the rate of change of the phase transition rate. These results provide important theoretical support for the design of SMA deep-sea actuators used in marine engineering applications.

Keywords: deep sea; shape memory alloy; logistic function; phase transformation kinetics; numerical fitting

Citation: Guo, J.; Pan, B.; Cui, W.; Hu, S. A Nonlinear Phase Transition Dynamic Model for Shape Memory Alloys Based Deep Sea Actuators. *J. Mar. Sci. Eng.* **2022**, *10*, 1951. https://doi.org/10.3390/jmse10121951

Academic Editors: Erkan Oterkus and José António Correia

Received: 17 October 2022
Accepted: 5 December 2022
Published: 8 December 2022

1. Introduction

The deep sea is located at a depth of more than 1000 m below the ocean's surface, and it accounts for about 90% of the ocean system. Due to its extreme environment of high pressure, low temperature, and no light, humans have minimal understanding of this part of the ocean. Nevertheless, the deep sea is the most important area for gathering oil, gas, minerals, and biological resources [1], and it also forms the core of what is called the "smart ocean." In recent years, with the development of emerging technologies, such as artificial intelligence and network cloud computing, in marine fields, the smart ocean has become another "peak" in the strategy for "caring about the ocean, understanding the ocean, and managing the ocean" [2]. The application of smart ocean engineering to the deep sea has revealed the limitations of traditional deep-sea drive systems, and new actuators based on smart materials are urgently required for new marine equipment that is miniaturized, intelligent, and refined.

Shape memory alloys (SMAs) are suitable for fulfilling these requirements. They are a new type of biofunctional material that integrate sensing, actuation, and control. At present, they are the most suitable smart material useful for practical applications of marine engineering [3,4]. In particular, titanium-nickel (Ti-Ni) SMAs are the most widely used and have the best thermomechanical properties. Accordingly, they have significant application potential not only in marine fields, but in aerospace, national defense, automotive, electromechanical, biomedical, and other fields as well [5]. Furthermore, Ti-Ni SMAs possess excellent shape memory effect (the maximum shrinkage strain of their one-way and two-way shape memory effects can reach 8% and 4.1%, respectively), superelasticity, corrosion resistance, and biocompatibility [6]. Moreover, the special phase composition (martensite phase and austenite phase) and crystal structure (crystalline martensite, detwinned martensite, and austenite) of SMAs results in a complex thermomechanical response to external stimulation. This structural relationship also exhibits nonlinear coupling. Currently, the mainstream SMA constitutive models are divided into macroscopic phenomenological models, microscopic mechanical models, and microscopic thermodynamic models. The macroscopic phenomenological models are considered to be the most valuable models for SMA structural analysis and engineering design because they have fewer parameters and their fitting and calculation methods are simple [7,8]. The most widely used models in this category are exponential function macromodels (e.g., the Tanaka and Boyd-Lagoudas models) and cosine function macromodels (e.g., the Liang–Rogers and Brinson models) [8–10]. The Brinson model subdivides martensite transformation into temperature-induced and stress-induced parts, which more accurately describes the relationship between stress, strain, temperature and martensite volume fraction, and effectively describes the martensite reorientation process. At present, it is a relatively complete phenomenological theory based constitutive model for actual engineering application. The main difference between these models is the assumed phase transition kinetic function. These traditional phase transition functions still have many common drawbacks, even though their fitting methods are simple [11–13]. On the one hand, the predicted thermomechanical responses at the beginning and end of the phase transition change abruptly, which is inconsistent with the smooth and gradual changes that are actually exhibited at the beginning and end of the phase transition. On the other hand, traditional phase transition kinetic models only focus on the phase transition and ignore the variable nature of its intermediate process. In order to avoid the fitting errors caused by these common drawbacks, it is necessary to improve the existing model in order to build a simple, accurate, and effective constitutive model that is useful for engineering applications, such as deep-sea actuators.

The logistic function is often used to describe the dynamic processes of systems to elucidate the "slow-fast-slow" system evolution law, and it has been improved many times to adapt to nonlinear growth modeling applications in different fields [14]. The logistic function is characterized by typical sigmoid nonlinear growth, which is consistent with the thermodynamic response characteristics of the SMA phase transition process [15]. In this study, based on the thermodynamic framework of the Boyd–Lagoudas constitutive model, the thermodynamic constitutive equation is extended to replace the traditional phase transition function with an improved logistic function. This serves to facilitate the construction of an SMA constitutive relation that can describe the slow and asymptotic phase in the initial and final regions of the entire phase transition as well as the transition rate variability. By comparing the numerical fitting results of the traditional Brinson model and the logistic constitutive model, the superior performance of the improved logistic function is verified. Lastly, the validity of the model is verified with experimental data.

The rest of this paper is organized as follows. Section 2 details the hydrostatic pressure dependence of the SMA phase transition process. In Section 3, the macroscopic phenomenological variable speed constitutive model is explained. Section 4 describes and discusses the results, and Section 5 draws the conclusions.

2. Hydrostatic Pressure Dependence of SMA Phase Transition Process

Hydrostatic pressure is an omnidirectional force that is exerted by a homogeneous fluid on an object, and it acts on all surfaces of the object with an equal vertical compressive stress. An increase in hydrostatic pressure reduces the volume of the object subject to that pressure but does not change its shape. The effect of omnidirectional hydrostatic pressure on SMA behavior is primarily reflected in the change of the material's elastic volume and does not involve plasticity [16]. When the volume of the SMA changes reversibly during the phase transition, the amount by which the hydrostatic pressure changes the phase transition temperature is independent of the phase transition mechanism and the distribution of the two phases at the phase transition temperature [17]. According to the Clausius–Clapeyron equation, the relationship between the omnidirectional hydrostatic pressure and the phase transition temperature can be expressed as

$$\frac{dP}{dT} = \frac{\Delta H}{T_0 \Delta V} = \frac{\Delta S}{\Delta V} \tag{1}$$

where P is the hydrostatic pressure, T is the phase transition temperature that occurs under the hydrostatic pressure; T_0 is the equilibrium temperature of the two phases, $T_0 = (M_s + A_s)/2$, where A_s and M_s are the characteristic temperatures of the SMA phase transition, respectively; ΔV is the change in volume that occurs during the SMA phase transition, ΔH is the latent heat of the phase transition, and ΔS is the change in entropy of the phase transition.

The temperature of the SMA phase transformation depends linearly on the external stress, and changes in the phase transition temperature do not depend on the macroscopic direction of the stress [18,19]. The rate of change of the equivalent hydrostatic stress as a function of phase transition temperature under multidimensional loading can be expressed as

$$\frac{dP}{dT} = C_i \tag{2}$$

where C_i is the rate of change of the stress of the martensite ($i = M$) or austenite ($i = A$) transformation (which is the slope of the stress-transition temperature curve). Integrating Equation (2) results in

$$T = \frac{P}{C_i} + T_0' \tag{3}$$

where T_0' is the transformation temperature under normal pressure.

Because the thermoelastic phase transition of dense solid SMA wire is almost volume-conserved, the percentage of the volume change caused by the phase transition under an external force is very small and almost negligible. Therefore, the effect of hydrostatic pressure on the transformation strain of solid SMA wire is negligible, i.e., the influence of omnidirectional hydrostatic pressure on the solid SMA constitutive model can be ignored [20,21].

3. Constitutive Model of Nonlinear Phase Transition

SMAs have special thermodynamic properties that are different from the general properties of metals, and their constitutive model exhibits a coupled nonlinear stress–strain–temperature relationship. Most of the engineering applications of deep-sea actuators are based on SMA wire [22,23]. The macroscopic phenomenological constitutive model based on thermodynamics and phase transition dynamics can facilitate simple and efficient simulations and predictions of the macroscopic thermodynamic behavior of SMAs, which is relevant for the design of intelligent actuators used in marine engineering applications.

3.1. *Thermodynamic Constitutive Equations of SMAs in the Deep Sea Environment*

In the deep sea environment, fully dense SMA wire is constrained by a multidimensional stress tensor, which is composed of two parts: the omnidirectional hydrostatic pressure tensor and the deviatoric stress tensor. As the elastic volume changes, the deviatoric stress tensor primarily causes the deformation of the material. Small-diameter and dense SMA wire used in deep-sea actuators undergoes very minimal volume changes due to the omnidirectional hydrostatic pressure, and the effect of the hydrostatic pressure on the phase transformation strain in SMAs is negligible [20,21]. Assuming SMA wire is isotropic, the multidimensional load state of the SMA wires in the deep sea environment can be simplified to one dimension.

According to Hartl et al. [24], the total strain in an SMA is the sum of the elastic strain, phase transformation strain, and thermal expansion strain. This can be expressed as

$$\varepsilon = \varepsilon^{e} + \varepsilon^{t} + \varepsilon^{\alpha} \tag{4}$$

where ε is the total strain, ε^{e} is the elastic strain, ε^{t} is the phase transformation strain, and ε^{α} is the thermal expansion strain.

Differentiating Equation (4), the Green–Lagrange strain increment form can be expressed as

$$d\varepsilon = d\varepsilon^{e} + d\varepsilon^{t} + d\varepsilon^{\alpha} \tag{5}$$

where $d\varepsilon^{e}$ is the elastic strain increment, $d\varepsilon^{t}$ is the transformation strain increment (which is related to the martensite volume fraction), and $d\varepsilon^{\alpha}$ is the thermoelastic strain increment.

According to the principles of elasticity and thermodynamics,

$$\begin{cases} d\varepsilon^{e} = d\sigma / D(\xi) \\ d\varepsilon^{\alpha} = \alpha dT \end{cases} \tag{6}$$

where σ is the stress, T is the temperature, ξ is the volume fraction of martensite, $D(\xi)$ is the elastic modulus of the SMA, and α is the thermal expansion coefficient of the SMA material.

Given that the transformation strain varies linearly with the volume fraction of the detwinned martensite, the maximum transformation strain is reached in the state in which 100% of the martensite is detwinned [25]. This can be expressed as

$$\varepsilon^{t} = \varepsilon_{L}\xi_{DM} \tag{7}$$

where ε_{L} is the maximum recoverable residual strain of the SMA.

Combining Equations (5)–(7), the SMA constitutive equation can be expressed as

$$d\varepsilon = \frac{d\sigma}{D(\xi)} + \varepsilon_{L} d\xi_{DM} + \alpha dT \tag{8}$$

Based on the crystal phase structure composition of SMAs, and the fact that the martensite contains twinned martensite and detwinned martensite variants, the total martensite volume fraction and elastic modulus of SMAs can be expressed as

$$\begin{cases} \xi = \xi_{DM} + \xi_{TM} \\ D(\xi) = D_{A} + \xi(D_{M} - D_{A}) = D_{A} + \xi(D_{TM} - D_{A}) + \xi_{DM}(D_{DM} - D_{TM}) \end{cases} \tag{9}$$

where ξ_{TM} and ξ_{DM} are the volume fractions of the twinned and detwinned martensite, respectively; D_{M} and D_{A} are the elastic moduli of the complete martensite and austenite phases, respectively, and D_{TM} and D_{DM} are the elastic moduli of the twinned and detwinned martensite, respectively.

Integrating Equation (8), the SMA constitutive equation in full form can be expressed as

$$\varepsilon - \varepsilon_{0} = \frac{\sigma}{D(\xi)} - \frac{\sigma_{0}}{D(\xi_{0})} + \varepsilon_{L}(\xi_{DM} - \xi_{DM0}) + \alpha(T - T_{0}) \tag{10}$$

where ε_{0}, σ_{0}, ξ_{DM0}, and T_{0} represent the initial values of the corresponding variables.

3.2. Dynamic Equation of Phase Transition Based on Logistic Function

The kinetic equation of the phase transformation characterizes the dependence of the martensite volume fraction on changes in the external loads (e.g., temperature and stress), and is the basis for the construction of the SMA constitutive model. The logistic function is a continuous, smooth, and strictly monotonic threshold function, and its mathematical formula is defined as [26]

$$f(x) = \frac{1}{1+e^{-x}} \tag{11}$$

where x is the independent variable.

The logistic function is characterized by typical sigmoid nonlinear growth [14]. The bell-shaped curve of its natural growth rate corresponds to the rate of change of the peak of the heat flow-temperature curve of SMAs, and the cumulative growth curve is very similar to the rate of change of the martensite volume fraction as a function of temperature, as shown in Figure 1. Figure 1a shows the curve of logistic function, and Figure 1b shows the derivative of logistic function.

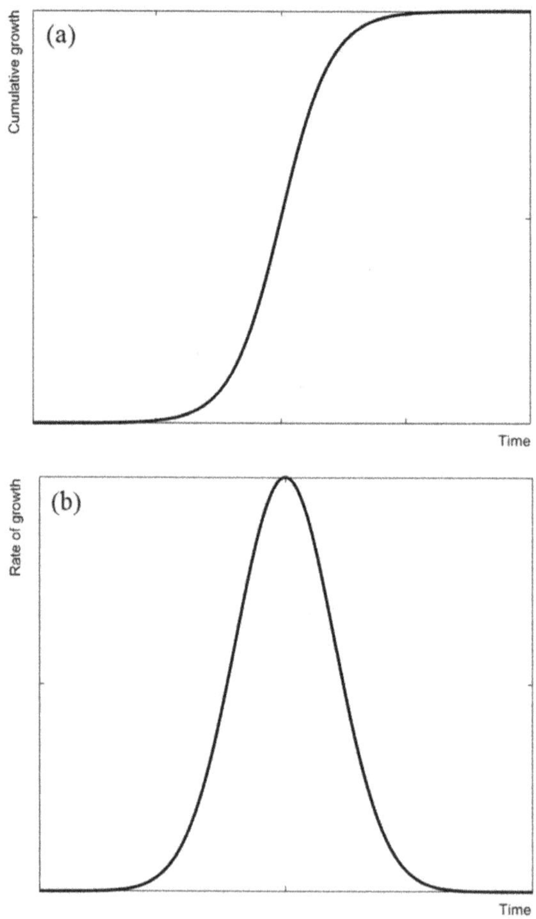

Figure 1. Typical (**a**) logistic function for the cumulative martensite volume fraction and (**b**) bell curves for the rate of change martensite volume fraction.

According to the differential scanning calorimetry (DSC) test curve of Ikuta et al. [19], the rate of change of the martensite volume fraction is a typical S-shaped nonlinear growth curve. On the basis of the phase transition kinetic model constructed by Ikuta et al., the description of the intermediate variable process using the phase transition rate coefficient was added. By replacing the traditional phase transition kinetic equation with the improved logistic function, the slow and gradual phase transition in the initial and final regions of the entire phase transition process could be described and the thermodynamic behavior of SMA with different phase transition rates could be expressed. This generated numerical simulation results that were more consistent with the experimental results. The shape memory effect has different transformation kinetic equations in the process of martensite forward and reverse transformation.

When $\dot{T} > 0$, the martensite volume fraction can be expressed as

$$\xi_{M\to A} = \frac{1}{1 + e^{\frac{ab}{A_f - A_s}\left(T - T_A - \frac{\sigma}{C_A}\right)}} \tag{12}$$

$$T_A = \frac{A_s + A_f}{2} \tag{13}$$

and when $\dot{T} < 0$, the martensite volume fraction can be expressed as

$$\xi_{A\to M} = \frac{1}{1 + e^{\frac{ab}{M_s - M_f}\left(T - T_M - \frac{\sigma}{C_M}\right)}} \tag{14}$$

$$T_M = \frac{M_s + M_f}{2} \tag{15}$$

where T_M and T_A are the average characteristic temperatures of the martensite and austenite phases under zero stress, respectively. M_s and M_f are martensite start and finish temperature, respectively, during cooling. A_s and A_f are austenite start and finish temperature, respectively, during heating. C_M and C_A are the rates of change of the stress of the forward and reverse martensite transformations (which are the slopes of the stress-transformation temperature curve), respectively, a is the phase transition rate coefficient, and b is a material constant related to the atomic ratio of Ni-Ti alloys (which can be obtained via a DSC test). When $\dot{T} = 0$, the SMA transforms from twinned martensite to detwinned martensite due to the stress. The detwinned martensite volume fraction can be expressed as

$$\xi_{DM} = \begin{cases} 0 & \sigma \le \sigma_s \\ \frac{1}{1+e^{\frac{ab}{\sigma_s - \sigma_f}(\sigma - \sigma_d)}} & \sigma_s < \sigma < \sigma_f \\ 1 & \sigma \ge \sigma_f \end{cases} \tag{16}$$

$$\sigma_d = \frac{\sigma_s + \sigma_f}{2} \tag{17}$$

where σ_s and σ_f are the critical stresses at the beginning and end of the martensite detwinning, respectively, and σ_d is the average value of the critical stress of the detwinned martensite phase in the zero stress state.

4. Results and Discussion

Based on the improved SMA constitutive model of the logistic function, the dynamic process of the SMA phase transition was fitted and analyzed. Numerical fitting and prediction were carried out through the shape memory effect under the two scenarios of a constant load and constant temperature, and the model was compared and verified by traditional constitutive data as well as literature test data.

4.1. Logistic Variable Speed Phase Transition Kinetics Fitting and Analysis

Under the constant load scenario, the initial stress was assumed to be 50 MPa and the temperature cycling range was 0–150 °C. The relationship between the martensite volume fraction and the temperature was numerically fitted using the Brinson model and the logistic variable speed constitutive model, and the results are shown in Figure 2. The fitted curves show that the dynamic equation of the logistic variable-speed transformation adjusted the shape of the curve using a single coefficient, and that it could flexibly adjust the rate of change of the transformation rate. The rate of change of the martensite volume fraction at a given temperature was not unique. However, for a given material parameter in the Brinson model, the rate of change of the martensite volume fraction was unique, the phase transformation rate was constant, and the intermediate process could not be flexibly adjusted. In contrast, the thermomechanical responses in the initial and final regions of the phase transition represented by the logistic variable-speed phase transition kinetic equation were completed in a smooth and gradual manner, while the phase transition region represented by the Brinson model was relatively sharp and abrupt. When the transformation rate coefficient was $a = 1$, the rates of change of the martensite volume fraction fitted by the logistic variable-speed transformation kinetic equation and the Brinson model were essentially the same except for the beginning and end of the transformation. As the phase transition coefficient increased, the curve became steeper in the phase transition region, and the phase transition responded more sharply in the initial and final regions. In summary, the improved variable-speed phase transition kinetic equation based on the logistic function could flexibly adjust the intermediate process of the phase transition and had superior adaptability compared to the traditional model.

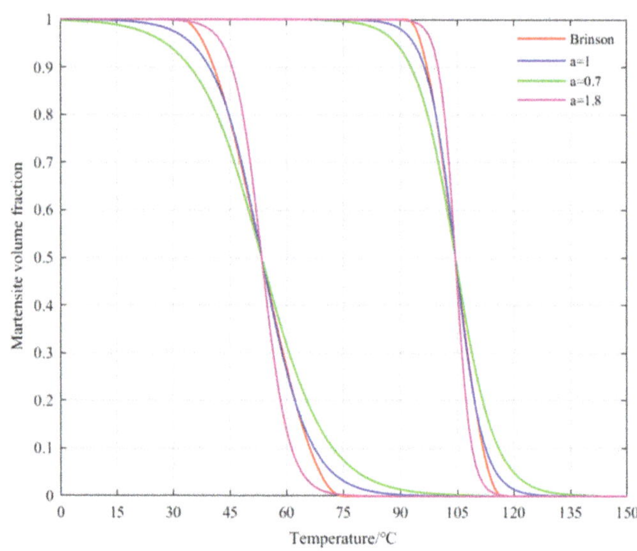

Figure 2. Relationship between the martensite volume fraction and the temperature for different transformation rate coefficients.

Figure 3 shows the numerical fitting results for the volume fraction rates of change for martensite and austenite with temperature cycling under a constant stress of 50 MPa. During the process of martensite forward and reverse transformation, the rate of change of the martensite volume fraction initially increased and then decreased as the temperature changed. The change was relatively gradual at the beginning and end of the phase transition, and it reached a peak in the central region of the phase transition. This was generally consistent with the trends in the rates of change of the endothermic and exothermic peaks in the heat flow-temperature curve measured by the DSC test. When the transformation rate coefficient $a = 1$, the logistic function basically agreed with the rate of change of the martensite volume fraction curve described by the transformation control equation in the form of a cosine function. As the phase transition rate coefficient increased, the range of the rate of change gradually increased, and the peak value that appeared increased gradually.

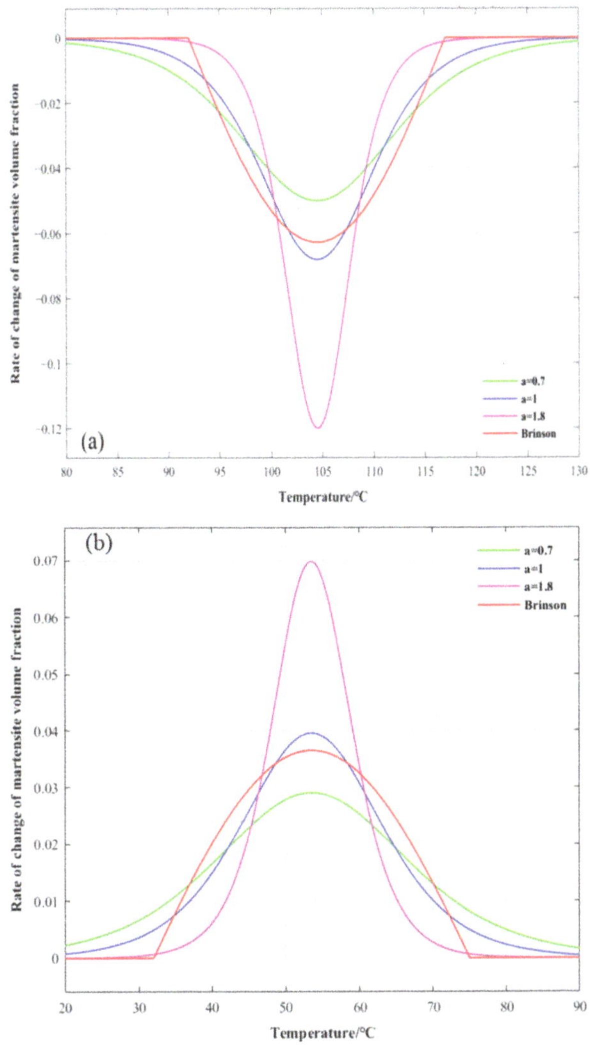

Figure 3. Relationship between the rate of change of the martensite volume fraction and the temperature for different transformation rate coefficients: (**a**) reverse martensitic transformation, (**b**) martensitic transformation.

4.2. Numerical Fitting and Analysis of Phenomenological Constitutive Model

The constitutive model improved by the logistic function could completely describe the thermodynamic response characteristics of the beginning, ending, and intermediate processes of the SMA phase transition. In order to verify the accuracy and validity of the model, the SMA shape memory effect under constant load and constant temperature was numerically fitted and predicted using MATLAB, and the fitting results were compared to the Brinson model and the experimental data. The material parameters involved in the fitting are shown in Table 1. The thermomechanical performance parameters refer to the data in [25].

Table 1. The transformation temperatures and thermomechanical properties of SMA.

Phase Transition Temperature	Thermal Coefficient	Elastic Modulus	Phase Transition Parameters	Maximum Residual Strain
M_f = 22 °C	$\alpha = 1 \times 10^{-5}$ °C^{-1}	D_{TM} = 22,800 MPa	C_M = 5 MPa/°C	ε_L = 0.072
M_s = 65 °C	b = 6.8	D_{DM} = 17,600 MPa	C_A = 5 MPa/°C	
A_s = 82 °C		D_A = 42,800 MPa	σ_s = 136 MPa	
A_f = 107 °C			σ_f = 161 MPa	

4.2.1. Different Load Cycles at Constant Temperature

Under different load conditions at constant low temperature, the initial crystal phase structure of the SMA was completely twinned martensite, and its initial conditions were $T_0 = -10$ °C, $\sigma_0 = 0$ MPa, $\varepsilon_0 = 0$, $\xi_{DM0} = 0$, $\xi_0 = 1$, and $a = 0.5$. When the twinned martensite was completely transformed into detwinned martensite, the end conditions were $\xi_{DM} = 1$, $\xi = \xi_{DM}$. Combining Equations (10) and (16), we can get the complete stress–strain relationship in the detwinning process. Figure 4 shows the numerical fitting of the stress–strain relationship of the SMA wire at constant temperature performed by the logistic variable-speed constitutive model and the Brinson model; this was also verified by the experimental data taken from the literature [25]. At constant temperature, the stress–strain relationship of the SMA under different loads presented three different response regions during the loading process. In the initial elastic region, the stress increased linearly with the strain. When the stress reached the critical stress, it entered the phase transformation plateau region; the twinned martensite began to transform into detwinned martensite, the strain increased significantly, and the stress remained stable. When the stress reached the critical stress, it entered the linear elastic region again, the twinned martensite was completely transformed into detwinned martensite, and the stress increased linearly with the strain again. During the unloading process, the martensite existed as a completely detwinned variant, and the stress decreased linearly with the strain. When the stress dropped to zero, a small part of the elastic strain recovered, and most of the transformation strain remained. When the temperature was raised again, the residual transformation strain recovered, thus realizing the shape memory effect. Figure 4 shows that the elastic moduli of the twinned martensite and detwinned martensite structures were not equal, and the fitting results of the Brinson model exhibited a certain deviation in the initial and final stages of the transformation and the unloading stage. The rate of change of the structural model at the beginning and end of the phase transition was relatively smooth and slow, and the fitted curve was more similar to the experimental data.

4.2.2. Different Temperature Cycles under a Constant Load

Under a constant load, the initial crystal phase structure of the SMA was fully twinned martensite, the temperature range was −20–200 °C, and the initial conditions were $T_0 = -20$ °C, $\sigma_0 = 100$ MPa ($\sigma_0 < \sigma_s$), $\varepsilon_0 = 0$, $\xi_{DM0} = 0$, $\xi_0 = 1$, and $a = 0.5$. Combining Equations (10), (12) and (14), we can get the relationship between strain and temperature of the complete temperature cycle. Figure 5 shows the numerical fitting of the strain-temperature relationship of the SMA wire under constant stress for the logistic model and the Brinson model. The logistic model was verified by the experimental data in the literature [25]. The results show that the strain–temperature relationship of the SMA under constant load had three different response regions: the pure martensite state, pure austenite state, and mixed transformation state. The SMA strain in the non-transition region slowly decreased and increased with increasing and decreasing temperature, respectively, until it became stable. The mixed phase transition region presented a nonlinear mutation, and the mutation point was basically consistent with the four phase transition temperature characteristic points (M_s, M_f, A_s, and A_f). The strain–temperature curves did not coincide during heating and cooling, and there was an obvious strain difference. Figure 5 shows that the strains of the three different response regions were all nonlinear with temperature, while the curves fitted by the Brinson model were all linear in the non-transition region and their transitions were too sharp at the beginning and end of the phase transition. Furthermore, they deviated by a certain amount from the test results. The fitting results of the improved phase transition model based on the logistic function were in good agreement with the experimental results. The nonlinear relationships in different regions were fitted completely, and the slow and gradual phase transition processes at the initial and final stages of the phase transition were accurately described. The small deviation between the heating fitting results and the experimental data was due to the fact that the temperature obtained from the experiment was 2 °C lower than the parameter value used in the numerical fitting.

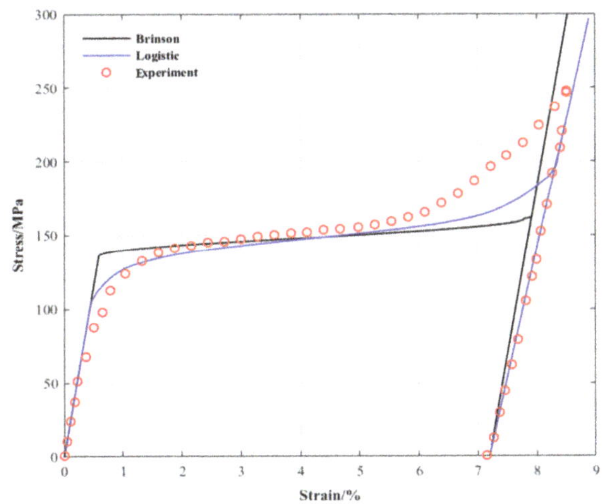

Figure 4. Comparison between the simulation and experimental results of the stress-strain relationship at −10 °C.

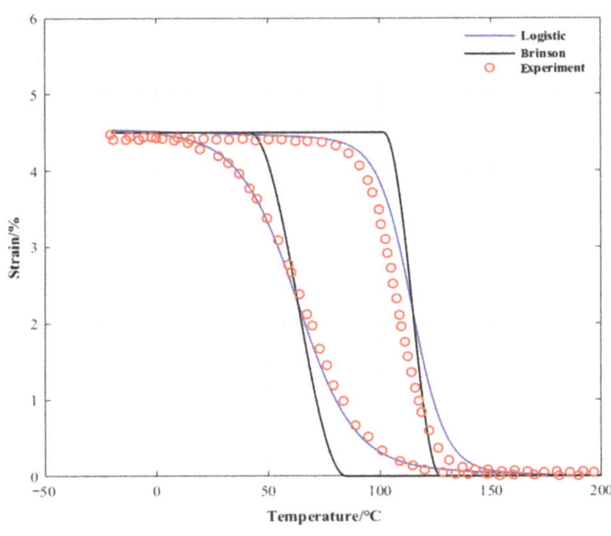

Figure 5. Comparison between the simulation and experimental results of the relationship between strain and temperature at 100 MPa.

4.3. Consideration of Marine Engineering Application

In addition to limit factors of traditional SMA actuators, there are very special aspects that need to be considered when SMA actuators are applied in deep sea extreme environment, including the conductivity of seawater, strong corrosion, low temperature, and high hydrostatic pressure.

4.3.1. Conductivity of Seawater

The biggest advantage of SMA is that it can integrate execution and sensing. Moreover, it can feedback the execution of the driver through its own resistance change. However, seawater is a multicomponent electrolyte solution, and the conductivity of different sea areas will be different due to the difference of temperature, salinity, and pressure, which will lead to a deviation of the closed-loop control method based on resistance feedback. Additionally, the conductivity of seawater will also cause the leakage of the SMA and connecting wires, which will inevitably lead to a loss of energy efficiency of the driver.

4.3.2. Strong Corrosive

Seawater is an excellent alkaline environment that can promote the formation of an electrochemical mechanism of corrosion and hydrolysis. When SMA is not energized, the stable and discontinuous adhesive passivation film formed on its surface can keep good corrosion resistance and biocompatibility of SMA. When the SMA is energized underwater, the seawater will be electrolyzed by electrical signals and hydrogen is produced. Ti based SMA are very sensitive to hydrogen, and sudden failure which is known as hydrogen embrittlement will happen. Therefore, anti-corrosion protection must be considered for deep sea SMA actuators, including soft sealing, oil-filled sealing, and insulating coating.

4.3.3. High Hydrostatic Pressure

The maximum hydrostatic pressure in an extreme deep-sea environment can reach 113 MPa, and the phase transformation strain of SMA with small diameter and compact structure is small under the hydrostatic pressure of 113 MPa. However, the damage to the porous structure caused by long-term electrolysis will gradually increase the influence of hydrostatic pressure on the SMA, which cannot be ignored.

4.3.4. Low-Temperature Environment

The thermodynamic performance of SMA is closely related to the ambient temperature. The maximum difference between the temperature of the deep seabed and the normal temperature is 10 times. The low temperature will accelerate the surface heat dissipation of SMA, which requires more driving thermal power. The thermodynamic performance parameters in the air can no longer be applied to the underwater environment. Therefore, the driving thermodynamic properties of SMA in deep-sea applications need to be further explored.

5. Conclusions

The traditional phenomenological model used to describe SMAs has the common drawbacks of abrupt transitions at the beginning and ending of the phase transition and an invariable speed in the phase transition process. The traditional phase transition models with constant phase transition speed, such as cosine functions, linear functions, and exponential functions, are suitable for single crystal SMAs, while polycrystalline SMAs require a more flexible variable speed phase transition function. The improved logistic function developed in this study is an excellent substitute for the traditional phase transition function.

The phenomenological variable-speed constitutive model expanded the thermodynamic constitutive equation through the decomposition of the phase transition strain, replaced the traditional phase transition function with an improved logistic nonlinear function, and increased the phase transition rate coefficient to systematically describe the variability of the intermediate phase transition process. Compared to the traditional model, this model has more advantages and better flexibility, and is more consistent with actual test results. Through an analysis of the fitting performance in multiple scenarios, it was found that the dynamic model of the variable-speed phase transition improved by the logistic function could adjust the shape of the curve through a single coefficient, and it could also perform simple and efficient fitting and prediction of the SMA shape memory effect with fewer parameters. Additionally, it accurately modeled the gradual changes in the initial and final regions of the phase transition and had the ability to flexibly adjust the rate of change of the phase transition rate. These results can be used to lay a theoretical foundation for the design of SMA deep-sea actuators used in marine engineering applications.

Author Contributions: Conceptualization, J.G., B.P. and W.C.; methodology, J.G.; validation, J.G., B.P., W.C. and S.H.; formal analysis, J.G. and B.P.; writing—original draft preparation, J.G.; writing—review and editing, J.G., B.P. and W.C.; project administration, S.H. All authors have read and agreed to the published version of the manuscript.

Funding: This research was funded by the Science and Technology Research Project of Social Development of Shanghai Science and Technology Commission, grant number 21DZ1205503.

Institutional Review Board Statement: Not applicable.

Informed Consent Statement: Not applicable.

Data Availability Statement: Not applicable.

Conflicts of Interest: The authors declare no conflict of interest.

References

1. Li, C. The potential and challenges of China's marine science and technology development. *People's Forum Acad. Front.* **2017**, *18*, 37–43.
2. Li, J.; Liu, F.; Ma, W. Development status of intelligent equipment industry for marine unmanned systems at home and abroad. *Ship Eng.* **2020**, *42*, 25–31.
3. Guo, J.; Pan, B.; Cui, W.; Hu, S.; Han, Z. Review of deep-sea actuators and marine bionic robots based on intelligent materials. *Ship Mech.* **2022**, *26*, 13.
4. Bai, Z. Summary of research progress and application of smart materials. *Mil. Civ. Dual-Use Technol. Prod.* **2020**, *3*, 15–20.
5. Guo, K.; Xia, P. Research progress of smart composite materials. *Funct. Mater.* **2019**, *50*, 4017–4022, 4029.
6. Hao, L. Research on Mechanical Properties of Shape Memory Alloy Bulge. Ph.D. Thesis, Nanjing University of Aeronautics and Astronautics, Nanjing, China, 2018.
7. Moumni, Z.; Zaki, W.; Nguyen, Q.S. Theoretical and Numerical Modeling of Solid–Solid Phase Change: Application to the Description of the Thermomechanical Behavior of Shape Memory Alloys. *Int. J. Plast.* **2008**, *24*, 614–645. [CrossRef]
8. Yang, J.; Huang, B.; Gu, X.; Wang, J.; Zhang, Y.; Zhu, J.; Zhang, W. Review of Mechanical Behavior and Application of Shape Memory Alloys. *J. Solid Mech.* **2021**, *42*, 345–375.
9. Liu, K. Design and Control Technology of Underwater Vector Deflection Actuator Based on Ni-Ti Shape Memory Alloy. Master's Thesis, South China University of Technology, Guangzhou, China, 2020.
10. Cisse, C.; Zaki, W.; Zineb, T.B. A review of constitutive models and modeling techniques for shape memory alloys. *Int. J. Plast.* **2016**, *76*, 244–284. [CrossRef]
11. Lagoudas, D.; Hartl, D.; Chemisky, Y.; Machado, L.; Popov, P. Constitutive model for the numerical analysis of phase transformation in polycrystalline shape memory alloys. *Int. J. Plast.* **2012**, *32–33*, 155–183. [CrossRef]
12. He, J.; Toi, Y. Improved constitutive modeling for phase transformation of shape memory alloys. *J. Solid Mech. Mater. Eng.* **2013**, *7*, 11–26. [CrossRef]
13. Lagoudas, D.C.; Entchev, P.B.; Popov, P.; Patoor, E.; Catherine Brinson, L.; Gao, X. Shape memory alloys, Part II: Modeling of polycrystals. *Mech. Mater.* **2006**, *38*, 430–462. [CrossRef]
14. Kucharavy, D.; Guio, R.D. Application of S-shaped curves. *Procedia Eng.* **2011**, *9*, 559–572. [CrossRef]
15. Guan, J.H.; Pei, Y.C.; Wu, J.T. A driving strategy of shape memory alloy wires with electric resistance modeled by logistic function for power consumption reduction. *Mech. Syst. Signal Process.* **2021**, *160*, 107839. [CrossRef]
16. Ding, Y. Research on the Multidimensional Constitutive Relation of Shape Memory Alloys. Master's Thesis, Xi'an University of Architecture and Technology, Xi'an, China, 2005.
17. Johari, G.P.; McAnanama, J.G.; Sartor, G. Effect of hydrostatic pressure on the thermoelastic transformation of Ni-Ti alloy and the entropy of transformation. *Philos. Mag. B* **1996**, *74*, 243–257. [CrossRef]

18. Wan, J.L.; Chen, Z.B.; Qin, S.J.; Shang, J. Effect of hydrostatic pressure on thermally induced phase transformation in NiTi alloy: A molecular dynamics study. *Comput. Mater. Sci.* **2018**, *153*, 119–125. [CrossRef]
19. Ikuta, K.; Tsukamoto, M.; Hirose, S. Mathematical model and experimental verification of shape memory alloy for designing micro actuator. In Proceedings of the IEEE Micro Electro Mechanical Systems, Nara, Japan, 30–2 January 1991; pp. 103–108.
20. Liu, B. Study on the Phase Transformation Mechanism and Mechanical Properties of Porous Shape Memory Alloys. Ph.D. Thesis, Beijing Jiaotong University, Beijing, China, 2013.
21. Zhu, X.; Dui, G. Modeling of superelastic–plastic behavior of porous shape memory alloys incorporating void shape effects. *Acta Mech. Solida Sin.* **2021**, *34*, 632–644. [CrossRef]
22. Sumoto, H.; Yamaguchi, S. A study on a control meth- od of artificial muscle using segmented binary control for an up-scaled fish type robot. In Proceedings of the 21st International Offshore and Polar Engineering Conference, Maui, HI, USA, 19–24 June 2011; pp. 223–229.
23. Angilella, A.J.; Gandhi, F.S.; Miller, T.F. Design and testing of a shape memory alloy buoyancy engine for unmanned underwater vehicles. *Smart Mater. Struct.* **2015**, *24*, 115018. [CrossRef]
24. Hartl, D.J.; Lagoudas, D.C. Thermomechanical Characterization of Shape Memory Alloy Materials. In *Shape Memory Alloys: Modeling and Engineering Applications*; Springer: Boston, MA, USA, 2008.
25. Li, J.; Yang, Z. Numerical analysis and modification of thermodynamic empirical constitutive model for shape memory alloys. *Mater. Rev.* **2021**, *35*, 18116–18123.
26. Kudryashov, N.A. Logistic function as solution of many nonlinear differential equations. *Appl. Math. Model.* **2015**, *39*, 5733–5742. [CrossRef]

Journal of
Marine Science and Engineering

MDPI

Article

Numerical Study on the Water Entry of a Freely Falling Unmanned Aerial-Underwater Vehicle

Liyang Dong [1,2,3], Zhaoyu Wei [1,*], Hangyu Zhou [1], Baoheng Yao [1,3] and Lian Lian [1,3,*]

1 School of Oceanography, Shanghai Jiao Tong University, Shanghai 200240, China
2 School of Naval Architecture & Ocean Civil Engineering, Shanghai Jiao Tong University, Shanghai 200240, China
3 State Key Laboratory of Ocean Engineering, Shanghai Jiao Tong University, Shanghai 200240, China
* Correspondence: wzhyu@sjtu.edu.cn (Z.W.); llian@sjtu.edu.cn (L.L.)

Abstract: The unmanned aerial–underwater vehicle (UAUV) is a new type of vehicle that can fly in the air and cruise in water and is expected to cross the free water surface several times to perform continuous uninterrupted observation and sampling. To analyze the hydrodynamic and motion characteristics of the vehicle, the whole water-entry process of a multi-degree-of-freedom UAUV with various velocity and pitch angle was investigated through a Reynolds-averaged Navier–Stokes method. The computational domain was meshed by trimmer cells. The relative movement between the vehicle and fluid domain was simulated using moving reference frame overset mesh to delineate the interaction region around vehicle body. To reduce the computational cost, adaptive mesh refinement and adaptive time-stepping strategy were used to capture the slamming pressure accurately with reasonable computational effort. First, convergence study is considered. Simulations of the vehicle with various initial velocities and different pitch angles were performed. The variable physical properties were analyzed, and detailed results through the time-varying force and velocity were shown. Initial velocity and pitch angle are found to significantly influence hydrodynamic behavior, including the time-varying force, while thickness ratio has a great impact on added mass and pressure. The results show that higher entry velocity results in greater peak vertical force. The transverse hydrodynamic characteristics for oblique water entry of the vehicle with varies pith angle are quite different.

Keywords: unmanned aerial–underwater vehicle; adaptive mesh refinement; water entry; hydrodynamic forces; freely falling

Citation: Dong, L.; Wei, Z.; Zhou, H.; Yao, B.; Lian, L. Numerical Study on the Water Entry of a Freely Falling Unmanned Aerial-Underwater Vehicle. *J. Mar. Sci. Eng.* **2023**, *11*, 552. https://doi.org/10.3390/jmse11030552

Academic Editor: Rafael Morales

Received: 2 February 2023
Revised: 25 February 2023
Accepted: 1 March 2023
Published: 4 March 2023

1. Introduction

The unmanned aerial vehicle (UAV) and the autonomous underwater vehicle (AUV) have become significant in joint multidomain marine operations, which have been widely favored by research communities. A single independent unmanned vehicle cannot satisfy the mission demand well, whereas it also is a daunting challenge to integrate various kinds of vehicles into one system properly. The unmanned aerial–underwater vehicle (UAUV) exhibits the characteristics of both the UAV and AUV, which meets demands for aerial and underwater tasks. The UAUV can fly in the air and cruise in water and is expected to cross the free water surface several times (Nicolaou [1]). The UAUV concept has been introduced as the LPL (Russian abbreviation for Flying Submarine) projects by the Soviet Union, which can navigate in either air or water (Petrov [2]). So far, UAUVs are still at the stage of prototype concept, and most work has been devoted to designing a configuration fitting for both air and water domain operations, and its irregular surface shape poses some challenges in the final manufacture. With their potential civil and military requirement, UAUVs have attracted more and more attention from researchers in recent years.

The complete motion process of the vehicle consists of three main stages: air flight, underwater navigation, and media crossing motion, including water entry and exit stages.

The wing-folding machinery for the vehicle is difficult to design and manufacture. The propulsion systems should be suitable for both water-to-air and air-to-water transition (Li et al. [3]). Among the key technologies for the development of UAUVs, the techniques for air-to-water transition are quite significant, especially for fixed-wing shaped UAUVs. During the transition, fixed-wing UAUVs always undergo a sharp change in hydrodynamic force and a complicated slamming phenomenon associated with cavity formation, splash, and free surface deformation, which may cause damage to the vehicles' components. For instance, the impact load may make UAUVs bounce on the water surface, damaging their components (Oumeraci [4]). Instability of the motion may also produce negative effects for the working conditions of the cruise. Furthermore, the hydrodynamic forces of lift and drag affect motion during the smoothing and consecutive transition from air flight to water cruise (Wei et al. [5]). To consider this aspect, the hydrodynamic characteristics of UAUVs were studied in the present work using the computational fluid dynamics (CFD) method.

The pioneer reports on the phenomenon of water impact by a rigid body were published by Von Karman [6], who gave an analytical expression for the impact force derived from the momentum theorem without considering the splash-up effect. Based on Von Karman [6], Wagner [7] derived a fair flat theoretical assumption and proposed a mathematical model that obtained satisfactory results. Based on an integro-differential equation introduced by Wagner [7], Dobrovol'skaya [8] extended the model and obtained self-similar solutions for water entry of a wedge at a constant speed. Toyama [9] presented typical solutions for asymmetrical water entry of unsymmetrical section bodies without flow separation, based on the potential flow introduced by Wagner [7]. Based on Dobrovol'skaya [8], Mei et al. [10] introduced a boundary element method (BEM). Early investigations on the problem of water entry were studied by many researchers (Vorus [11], Semenov & Iafrati [12], Xu & Wu [13], Han & Sun [14]) using potential flow and boundary elements theory, and variations in pressure distribution were also obtained. Based on the accurate surface shape of the fluid-structure coupling, Wang et al. [15] established an oblique water-entry impact model coupled with ballistic and dynamic models. However, the above methods ignore some physical properties, such as viscosity, vortex, and cavitation of the fluid.

With the rapid development of high-performance computational technology in recent years, the parallelization of computation efficiency and accuracy has increased so that complicated coupling problems are expected to be solved (Afzal, 2017 [16]). Qiu et al. [17] studied the water entry process of bodies of revolution by the CFD analysis software FLUENT with user-defined functions (UDFs) and a MIXTURE process model. Based on the Reynolds-averaged Navier–Stokes (RANS) method, Guo et al. [18] simulated the planned forced landing of a transport aircraft on the water free surface and studied the water-entry impact forces and the proper initial angle. Shi et al. [19] designed a buffer cap for an underwater vehicle and analyzed the influence of various parameters on the water-entry process. Chen et al. [20] presented a numerical study of the launching and recovery process for an autonomous underwater helicopter from a research vessel using the commercial software STAR-CCM+, showing that the optimal water-entry angle was 75°, which contributes to shape optimization. Du et al. [21] studied the influence of geometrical shapes on cavity evolution, flow fields, and vortex structure evolution during water-entry processes by using experimental and OpenFOAM platforms.

By simulating the hydrodynamics of water entry, some further applications about the vehicles have also been studied by researchers. Based on the motions and forces result by numerical simulations, identification of vehicle parameters can be done thereafter. Wu et al. [22] established the dynamic model of a freely falling UAUV immerging into the water and obtained the satisfactory trajectory using a cuckoo search (CS) optimization algorithm. Wu et al. [23] also studied the cross-domain trajectory for an aerial–aquatic coaxial eight-rotor vehicle, considering the influence of navigation error. On the other hand, some control could complete the control task of the unmanned vehicle system, without a priori knowledge or an identification process. Baldi & Roy [24] designed an autopilot framework, which needs no knowledge of the vehicle dynamics and environment messages. Embedding adaptation mechanisms in guidance law, Wang & Roy [25] proposed an adaptive vector field guidance method for UAVs, where a priori knowledge is also not required, and obtained fine results in the tests on the hardware platforms.

When a fixed-wing UAUV enters the water, the impact always induces a series of problems. This work considers the hydrodynamic problems of UAUVs using the CFD method. In contrast to the regular shapes, such as wedges, cones, and circulars mentioned above, vehicles have curvy surfaces and relatively complicated shapes, which makes the problem no longer linear. The volume of fluid (VOF) method is utilized to numerically capture the free surface profiles (Wang et al. [26]), and overset mesh technology is utilized to achieve the interaction between the background domain and the body region. This technology has an instructive significance in practical engineering applications that provides a reference in the dynamic modeling of cross-domain bodies. The present study is motivated by the need to evaluate the impact loads that may induce damage to the structure or bouncing of the falling object.

2. Materials and Methods

For freely falling bodies, it is also common to assume that the motions can be decoupled into longitudinal and lateral motions. In this study, it is speculated that the vehicle enters the water in a vertical plane, as shown in the sketch in Figure 1. It is of interest to analyze the slamming force and the motion responses of the vehicle.

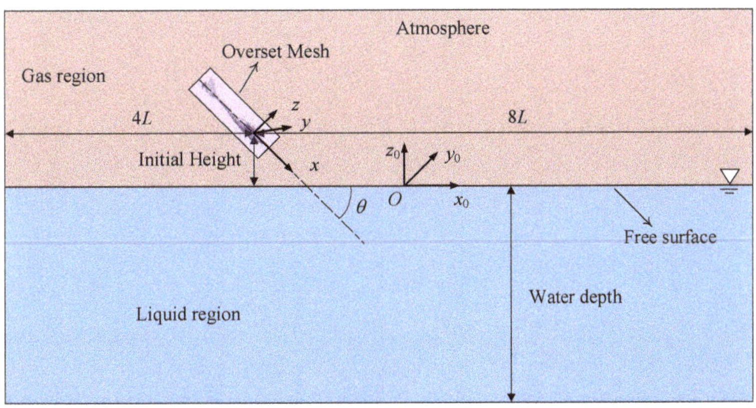

Figure 1. The sketch of the problem definition.

2.1. Governing Equations and Turbulence Model

As the velocity of the vehicle is much lower than the acoustic speed, the compressibility of water can also be ignored. Without energy exchange, the numerical solution only follows the conservation equation of mass and continuum:

$$\frac{\partial \rho}{\partial t} + \frac{\partial (\rho u_i)}{\partial x_i} = 0 \tag{1}$$

$$\frac{\partial (\rho u_i)}{\partial t} + \frac{\partial (\rho u_i u_j)}{\partial x_j} = -\frac{\partial p}{\partial x_i} + \frac{\partial \rho (\tau_{ij})}{\partial x_j} + \rho g_j \tag{2}$$

where x_j are the Cartesian coordinates; u_i and u_j are the Cartesian components of the time-averaged velocity ($i, j = 1, 2, 3$), p is mean pressure, ρ is the fluid density, g denotes the body forces due to the gravitational acceleration, and τ_{ij} is the viscous stress tensor belonging to a new unknown in which the equations are not closed.

To close the Reynolds stresses in the equations, the RANS has been modeled with the Shear-Stress Transport (SST) k-ω model (Menter [27]). Wilcox derived a 2-equation k-ω

turbulence model for the correlation between the transport equations for 2 parameters: the turbulent kinetic energy k and the specific dissipation rate $\omega = \varepsilon/C_\mu k$, which are obtained from the following transport equation:

$$\frac{\partial}{\partial t}(\rho k) + \frac{\partial}{\partial x_i}(\rho k u_i) = \frac{\partial}{\partial x_j}\left(\Gamma_k \frac{\partial k}{\partial x_j}\right) + \widetilde{G_k} - Y_k \tag{3}$$

and:

$$\frac{\partial}{\partial t}(\rho \omega) + \frac{\partial}{\partial x_i}(\rho \omega u_j) = \frac{\partial}{\partial x_j}\left(\Gamma_\omega \frac{\partial \omega}{\partial x_j}\right) + \widetilde{G_\omega} - Y_\omega + D_\omega \tag{4}$$

where Γ_k and Γ_ω are the effective diffusivities for the SST k-ω model, respectively:

$$\Gamma_k = \mu + \frac{\mu_t}{\sigma_k} \tag{5}$$

$$\Gamma_\omega = \mu + \frac{\mu_t}{\sigma_\omega} \tag{6}$$

where σ_k and σ_ω represent the turbulent Prandtl numbers for kinetic energy k and the specific dissipation rate ω, respectively. The turbulent viscosity μ_t is computed as:

$$\mu_t = \frac{\rho k}{\omega} \frac{1}{\max\left[\frac{1}{a^*}, \frac{SF_2}{a_1 \omega}\right]} \tag{7}$$

$$\sigma_k = \frac{1}{\frac{F_1}{\sigma_{k,1}} + \frac{(1-F_1)}{\sigma_{k,2}}} \tag{8}$$

$$\sigma_\omega = \frac{1}{\frac{F_1}{\sigma_{\omega,1}} + \frac{(1-F_1)}{\sigma_{\omega,2}}} \tag{9}$$

where S is the modulus of the mean rate-of-strain tensor. The term $\widetilde{G_k}$ represents the production of turbulence kinetic energy, which is defined as:

$$\widetilde{G_k} = \min(G_k, 10\rho\beta^* k\omega) \tag{10}$$

To evaluate the term G_k consistent with the Boussinesqu hypothsesis:

$$G_k = \mu_t S^2 \tag{11}$$

The term G_ω is defined as:

$$G_\omega = \frac{a}{v_t}\widetilde{G_k} \tag{12}$$

and D_ω is defined as:

$$D_\omega = 2(1 - F_1)\rho \frac{1}{\omega \sigma_{\omega,2}} \frac{\partial k}{\partial x_j} \frac{\partial \omega}{\partial x_j} \tag{13}$$

This turbulence model provides accurate and reliable solutions for a wide variety of flows, including adverse pressure gradient flow, effectively blending the robust and accurate formulation of the model in the far field [28]. For the fully developed turbulent flow, the higher accuracy of the k-ω model helps in predicting characteristics of the flow field and tracking streamline distribution during the whole water-entry process. More details on the closure coefficients and auxiliary relationships in the above equations were presented in previous papers by Wilcox [29,30].

2.2. Treatment of Free Surface

For multiphase analyses, the Eulerian-based VOF method developed by Hirt and Nichols [31] is applied to capture the interface in multiphase flows. This method is based on the rationale of the determination of air-water interface by calculating the volume function of the air and water phases within the grids [32]. In this work, the free surface is computed through the VOF method. The VOF formulation is set to be unsteady implicit in this study. Throughout the computational domain, air above the free surface of the calm water is selected as the gas phase, while the water is selected as the liquid phase. In the beginning step, flat wave has been selected.

To simulate the time-varying process of a vehicle submerging into the water, the overset mesh technique is used to track the motion of the vehicle region. This technique utilizes a set of Chimera embedding grids to discretize complex regions into simpler meshes [33]. The scheme is to transfer physical quantities between the multiblock mesh systems by interpolations in which the solution from one grid is linked to the solution on the embedding grids.

For free surface problems and overset mesh interaction problems, the implementation of grid refinement plays an important role in the accuracy of numerical solutions. Several arbitrarily shaped volumetric control regions were previously adopted to refine some meshes near the body surface and the free surface with a larger curvature. In order to balance accuracy and efficiency during the simulation, the AMR method, which adjusts the cell grid adaptively, is developed in the STAR-CCM+. To improve the stability and the accuracy of the results during the simulation, the adaptive mesh refinement (AMR) technology is used to increase the convergence rate and reduce calculation errors and [34]. An initial uniform mesh is generated at the beginning, and the local mesh near the specified area will be redefined adaptively in next iteration step. The accuracy error of numerical solutions is the criterion for adjusting each cell node during the iteration. In the regions with higher nonlinearity, the distribution of gird nodes will be denser to increase computational accuracy using the proposed method. In order to decrease computational burden, it will be sparser in the regions with lower nonlinearity in the meantime.

Adaptive time-stepping strategy is also used to reduce run time of the simulation and attain a specified temporal resolution. In the iteration process, this strategy first calculates local time step automatically, based on the free surface Courant-Friedrichs-Lewy (CFL) condition. Then, the global time step is obtained based on Courant number weight using the rate of change of volume fraction. This adjustment is particularly useful for cases with large variations of flow topology or large varying time scales of the physics [34].

2.3. Numerical Procedure

The motion of the body interacts with the unknowns of flow characteristics, so it should be decoupled. With the dynamic fluid body interaction (DFBI) module superposed rotation and translation motion, the motions of the vehicle can be obtained at each time step. First, the pressure and shear on each boundary are obtained through Equations (1) and (2) above, and the resultant forces and moments acting on the body surface can also be calculated by integration, according to the related kinetic parameters. At this time, the accelerations can be obtained through the governing equations of the 6 degrees of freedom (6-DOF) motion solver. Once the new position and orientation of the body are updated over time, the body surface will in turn act on the nearby fluid domain. Then, the real-time physical state of the body is computed again.

The simulations are conducted using the CFD software STAR-CCM+ (version 17.02) to numerically simulate the problem. The solution method for solving the pressure-velocity coupled equation system in multiphase flow is the SIMPLE scheme algorithm. In discretization of volume fraction and turbulent kinetic energy, 1st-order upwind is selected until the residuals converge and then are turned into 2nd-order backward Euler scheme to improve precision of the simulation. The implicit unsteady algorithm is used for velocity–pressure coupling. The same numerical scheme had been used to investigated the oblique water entry of projectiles (Song & Duan [35]). The gravity acceleration in this case is set in negative z_0-direction. A robust automated prism layer meshing algorithm is also used on the vehicle surface, with two-layer all y+ wall treatment, to better capture the boundary layer.

3. Test Object and Computational Mesh

In this section, studies on the water-entry process of UAUVs is carried out. This vehicle can be carried by a supporting vessel and transit between two phases smoothly. Similar to AUVs and UAVs, UAUAVs can cruise underwater and fly in the air.

3.1. Description of the UAUV

As shown in Figure 2, the main stages of the complete operation are taking off, climbing, aerial flight, diving, water entry, underwater cruise, floating, and water exit. In an ocean survey, when the supporting vessel arrives at the ingress area, the vehicle is launched from the deck. Thereafter, the vehicle flies to the wrecked area and surveys aerially above the water surface. Then, the vehicle immerges into the water to investigate the region of interest. When the underwater cruise is completed, the vehicle floats up by the propellers. The stream flow produced by the propellers and the lift forces generated by the wings can also help the UAUV convert into flight state smoothly. Upon water exit, the UAUV communicates with the supporting vessel and transmits key data for making timely decisions afterward. Then, the UAUV could continue aerial or underwater observation missions. This vehicle has a real-time ability to exchange information and achieve extensive range of both aerial and underwater investigations such as relay communication, marine environment reconnaissance, seabed detection, and rescue applications.

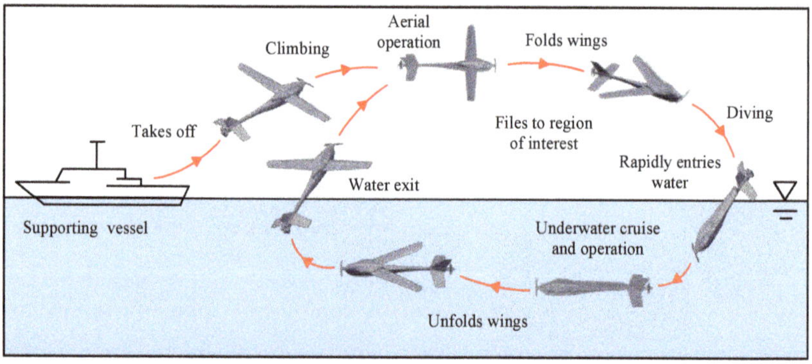

Figure 2. Schematic sketch of operational regimes of the UAUV.

Definition of the main parameters are given in Table 1, and the configuration of the UAUV is shown in Figure 3. Figure 3a–c shows three views of the UAUV and Figure 3d shows the stereogram view of the UAUV. The fixed wing is based on an ultrathin RSG-28 airfoil commonly used in UAVs, which remarkably saves space to facilitate folding. Two small winglets are implemented to improve the kinetic stability of the vehicle, ensuring aerodynamic efficiency during air flight. An aerodynamic propeller is adopted on the vehicle's fore, and a hydrodynamic propeller is adopted on the aft.

Table 1. Parameters of the object.

Main Parameters	Symbol	Unit	Quantitative Values
Length of the vehicle	L	m	1.62
Span of the wing (UAV moed)	b	m	1.72
Width of the vehicle (AUV mode)	B	m	0.23
Weight of the vehicle	W	kg	7.86 kg
Pitch angle of the vehicle	θ	°	variable

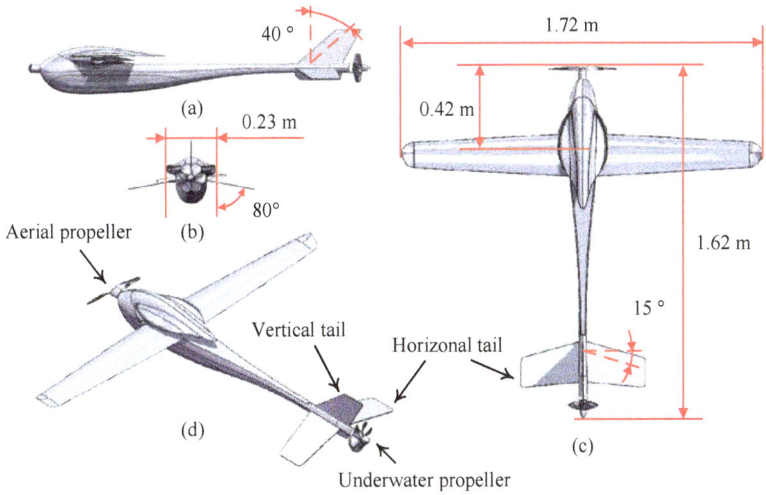

Figure 3. Prototype configuration for the UAUV: (**a–c**) three views; (**d**) stereogram view.

The equilibrium of the buoyancy force and gravity influences the stability of underwater vehicles. The relative mass density of a vehicle determines its negative, positive, or neutral buoyancy. Most underwater vehicles are designed as neutrally or positively buoyant so that they can float passively in case of an accident. In the simulations, the UAUV is set as neutrally buoyant.

3.2. Coordinate Systems and Dimensionless Quantities

To describe the hydrodynamic forces and motions of the vehicle, there are two right-handed Cartesian coordinate systems as shown in the sketch in Figure 4. In this study, the vehicle changes its wing span and orientation before water entry, and the simulated object is the UAUV under the folded state (AUV mode). In order to display the direction of the y axis more clearly, the wings are unfolded as in Figure 4. O-$x_0y_0z_0$ is the inertial Earth-fixed coordinate system, and fixed in the space is defined with O arranging at the center of gravity (CG) of the vehicle; the x_0 axis coincides with the mean water surface, and the z_0 axis points upward. In addition, a moving coordinate system of O-xyz fixed on the vehicle is defined, with its horizontal axis Ox and vertical axis Oz aligned at the right angle and along the moving direction of the vehicle, respectively.

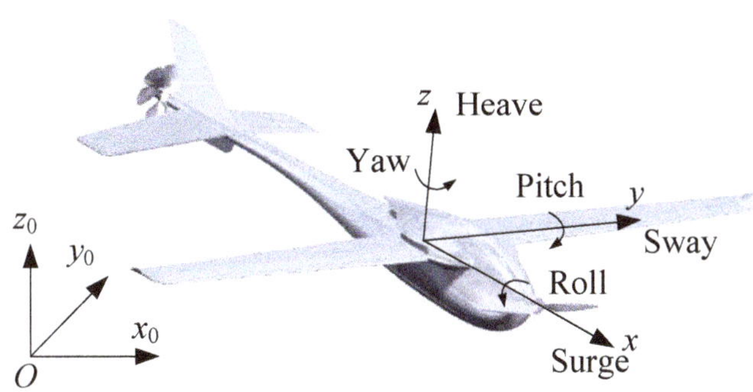

Figure 4. Schematic diagram of the Earth-fixed coordinate and body-fixed coordinate system of the UAUV.

To simplify the problem, the UAUV moves only in the x_0-O-z_0 plane and the out-of-plane motions are neglected, which indicates the motion plane of the vehicle being normal to the y_0 axis. As the vehicle entered the water, the body-fixed coordinate system also moved in the x_0-O-z_0 plane. In this study, the simulated object is the UAUV under the folded state. The simulated forces and moments exerted on the vehicle in 3-DOF are estimated in the O-xyz coordinate including the surge force, force heave, and pitch moment, designated as X, Z and M, respectively. To simplify the problem, the aerial propeller and the underwater propeller are also neglected, because the underwater and aerial propellers will stop rotating. The water-entry process can be regarded as freely falling.

3.3. Computational Domain and Boundary Conditions

To solve the integral Equations (1) and (2) numerically, the boundaries are represented in spatial and temporal discretized form. A mesh strategy with high computational accuracy and fast efficiency should be selected for simulation calculation. The sketch of the computational domain is shown in Figure 5a. The surface remesher is used to generate the volume mesh for a trimmed nonstructured mesh, which has better geometric adaptability. There is an overset mesh block around the free surface, moving in the vertical plane, as shown in Figure 5b. Data exchange between two blocks is realized by the least square interpolation. The DFBI solver should been created, and it is necessary to specify the manner in which the vehicle body can move freely. In this study, the vehicle body is set free to translate in the x axis (lateral direction) and in the z axis (transverse direction) and to rotate about the y axis (pitch). Overset AMR and free surface AMR are utilized during the simulation process.

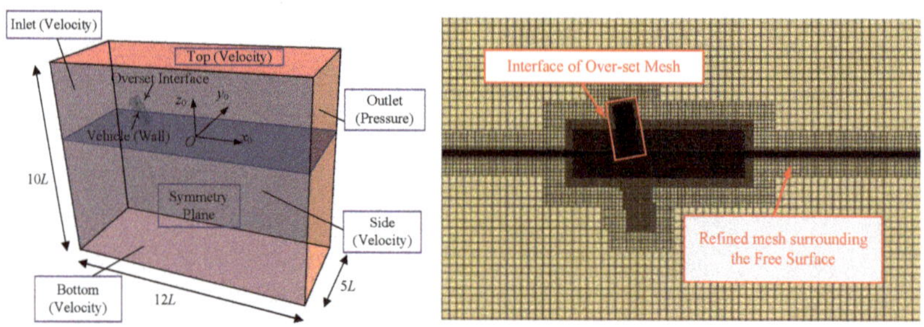

(**a**) (**b**)

Figure 5. The sketch of CFD settings: (**a**) configuration of the boundary conditions; (**b**) mesh of computational domain.

During the CFD simulation, a rectangular computational domain region of $12L \times 10L \times 5L$ is selected, as shown in Figure 5a. Due to the symmetry of the flow field, a symmetric surface treatment is applied for the computational domain. The vehicle surface is defined as moving walls with certain velocities. For inlet, the velocity inlet boundary is set with a specification of inlet velocity magnitude and specified at $4L$ upstream of the vehicle. The pressure outlet with specified pressure is specified at $8L$ downstream of the vehicle. After the vehicle model is released, it falls freely due to Earth's gravity with an initial velocity. The water depth is set as $6L$, and the initial height of the vehicle is specified as 0.5 m above the mean water. The y+ values should be larger than 30 to ensure accuracy for the SST k-ω model [36].

4. Convergence Study

This section describes the convergence study conducted to make sure the appropriate time step and mesh size are chosen so that the whole process of the UAUV entering the water can be solved. To test the numerical error, convergence studies were conducted for the UAUV to evaluate the influences of mesh size dl and time step Δt on the simulation results. The mesh convergence study was carried out with the *dl* of the fine grid to 20 mm; the *dl* of the other two meshes were 25.19 and 31.75 mm. The initial mesh number N_T was scaled by a factor of 2.0 with three typical sizes. Therefore, three meshes were evaluated with a grid refinement ratio of $r_G = \sqrt[3]{2}$, In each mesh system, the same basic length scale was applied to both background mesh and overset mesh.

Three groups of elements were chosen to carry out the grid uncertainty analysis, as shown in Table 2. The time interval of the grids changed with the decrease of mesh number due to the refinement *dl*. The medium grid illustrated in Figure 6a,b contains a total of 2,451,104 cells. The overset domain contains about 80% of all meshes. The cell parameters should be decided with caution, as this involves much computational resources. Meanwhile, most meshes of the background were near the free surface to obtain a convergent profile. The meshes around the block interface should be similar in size to ensure efficient data transfer through the interface. For better computational convergency, the grid size and discrete scheme should be approximate between the background and the overset region.

Table 2. Detailed information on the initial mesh systems.

Mesh	Basic Mesh Size *dl* (mm)	Time Interval Δt (s)	Initial Mesh Number N_T
Coarse	31.75	0.002	1225 k
Medium	25.19	0.001	2451 k
Fine	20	0.0005	4902 k

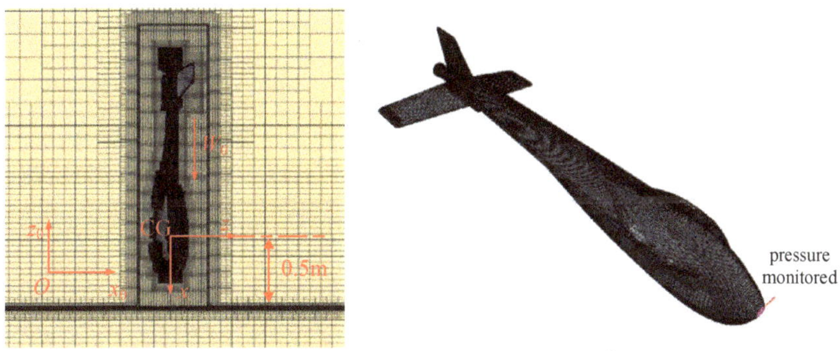

(a) (b)

Figure 6. The sketch of the CFD mesh generations: (**a**) cross-view details; (**b**) mesh partition of the UAUV surface.

For a vehicle with a smooth fore, the stagnation point is near the fore and may move toward the upstream direction. According to Bernoulli's equation, the fore of the vehicle is generally exposed to the pressure peak, which can be taken into consideration. Therefore, the maximum pressure at the fore of the vehicle is defined as a representative pressure, as shown in Figure 6b. In the simulations, the vertically deposited UAUV with initial pitch angle $\theta_0 = 90°$ falls freely and vertically with $W_s = 4.0$ m/s, where W_s is the initial vertical water-entry velocity in the Earth-fixed coordinate. For the vertical water-entry problem, the vertical velocity W equals the transverse velocity u in the body-fixed coordinate.

The STAR-CCM+ predicted pressure at the fore and the vertical force Z_0 (in z_0-direction) as shown in Figure 7a,b, respectively. There are some differences between the fine mesh and the other two meshes, especially in pressure. Too coarse a mesh may cause computation divergency, because the fluid domain is discretized into unstructured trimmer cells in Star-CCM+ simulations. The basic mesh size and the time step in the third mesh could capture key features of the flow field better. However, the fine mesh system involved too much computational resources and CPU time. The hydrodynamic performance of the vehicle is one of the major focuses in present study, and the medium mesh system is enough to ensure the fidelity of hydrodynamic forces and motion during the numerical procedure, occupying fewer computational resources. Empirically based on extensive test simulation, the second mesh fits due to the compromise of computation efficiency and accuracy. The medium mesh system spatial size grid is shown in Figure 6.

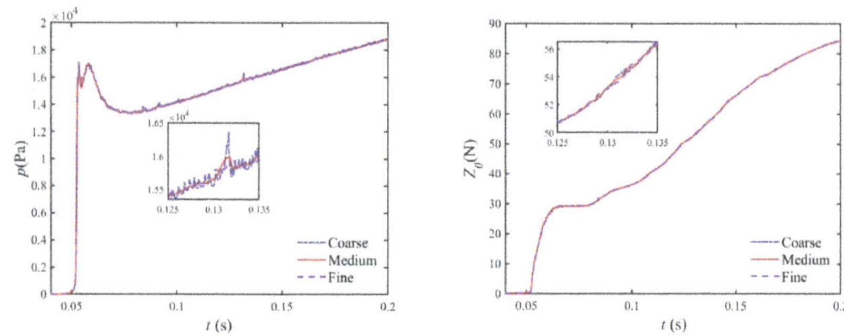

(a) **(b)**

Figure 7. Convergent analysis of the time history curves of (**a**) pressure at the fore; (**b**) vertical force for the water entry of UAUV with different mesh systems.

5. Numerical Results and Discussions

This section computes more results of UAUVs entering the water using the present method and considers the effects of two main parameters: initial vertical velocity W and initial pitch angle θ. A series of numerical simulations are presented in this section. The initial time step $\Delta t = 0.001$ s and the basic mesh size $dl = 25.19$ mm are chosen for all the cases in this section. The time step and mesh are automatically adjusted during the course of calculations, followed by an adaptive time-stepping strategy and AMR technology.

5.1. Effect of Initial Vertical Velocity

This section considers four conditions in which the UAUV falls freely into the water at initial velocities of $W_s = 2.0, 4.0, 8.0$, and 12 m/s. According to the results, the hydrodynamic forces acting on the approximately symmetric vehicle is approximately along the vertical axis during the whole water entry, and the horizontal (in x_0-direction) motion is somewhat smaller than the vertical (in z_0-direction) motion. Thus, the horizontal motion can be ignored here, so this section lists only the results of the vertical motion. For the sake of comparison, the Froude number Fr can be defined as

$$Fr = \frac{V_z}{\sqrt{gL}} \tag{14}$$

where g is the gravity acceleration and V_z is the vertical velocity in z_0-direction.

In some water-entry problems, a time-varying and deformable cavity may form while the body is immersing into the water (Birkhoff & Zarantonello [37]). To study the influence of initial velocity, the simulated free surface profiles over the UAUV at $W_s = 4.0$ m/s and 12.0 m/s at each moment are shown in Figures 8 and 9, respectively. The deformed free surface finally tends to move downward due to the acting force from the vehicle surface and the restoring force of the gravity effect. As x position increases, the evolution of water first increases sharply and then gradually falls to zero far away from the vehicle due to the infinity boundary condition.

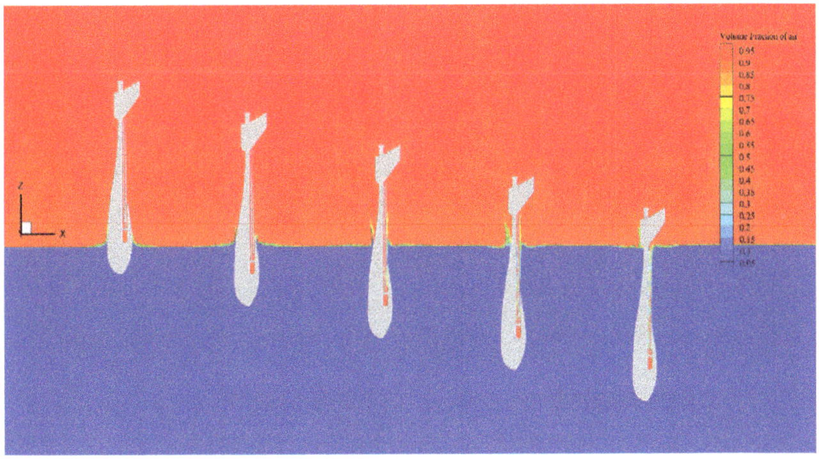

(a) (b) (c) (d) (e)

Figure 8. Free surface profiles of the vehicle entering water at W_s = 4.0 m/s at different times: (**a**) t = 0.10 s, (**b**) t = 0.15 s, (**c**) t = 0.20 s, (**d**) t = 0.25 s, and (**e**) t = 0.30 s.

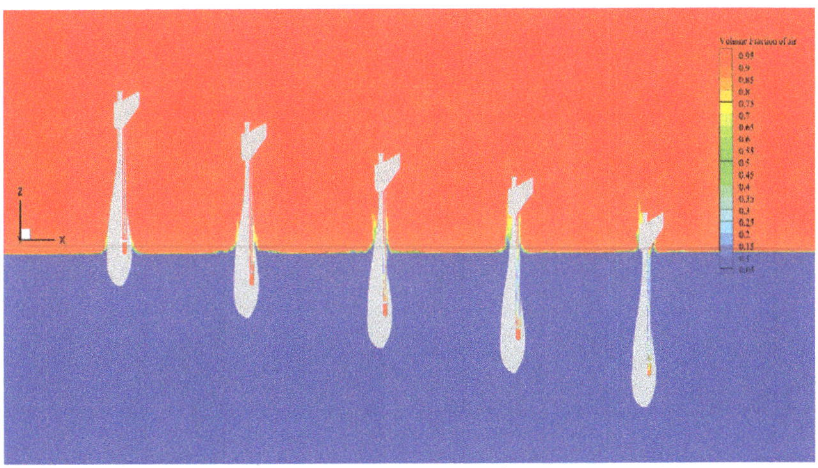

(a) (b) (c) (d) (e)

Figure 9. Free surface profiles of the UAUV entering water at W_s = 12.0 m/s at different times: (**a**) t = 0.04 s, (**b**) t = 0.06 s, (**c**) t = 0.08 s, (**d**) t = 0.08 s, and (**e**) t = 0.12 s.

Figure 10a shows the jet flow around the intersection area between the water and the vehicle. The flow separation depends on the Froude number and the shape of body (Gurevich [38]). In some cases, the water would separate from the body in the absence of fluid gravity, and the water would break up into spray, as shown in Figure 10a. Streamlines around and velocity magnitude distributions at t = 0.15 s are shown in Figure 10b. Because of gravity, the fluid particles attached to the body surface were pulled down eventually, and the attached water could not be sustained indefinitely. Fluid/structure interaction may also make the flow more turbulent.

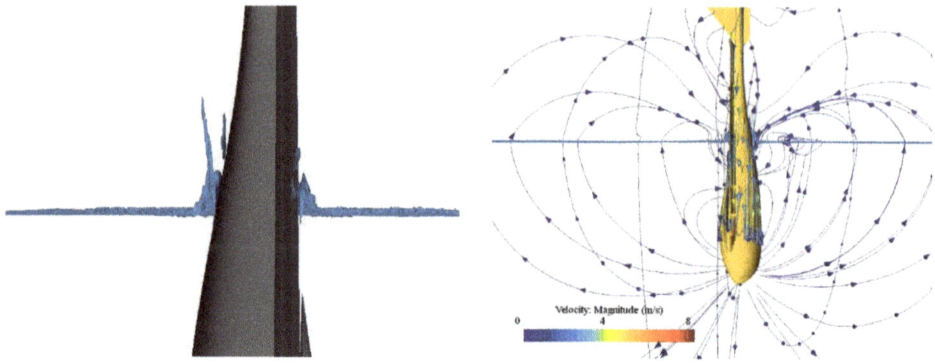

(**a**) (**b**)

Figure 10. Profiles for (**a**) jet flow and (**b**) streamlines around the UAUV with $W_s = 4.0$ m/s at $t = 0.15$ s.

The characteristics of the predicted vertical velocity, acceleration, and hydrodynamic force are analyzed against a submergence parameter s. The depths curves of the vertical translation s and vertical velocity (in the Earth-fixed coordinate) are given in Figure 11a,b. While impacting the water, the vehicle decelerates due to the slamming force exerted on it. Through the comparison of different W_s, it was found that vertical translation varies faster with larger vertical velocity.

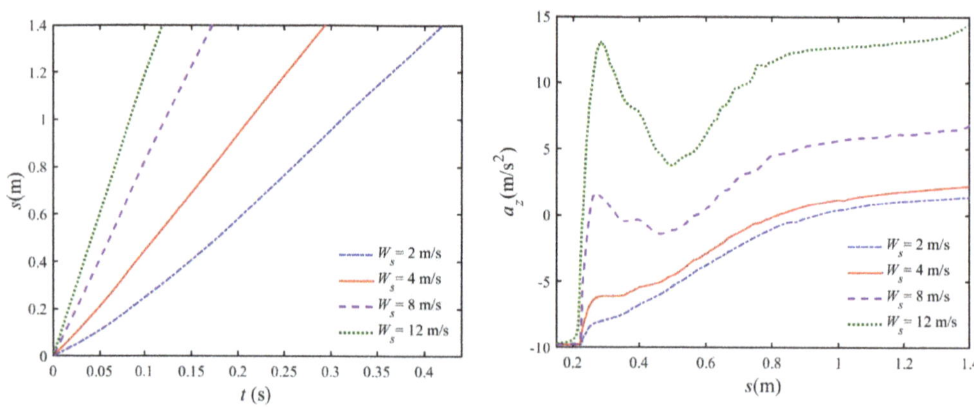

(**a**) (**b**)

Figure 11. Comparison of the depth curve of (**a**) vertical displacement and (**b**) vertical velocity for the water entry of UAUV with different initial velocities.

The velocity variation is highly correlated with the real impact velocity W_0, which is the velocity magnitude when the fore end of the vehicle first touches the water. Due to the gravity effect, W_0 may be larger than the intended initial speed. Table 3 shows the comparison of the velocity variation at certain positions. W_1 is the velocity magnitude at $s = 0.8$ m, and W_2 is the velocity magnitude at the final distance of $s = 1.4$ m. $W_0\% = (W_0 - W_s)/W_s$, $W_1\% = (W_1 - W_s)/W_s$, and $W_2\% = (W_2 - W_s)/W_s$ are the velocity variation percentages relative to W_s. Both vertical velocity reductions $W_1\%$ and $W_2\%$ become less for the vehicle with larger W_s. This is because in the vehicle with larger velocity the interaction of fluid/structure impact becomes more significant, thus making kinetic energy dissipate faster.

Table 3. Comparison of the velocity variation of the UAUV with various W_s.

Initial Velocity	W_0 [m/s]	W_1 [m/s]	W_2 [m/s]	$\Delta W_0\%$	$\Delta W_1\%$	$\Delta W_2\%$
W_s = 2.0 m/s	2.885	3.766	3.667	44.24	88.28%	83.37%
W_s = 4.0 m/s	4.486	4.985	4.814	12.16	24.63%	20.35%
W_s = 8.0 m/s	8.248	8.219	7.784	3.11	2.73%	−2.70%
W_s = 12.0 m/s	12.148	11.793	11.117	1.24	−1.73%	−7.36%

Figure 12 compares the monitored pressure at the fore of the vehicle. Undoubtedly, the pressure in the initial stage is mainly caused by the slamming, which is closely associated with the local deadrise angle (Wu et al. [39]). It can be found from the vehicle profile that the normal deadrise angle increases rapidly and then decreases gradually as the vehicle becomes submerged in water. There is a sharp peak pressure during the early stage, which can be defined as the slamming pressure. After a sharp increase, the pressure gradually decreases and then smoothly increases in the later stage. This is because the effect of fluid gravity becomes more significant in the water-entry process.

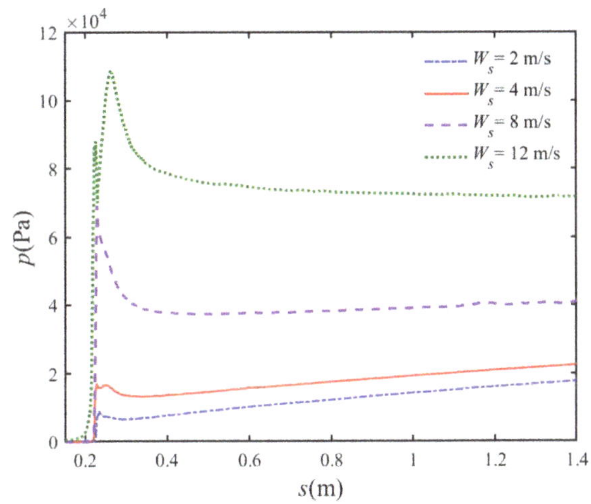

Figure 12. Comparison of the time history curve of monitored pressure for water entry of UAUV with various initial velocities.

The vertical force and vertical acceleration for different initial velocities are given in Figure 13a,b. Here, the vertical force Z_0 equals the transverse force in the body-fixed coordinate. As can be seen, the simulated acceleration curves from the four cases coincide with the force curves for the same case during the whole water-entry process. In addition, pressure and force with higher W_s are entirely larger than with lower W_s. Notably, the force Z_0 is always positive throughout the whole water entry due to the slamming effect. This may be because more active turbulence dynamics are caused by larger induced velocity. When s is small, the local deadrise angle is smaller, thus leading to higher pressure peaks and high loads. The slamming force is normally proportional with quadratic velocity as indicated by Von Karman [6]. After the peak slamming force, the vertical hydrodynamic force decreases rapidly until it drops to the minimum positive value and then climbs up again, which can also be verified by the results of the acceleration. According to its contributing factor, the total hydrodynamic force can be divided into viscous force and pressure force. In the early water-entry stage, the pressure force occupies the main component (Xiang & Wang et al. [40]), and the hydrodynamic forces in the late stage are essentially buoyancy forces. At the same time, when Froude is low, the growth trend is faster because there is a more obvious gravity effect in the late stage.

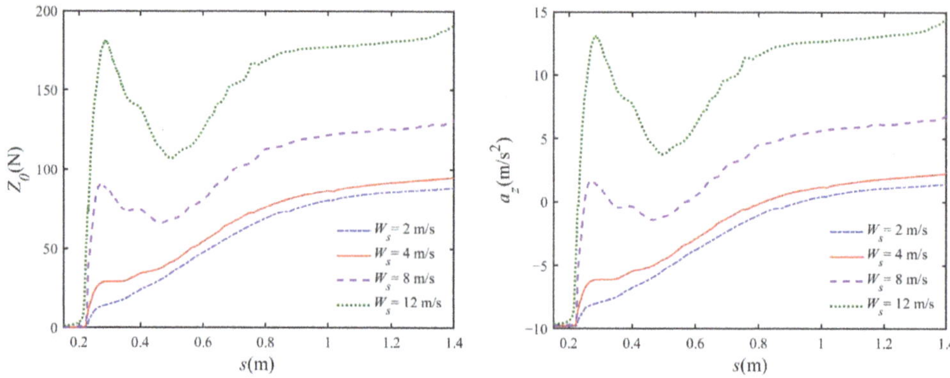

(**a**) (**b**)

Figure 13. Comparison of the displacement curve of (**a**) vertical force and (**b**) vertical acceleration for the water entry of UAUV with different initial velocities.

The acceleration histories can also illustrate the trend of hydrodynamic force. In a physical meaning, when the slamming velocity decreases, the Froude number also decreases, and the influence of gravity turns into remarkable gradually. Before the vehicle touches the water, the vertical acceleration approximately equals the gravity acceleration g for the effect of freely falling. For lower W_s, the curve would not be consistent with the results of larger W_s. During the initial water-entry stage, the vertical acceleration suddenly varies to a positive peak. After the vehicle submerges, the sign of the vertical acceleration is always positive because of buoyancy, resistance, and slamming force.

5.2. Effect of Initial Pitch Angle

Except for vertical water entry, it is of extensive interest to simulate the oblique water entry of a body with an initial pitch angle. As an important physical parameter, the difference of a vehicle's inclination has a significant influence on the media crossing motion. The water-entry velocity is u_s = 4.0 m/s, and the initial pitch angles θ_0 are 90° (reference case), 80°, 70°, and 60°, while holding other variables the same. The direction of initial water-entry velocity u_s is along the axis of the vehicle as shown in Figure 14a. Here, angle of inclination (AOI) is defined as the complementary angles of the pitch angles, which are 0°, 10°, 20°, and 30°, correspondingly. In this section, further results on horizontal motion and hydrodynamic force will be discussed.

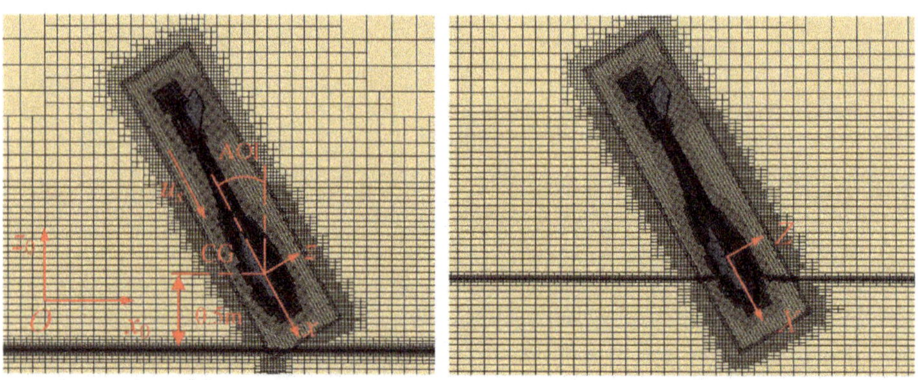

(a) (b)

Figure 14. The sketch of the CFD mesh partition of (**a**) initial time and (**b**) running time of the oblique water entry.

Figures 15–17 show snapshots of the free surface profiles when the UAUV enters the water with different θ_0. With time, the free surface first moves upward, with the most dramatic nonlinear effect observed near the jet root with the curviest free surface. As seen in the figures, flow separation occurs more violently for the vehicle with higher horizontal velocity. It can also be found that gravity would eventually pull the water down at this velocity as discussed in Section 5.1.

Figures 18a and 18b, respectively, give the time-varying moment and pitch angle of the vehicle. The net rotation of the vehicle with θ_0 = 60°and 70° is much larger than the other two cases, as shown in Figure 18b. The initial pitch angle affects the horizontal translation and rotation of the vehicle as it enters the water for a longer time. A larger righting moment exerted on the vehicle accelerates its rotation and restores its orientation. Figures 19a and 19b, respectively, gives the horizontal and vertical translation of the vehicle with different initial pitch angles. With time, the vehicle penetrates the water surface gradually. Transverse resistance is also exerted on the vehicle, which results in horizontal translation. The trajectories depend on both the direction of velocity and the inclination state. In Figure 19b, the curve of θ_0 = 80° is close to the reference case of vertical water entry.

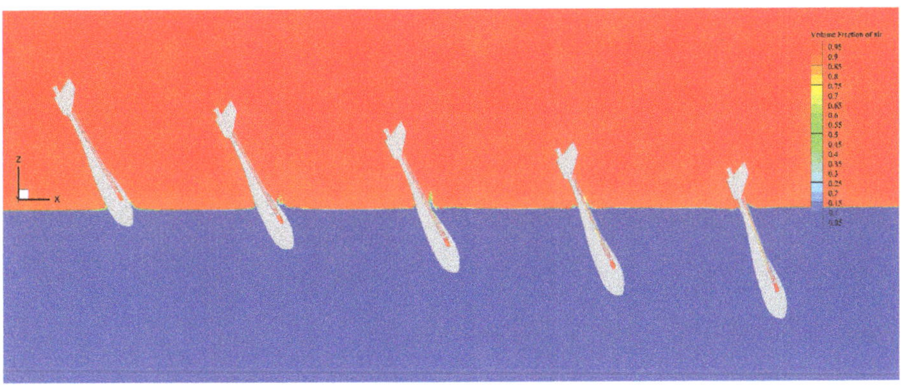

(a) (b) (c) (d) (e)

Figure 15. Free surface profiles of the vehicle entering water at θ_0 = 60° at different times: (**a**) t = 0.10 s, (**b**) t = 0.15 s, (**c**) t = 0.20 s, (**d**) t = 0.25 s, and (**e**) t = 0.30 s.

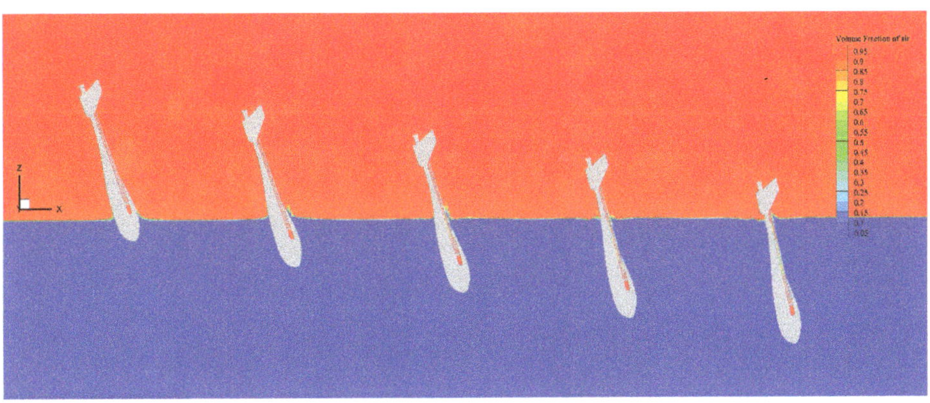

(a) (b) (c) (d) (e)

Figure 16. Free surface profiles of the vehicle entering water at θ_0 = 70° at different times: (**a**) t = 0.10 s, (**b**) t = 0.15 s, (**c**) t = 0.20 s, (**d**) t = 0.25 s, and (**e**) t = 0.30 s.

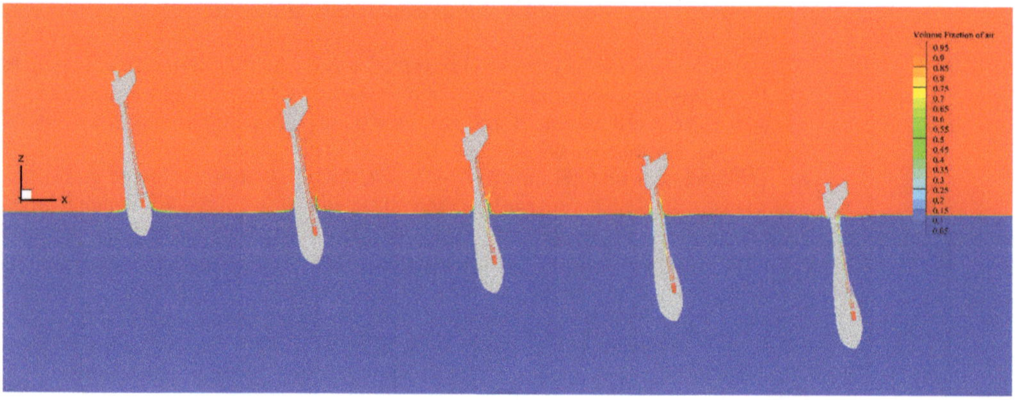

(**a**) (**b**) (**c**) (**d**) (**e**)

Figure 17. Free surface profiles of the vehicle entering water at $\theta_0 = 80°$ at different times: (**a**) $t = 0.10$ s, (**b**) $t = 0.15$ s, (**c**) $t = 0.20$ s, (**d**) $t = 0.25$ s, and (**e**) $t = 0.30$ s.

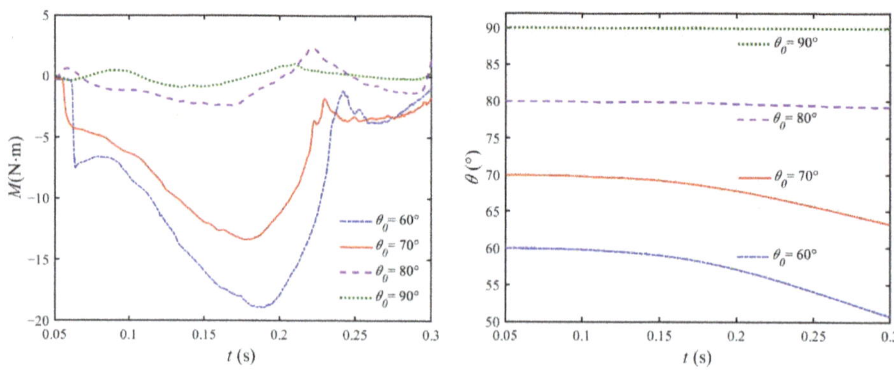

(**a**) (**b**)

Figure 18. Comparison of time history curve of (**a**) moment and (**b**) pitch angle for the water entry of UAUV with different initial pitch angles.

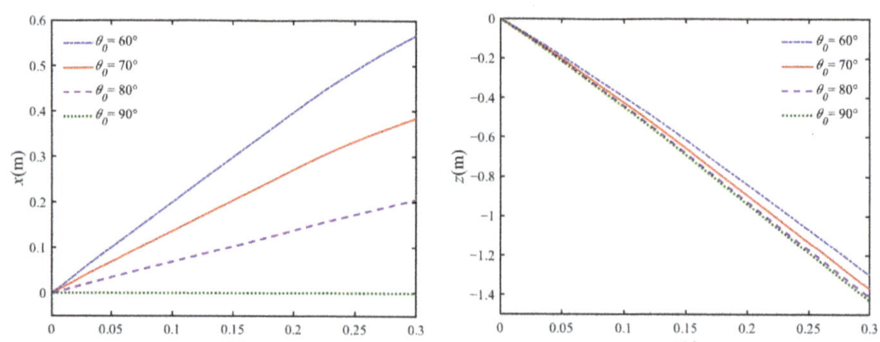

(**a**) (**b**)

Figure 19. Comparison of time history curve of (**a**) horizontal translation and (**b**) vertical translation for the water entry of UAUV with different initial pitch angles.

Figure 20 compares the time history curve of transverse and lateral forces in the body-fixed coordinate, whose direction is shown in Figure 14b. The simulated lateral force curves from all cases show similar trends, with a gentle peak in the initial stage due to the slamming effect. In the early stage of water entry, the vertical slamming peak of the transverse lateral force is sharper and slightly increases correspondingly with the increasing of inclination. Transverse force not only depends on the entry velocity but is also related to the projected wetted surface and volume. The associated wetted area of the vehicle is affected by both the pitch angle and penetration depth. When the vehicle touches the water surface, the pressure difference between the two sides results in a moment in the y_0 direction, resulting in the vehicle's rotation. As the vehicle penetrates obliquely deeper, the transverse wetted area increases, resulting in a larger drag force in the vertical direction as shown in Figure 20a. Therefore, the inclination increases due to the wetted surface area increase. Then, as result of the rotation, the associated wetted area in the horizontal direction also decreases compared with the vertical water-entry case, which causes a smaller vertical hydrodynamic force. The vertical slamming peak force slightly increases correspondingly with the increasing of inclination, and it can be inferred that vertical water entry mildly helps reduce the early vertical slamming force for a slender vehicle. As the vehicle penetrates deeper, its vertical projected wetted area constantly increases. However, for vertical water entry of $\theta = 90°$, the curve would not be consistent with the results of oblique water entry. In the early period, the difference in vertical force become less obvious after a numerical transition period, and then the gap increases faster in a later period.

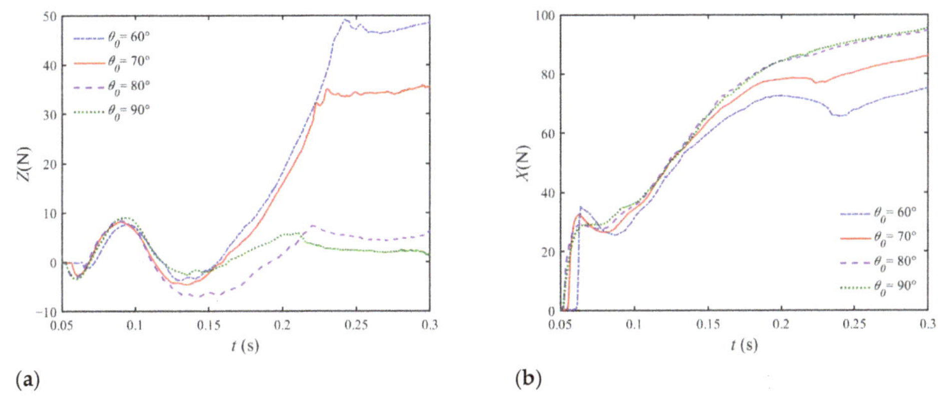

Figure 20. Comparison of time history curve of (**a**) transverse force and (**b**) lateral force for the water entry of UAUV with different initial pitch angles.

Figure 21 compares the simulated horizontal and vertical velocity history curves of the vehicle, in the O-$x_0y_0z_0$ coordinate. As the vehicle enters the water for a longer time, it obviously begins moving in a transverse direction as shown in Figure 21a. It can be noticed that the final velocity at t = 0.3 s significantly decreased in contrast to u_s. The curve of the horizontal velocity changes less and gradually becomes close to linear. In addition, it can be clearly found in Figure 21b that the turning points of the curve are the corresponding moment when the vehicle touches the water surface. In the beginning, the lateral hydrodynamic force is still developing, so the vehicle does not move in the x_0 direction. In addition, the development of transverse hydrodynamic force is not unbalanced enough. Similarly, for the water entry of the cylinder, vortex is only induced after the initial 'warm-up' stage (Xiang & Wang et.al [36]). As shown in Figures 15–17, it can be observed that more flow separation occurs for the oblique water-entry cases in the later stage. More energy transfers to the water because of horizontal velocity, generating more water splashes. On the other hand, the pressure difference between the two sides of lager inclination is larger, resulting in a much larger horizontal hydrodynamic force. Thus, more air is trapped on the downstream side, which can also be observed in Figures 15–17. Compared with the oblique water-entry cases, the time history curve of transverse force for the vertical water entry is much flatter in the later stage. In the early stage, the pitch angle has less effect on both transverse force and lateral force. Over time, the pitch angle produces more effect on the hydrodynamic force, especially transverse force. This may be induced by the asymmetric shedding vortex of the body with higher inclination angle.

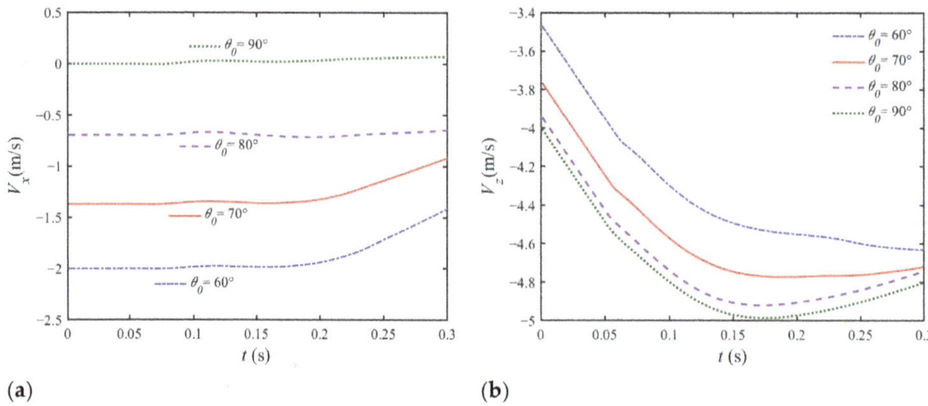

Figure 21. Comparison of time history curve of (**a**) horizontal velocity and (**b**) vertical velocity for the water entry of UAUV with different initial pitch angles.

6. Conclusions

In this work, the water-entry process of a multi-DOF vehicle have been simulated based on the RANS method with AMR technologies. The flow field and structural body motion are solved simultaneously using a DFBI solver. Numerical convergency with respect to mesh sizes and time step was verified, and more simulations were performed. Parametric studies have been conducted in addition to factors such as the vertical velocity and pitch angle of the UAUV. Flow separation occurs around the vehicle body, and some jet flow could be observed. The comparison of the time-varying velocity, pressure, force, and acceleration revealed that the vertical velocity exhibits a significant effect on the hydrodynamics of the vehicle. The results show that the slamming pressure and force occurred at the early stage. In the oblique water-entry cases, the velocity and force of the vehicle in each direction were analyzed deeply. It was found that the pitch angle significantly affects the motion and force in the later stage of water entry. The effect of initial inclination has been conducted in which the horizontal velocity decreases with a larger pitch angle. The influence of initial pitch angles on the horizontal force was more significant in the later stage of water entry. The numerical results indicated that the vertical hydrodynamic force was mainly affected by vertical velocity, and the contributions of pitch angle are less significant.

In the simulated oblique entry cases, some flow separation would occur near the vehicle body on the downstream side, around where cavities may occur. The studied vehicle is slender and symmetric, with its head entering the water vertically. Based on the present study, further investigations will be required with ventilation, air cavity phenomena, and initial air cushion.

Author Contributions: Conceptualization, Z.W. and L.L.; methodology, L.D. and Z.W.; software, L.D. and H.Z.; validation, L.D. and H.Z.; formal analysis, B.Y.; investigation, Z.W.; writing—original draft preparation, L.D.; writing—review and editing, L.D., Z.W and L.L.; project administration, L.L. All authors have read and agreed to the published version of the manuscript.

Funding: This research was funded by the project of the Shanghai Committee of Science and Technology, grant number 20dz1206600.

Institutional Review Board Statement: Not applicable.

Informed Consent Statement: Not applicable.

Data Availability Statement: Not applicable

Conflicts of Interest: The authors declare no conflict of interest.

References

1. Nicolaou, S. *Flying Boats & Seaplanes: A History from 1905*; Bay View Books Ltd.: Bideford, UK, 1996.
2. Petrov, G. Flying submarine. *J. Fleet.* **1995**, *3*, 52–53.
3. Li, Y.; Feng, J.; Hu, J.; Yang, J. Research on the motion characteristics of a trans-media vehicle when entering water obliquely at low speed. *Int. J. Nav. Archit. Ocean Eng.* **2018**, *10*, 188–200. [CrossRef]
4. Oumeraci, H. Review and analysis of vertical breakwater failures—Lessons learned. *Coast. Eng.* **1994**, *22*, 3–29. [CrossRef]
5. Wei, Z.; Teng, Y.; Meng, X.; Yao, B.; Lian, L. Lifting-principle-based design and implementation of fixed-wing unmanned aerial–underwater vehicle. *J. Field Robot.* **2022**, *39*, 694–711. [CrossRef]
6. Von Karman, T. *The Impact on Seaplane Floats during Landing*; Tech. Rep. Arch. Image Librar: Washington, DC, USA, 1929; Available online: https://digital.library.unt.edu/ark:/67531/metadc54062/ (accessed on 15 March 2019).
7. Wagner, H. Uber Stoss—und Gleitvorgange an der oberflacke flussigkiten. *ZAMM* **1932**, *4*, 193–235. [CrossRef]
8. Dobrovol'skaya, Z.N. On some problems of similarity flow of fluid with a free surface. *J. Fluid Mech.* **1969**, *36*, 805–829. [CrossRef]
9. Toyama, Y. Two-dimensional water impact of unsymmetrical bodies. *J. Soc. Nav. Arch. Jpn.* **1993**, *173*, 285–291. [CrossRef]
10. Mei, X.; Liu, Y.; Yue, D.K. On the water impact of general two-dimensional sections. *Appl. Ocean Res.* **1999**, *21*, 1–15. [CrossRef]
11. Vorus, W.S. A flat cylinder theory for vessel impact and steady planing resistance. *J. Ship Res.* **1996**, *40*, 89–106. [CrossRef]
12. Semenov, Y.A.; Iafrati, A. On the nonlinear water entry problem of asymmetric wedges. *J. Fluid Mech.* **2006**, *547*, 231–256. [CrossRef]
13. Xu, G.; Wu, G.; Duan, Y. Axisymmetric liquid block impact on a solid surface. *Appl. Ocean Res.* **2011**, *33*, 366–374. [CrossRef]
14. Han, B.; Peng, Y.; Li, H.; Liu, S.; Sun, S.; Shan, Y.; Sun, Z. Numerical investigations of a 2D bow wedge asymmetric free-falling into still water. *Ocean Eng.* **2022**, *266*, 112905. [CrossRef]
15. Wang, Y.; Shi, X.; Wang, P. Modeling and simulation of oblique water-entry of disk Ogive. *Torpedo Technol.* **2008**, *16*, 14–17.
16. Afzal, A.; Ansari, Z.; Faizabadi, A.R.; Ramis, M.K. Parallelization Strategies for Computational Fluid Dynamics Software: State of the Art Review. *Arch. Comput. Methods Eng.* **2017**, *24*, 337–363. [CrossRef]
17. Qiu, H.; Yuan, X.; Wang, Y.; Liu, C. Simulation on impact load and cavity shape in high speed vertical water entry for an axisymmetric body. *Torpedo Technol.* **2013**, *21*, 161–164.
18. Guo, B.; Liu, P.; Qu, Q.; Wang, J. Effect of pitch angle on initial stage of a transport airplane ditching. *Chin. J. Aeronaut.* **2013**, *26*, 17–26. [CrossRef]
19. Shi, Y.; Pan, G.; Huang, Q. Water entry impact cushioning performance of mitigator for AUV. In *Oceans 2017-Aberdeen*; IEEE: Aberdeen, UK, 2017.
20. Chen, C.; Wang, T.; Feng, Z.; Lu, Y.; Huang, H.; Ji, D.; Chen, Y. Simulation research on water-entry impact force of an autonomous underwater helicopter. *J. Mar. Sci. Technol.* **2020**, *25*, 1166–1181. [CrossRef]
21. Du, Y.; Wang, Z.; Wang, Y.; Wang, J.; Qiu, R.; Huang, C. Study on the cavity dynamics of water entry for horizontal objects with different geometrical shapes. *Ocean Eng.* **2022**, *252*, 111242. [CrossRef]
22. Wu, Y.; Li, L.; Su, X.; Gao, B. Dynamics modeling and trajectory optimization for unmanned aerial-aquatic vehicle diving into the water. *Aerosp. Sci. Technol.* **2019**, *89*, 220–229. [CrossRef]
23. Wu, Y.; Li, L.; Su, X.; Cui, J. Multi-phase trajectory optimization for an aerial-aquatic vehicle considering the influence of navigation error. *Eng. Appl. Artif. Intel.* **2020**, *89*, 103404. [CrossRef]
24. Baldi, S.; Roy, S.; Yang, S. Towards adaptive autopilots for fixed-wing unmanned aerial vehicles. In Proceedings of the 2020 59th IEEE Conference on Decision and Control (CDC), Jeju, Korea, 14–18 December 2020; IEEE: New York, NY, USA, 2020.
25. Wang, X.; Roy, S.; Farì, S.; Baldi, S. Adaptive Vector Field Guidance Without a Priori Knowledge of Course Dynamics and Wind. *IEEE/ASME Trans. Mechatron.* **2022**, *27*, 4597–4607. [CrossRef]
26. Wang, Z.; Stern, F. Volume-of-fluid based two-phase flow methods on structured multiblock and overset grids. *Int. J. Numer. Methods Fluids* **2022**, *94*, 557–582. [CrossRef]
27. Menter, F.R. Two-equation eddy-viscosity turbulence models for engineering applications. *AIAA J.* **1994**, *32*, 1598–1605. [CrossRef]
28. Durbin, P.A.; Reif, B.P. *Statistical Theory and Modeling for Turbulent Flows*; John Wiley & Sons: Chichester, West Sussex, UK, 2011.
29. Wilcox, D.C. *Turbulence Modeling for CFD*; DCW Industries: La Canada, CA, USA, 1998.
30. Wilcox, D.C. Formulation of the k-w turbulence model revisited. *AIAA J.* **2008**, *46*, 2832–2838. [CrossRef]
31. Hirt, C.W.; Nichols, B.D. Volume of fluid (VOF) method for the dynamics of free boundary. *J. Comput. Phys.* **1981**, *39*, 201–225. [CrossRef]
32. Nova Science Publishers. *Multiphase Flow Research*; Nova Science Publishers: New York, NY, USA, 2009.
33. Zahle, F.; Sørensen, N.N.; Johansen, J. Wind turbine rotor-tower interaction using an incompressible overset grid method. *Wind Energy Int. J. Prog. Appl. Wind Power Convers. Technol.* **2009**, *12*, 594–619. [CrossRef]

34. Siemens Digital Industries Software. *STAR-CCM+ User Guide, Version.17.02*; Siemens PLM Software: Plano, TX, USA, 2022; pp. 6882–7019.
35. Song, Z.; Duan, W.; Xu, G.; Zhao, B. Experimental and numerical study of the water entry of projectiles at high oblique entry speed. *Ocean Eng.* **2020**, *211*, 107574. [CrossRef]
36. Qiu, S.; Ren, H.; Li, H. Computational Model for Simulation of Lifeboat Free-Fall during Its Launching from Ship in Rough Seas. *J. Mar. Sci. Eng.* **2020**, *8*, 631. [CrossRef]
37. Birkhoff, G.; Zarantonello, E. *Jets, Wakes, and Cavities*; Elsevier: New York, NY, USA, 1960; p. 261.
38. Gurevich, M. *The Theory of Jets in an Ideal Fluid*; Elsevier: Amsterdam, The Netherlands, 2014.
39. Wu, G.; Sun, H.; He, Y. Numerical simulation and experimental study of water entry of a wedge in free fall motion. *J. Fluids Struct.* **2004**, *19*, 277–289. [CrossRef]
40. Xiang, G.; Wang, S.; Soares, C.G. Study on the motion of a freely falling horizontal cylinder into water using OpenFOAM. *Ocean Eng.* **2020**, *196*, 106811.1–106811.13. [CrossRef]

Journal of
Marine Science and Engineering

MDPI

Article

Analysis of the Descent Process and Multi-Objective Optimization Design of a Benthic Lander

Qiao Zhang [1,†], Chunming Dong [2,†], Zongze Shao [2,*] and Donghui Zhou [1,*]

1 School of Mechanical Engineering, Hangzhou Dianzi University, Hangzhou 310018, China
2 Key Laboratory of Marine Biogenetic Resources, Third Institute of Oceanography, Ministry of Natural Resources, Xiamen 361005, China
* Correspondence: shaozz@163.com (Z.S.); dhzhou@hdu.edu.cn (D.Z.)
† These authors contributed equally to this work.

Abstract: The growing need for deep-sea biological research and environmental monitoring has expanded the demand for benthic landers. Compared with remotely operated vehicles (ROVs) and autonomous underwater vehicles (AUVs), benthic landers can reduce overall operation cost and also possess longer endurance. Configuring a suitable descent velocity is important for benthic lander designs, helping them avoid retrieval failure and improve sea trial efficiencies. In this study, an effective scheme for the configuration and optimization of a self-developed benthic lander was outlined. First, the structural characteristics of the benthic lander were analyzed, and then a dynamic model was established. Second, the hydrodynamic coefficients of the benthic lander during its descent process were calculated using computational fluid dynamics (CFD) methods. Third, the MATLAB Simulink simulation environment was used to solve the dynamic model, and then the multi-objective optimization algorithm was introduced for the optimization design. Finally, the model was validated based on sea trial data, which demonstrated that the designed configuration and optimization scheme were correct and efficient. Collectively, this work provides a useful reference for the rational configuration and practical application of benthic landers.

Keywords: benthic lander; dynamic modeling; computational fluid dynamics (CFD); optimization design

Citation: Zhang, Q.; Dong, C.; Shao, Z.; Zhou, D. Analysis of the Descent Process and Multi-Objective Optimization Design of a Benthic Lander. *J. Mar. Sci. Eng.* **2023**, *11*, 224. https://doi.org/10.3390/jmse11010224

Academic Editor: Ernesto Weil

Received: 21 November 2022
Revised: 6 January 2023
Accepted: 13 January 2023
Published: 15 January 2023

1. Introduction

The deep sea is rich in energy, minerals and biogenetic resources; thus, it has significant economic value [1,2]. The impact of human activities on the marine environment is continually increasing [3,4], especially in offshore areas, but the response of the deep-sea to human activities remains unclear. In recent years, a large number of underwater vehicles have been used in marine climate change exploration, deep-sea observation, bioprospecting, mineral resource development and other fields. However, the mobility of conventional ROVs is severely limited due to the connection of the umbilical cable, while the cost of ROV sea trials is relatively high [5,6]. Although AUVs compensate for the abovementioned ROV shortcomings, they have poor endurance [7]. In contrast, benthic landers are playing an increasingly important role due to their relatively low cost, longer endurance and reusability.

Benthic landers have been widely utilized in in situ studies of deep-sea ecology, resources and environments. In 1975, Smith et al. designed a free vehicle respirometer (FVR) that could in situ measure the oxygen consumption of benthic communities relative to abyssal depths [8]. The other two types of benthic landers—the Autonomous Lander for Biological Experiments (ALBEX) [9,10] and the deep-sea benthic environmental observation system (Benvir) [11]—were developed for studies on deep-sea sediment community oxygen consumption. With the development of marine science and engineering technology, benthic landers can be equipped with various types of sampling modules for the collection of deep-sea seawater, sediment, microfauna and macrofauna. In 2009, the sediments incubated in the benthic lander named FLUFO [12] contained seep-associated fauna, and this study reveals the relationship between methane emissions from sediments with the seep-associated fauna. In the same year, a project called the Hadal Environment and Educational Program (HADEEP) was launched; it aimed to examine hadal biology and extend ecology research [13]. In this project, benthic landers were equipped with biological traps and seawater samplers to collect hadal samples. Thereafter, two function-similar benthic landers—a 7000-m-depth-rated lander (HaiJiao) [14] and a full-ocean-depth-rated lander—were separately produced by the Shenyang Institute of Automation of Chinese Academy of Sciences and the University of California San Diego (UCSD) [15]. Recently, Wei et al. successfully mounted their newly designed multiple in situ nucleic acid collections (MISNAC) on a benthic lander (Phoenix) to collect deep-sea microbial nucleic acid samples under in situ conditions [16]. In our previous study, a novel genus-level deep-sea bacterial species was isolated from an in situ enriched consortium collected by a benthic lander [17], and the lander used there is the prototype of the lander intended for optimization in this study. Benthic landers also have important applications in deep-sea acoustics [18], geohazards [19] and ocean-bottom seismometers (OBS) [20]. Collectively, in order to have an intuitive understanding of the background, onboard sensors and sampling modules of related landers, some typical cases of deep-sea environmental and biological studies using benthic landers are summarized in Table 1.

Table 1. Summary of typical benthic landers used in in situ studies of deep-sea ecology, resources and environments.

	FVR	ALBEX	FLUFO	HADEEP	Benvir	HaiJiao	UCSD-Lander
Background							
Country	U.S.A.	Netherland	Germany	Japan and UK	China	China	U.S.A.
Time	1975	1998	2009	2009	2009	2015	2019
Sensor							
CTD		✓	✓	✓	✓	✓	
Dissolved Oxygen (DO)	✓	✓			✓	✓	
PH					✓		
Sampling modules							
Camera	✓	✓	✓	✓		✓	✓
Seawater	✓			✓		✓	✓
Sediments		✓	✓			✓	
(Microbe Enrichment)						✓	
Biological trap				✓		✓	✓

In recent years, the frame structure of the benthic lander has also been greatly developed. The early lander mainly comprised a tripod frame [8,9,13] was constructed around a centrally located acoustic release, similarly to the FVR lander; on top of the three legs is an aluminum plate. The BIGO [21,22] lander also has a tripod frame, but a cuboid frame is added on the top of the floatation sphere array of the lander to carry more scientific devices. The three legs make up the simplest stable structure, and the plate provides a larger contact area to avoid retrieval failure caused by excessive subsidence depth with the seafloor. Then, the three legs gradually developed into four legs [23–25] for higher stability, which greatly increased the ability of the lander to carry scientific devices, similarly to the RAP2 [26] lander. The four-legged Gothenburg [27] lander adopted a two-frame solution, and this was significant because it allowed for the flexibility to deploy only the inner-frame from small vessels. However, landers used for ocean-bottom seismic research usually have flat boards [28] or annular structures [20,29].

The control of descent velocity is a key concern in the design and application of benthic landers. After the benthic lander separated from the ship, its descent velocity was mainly determined by the system's configuration [30]. When the descent velocity is too low, the efficiency of the sea trial will be reduced, and the lander may also experience uncontrolled lateral drift driven by ocean currents. Simultaneously, excessive descent velocities will also lead to a series of problems, such as lander damage or sinking too deep into the seabed sediment, which may all lead to a failure in lander retrieval. To solve these problems, Mortensen et al. [31] designed crossed wings to control the descent velocity of the lander. However, the velocity was too low, which resulted in lower efficiency and higher costs in sea trials. Jun [32] and Gang et al. [33] controlled the descent velocity by integrating a hydrofoil into the lander, but the effectiveness needs to be further verified. Therefore, achieving a more direct control scheme over the descent process is a problem that needs to be considered when designing a benthic lander.

Based on the above analysis, an effective scheme for the configuration and optimization of a self-developed benthic lander was outlined in this study. This scheme started with dynamic modelling and hydrodynamic coefficients calculation. Second, the CFD methods and multi-objective optimization algorithms were used to comprehensively study the influence mechanism of several variables on the lander's descent velocity, including the number of floatation spheres, the mass of the weight stack and the contact area of the weight stack. Finally, the mutual restriction relationship between the lander's descent time and subsidence depth was revealed. This study provides a useful reference for the rational configuration and practical application of a benthic lander.

2. Model Establishment of the Benthic Lander

2.1. Frame Structure and Working Principle of the Benthic Lander

The lander mainly comprises three parts: an upper floatation sphere array that provides positive buoyancy, a frame that supports the scientific device and a lower weight stack that provides descending gravity (see Figure 1a). The floatation sphere array is connected in series with the fiber rope, and each floatation sphere (model: NMS-FS-6700-17RO, Nautilus Marine Service GmbH) provides 26 kg of positive buoyancy; moreover, the number of floatation spheres is configurable depending on the weight of the scientific device. The frame is made of titanium alloys. Loaded within the frame is a syntactic foam that provides positive buoyancy and structural support to the frame, and on top of the frame is a lifting bail that facilitates the process of deployment and retrieval. At the center of the frame is an acoustic release (model: OCEANO 2500S Light, iXblue, Paris, France) that is responsible for releasing the disposable weight stack. The scientific device includes the microbe enrichment device [34] array, which is the main usage of the lander, and a dual-axis inclinometer sensor (model: LCT100A, Guigangshi Lecheng Information Technology Co., Ltd., Guigang, China) with an angle resolution of 0.0001° and a 1–100 Hz configurable output frequency. The weight in water of the frame (including foam, release and science devices without the weight stack) is −85 kg, which was obtained from the water tank experiment (see Figure A1). Based on the above explanation of the lander's frame structure, the working principle of the benthic lander is described as follows: At the end of the experiment, an acoustic remote-control unit (model: TT801, iXblue, Paris, France) is used to release the hook of the acoustic release; see Figure 1b. Then, the weight stack and bearing bracket (see Figure A2) are automatically separated from the frame, and the positive buoyancy provided by the floatation sphere array and syntactic foam allows the frame to ascend from the seafloor.

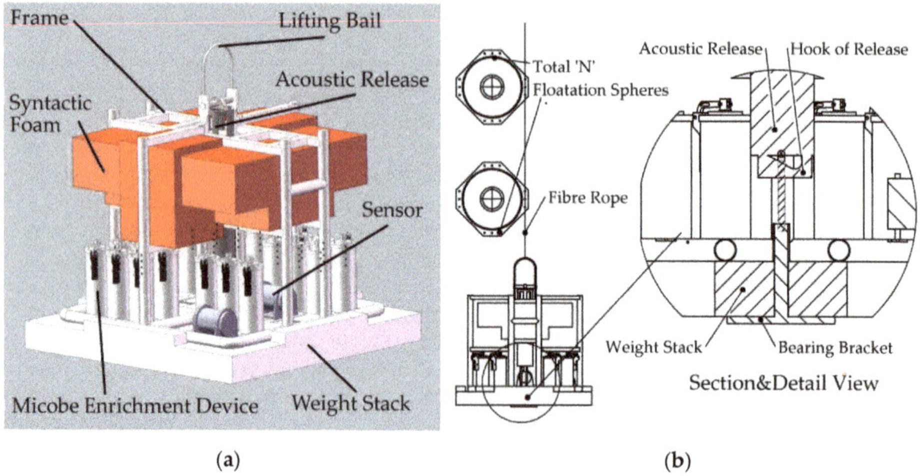

Figure 1. (**a**) The overall structure of the benthic lander; (**b**) the working principle of the benthic lander.

2.2. Benthic Lander Dynamic Modeling

The dynamic model of an underwater vehicle in six degrees of freedom can be described as a nonlinear equation of motion in matrix form [35]:

$$\mathbf{M}\dot{\boldsymbol{\upsilon}} + \mathbf{C}(\boldsymbol{\upsilon})\boldsymbol{\upsilon} + \mathbf{D}(\boldsymbol{\upsilon})\boldsymbol{\upsilon} + \mathbf{g}(\boldsymbol{\eta}) = \boldsymbol{\tau} \tag{1}$$

$$\dot{\boldsymbol{\eta}} = \mathbf{J}(\boldsymbol{\eta})\boldsymbol{\upsilon} \tag{2}$$

where $\mathbf{M} = \mathbf{M_{RB}} + \mathbf{M_A}$ is the inertia matrix for the rigid-body mass matrix and added mass matrix, respectively;

$\mathbf{C}(\boldsymbol{\upsilon}) = \mathbf{C_{RB}}(\boldsymbol{\upsilon}) + \mathbf{C_A}(\boldsymbol{\upsilon})$ is the Coriolis and centripetal matrix for rigid-body mass and added mass, respectively;

$\mathbf{D}(\boldsymbol{\upsilon}) = \mathbf{D_{quad.}}(\boldsymbol{\upsilon}) + \mathbf{D_{lin.}}$ is the quadratic and linear damping matrix, respectively;

$\mathbf{g}(\boldsymbol{\eta})$ is the hydrostatic restoring force matrix resulting from the different positions of the center of buoyancy and the center of gravity;

$\boldsymbol{\tau}$ is the sum of the acting forces and moments produced by the propulsion system of the vehicle, which is equal to zero in this case because the lander is a freely descending vehicle;

$\boldsymbol{\upsilon}$ is the generalized velocity vector; accordingly, $\dot{\boldsymbol{\upsilon}}$ is the generalized acceleration vector;

$\boldsymbol{\eta}$ denotes the generalized position coordinates, and $\mathbf{J}$ is the rotation matrix for the transformation from the body-fixed frame to the earth-fixed frame. Note: $\mathbf{M}, \mathbf{C}(\boldsymbol{\upsilon}), \mathbf{D}(\boldsymbol{\upsilon}) \in R^{6x6}$, $\mathbf{g}(\boldsymbol{\eta}), \boldsymbol{\tau}, \boldsymbol{\upsilon}, \boldsymbol{\eta} \in R^{6x1}$.

Due to the relatively low descent velocity (please refer to the data presented in "6 Sea Trials".), the Coriolis and centripetal matrix can be ignored, and its symmetrical structure design means that the lander is relatively stable with respect to its attitude (please refer to the data presented in "3.2. Calculation of Damping Force Coefficients".); thus, the quadratic and linear viscous coupling coefficients between the degrees of freedom are small and can be ignored. Finally, the descent process of the next analysis was only carried out for the lander's heave degree of freedom, and the model described in Equations (1) and (2) can be simplified as follows:

$$(M_{RB} + M_A)\dot{\upsilon} + (D_{quad.}(\upsilon) + D_{lin.})\upsilon = -g \tag{3}$$

where M_{RB} is the mass of the lander; $D_{quad.}$ and $D_{lin.}$ are the quadratic and linear damping coefficients, respectively; υ is the descent velocity in the heave degree of freedom; $-g$ is the net gravity for the free descending vehicle (lander). Moreover, gravity is composed of three parts: weight in seawater of the weight stack (positive), the buoyancy of the floatation sphere array (negative) and the buoyancy of the frame (negative).

3. Calculating Hydrodynamic Coefficients

3.1. Hydrodynamic Modeling

The benthic lander is an open frame structure with different configurations of the scientific device, which increases the difficulty of solving the hydrodynamic coefficients in Equation (3). The Favre-averaged Navier–Stokes (FANS) equation is a widely used method for hydrodynamic simulations of underwater vehicles with turbulent flows. In this study, the CFD software add-in, SOLIDWORKS Flow Simulation, is used to implement the FANS to compute the hydrodynamic coefficients. Assuming that the fluid comprises continuous media and does not exhibit energy conservation, the conservation laws for mass and angular momentum can be written in the conservation form as follows [36]:

$$\frac{\partial \rho}{\partial t} + \frac{\partial}{\partial x_i}(\rho u_i) = 0 \tag{4}$$

$$\frac{\partial \rho u_i}{\partial t} + \frac{\partial}{\partial x_j}(\rho u_i u_j) + \frac{\partial p}{\partial x_i} = \frac{\partial}{\partial x_j}\left(\tau_{ij} + \tau_{ij}^R\right) + S_i \; i = 1,2,3 \tag{5}$$

where u is the fluid velocity, and ρ is the fluid density. Moreover, S_1 denotes a mass-distributed external force per unit mass due to a porous media resistance, and S_2 denotes buoyancy ($S_2 = -\rho g_i$, where g_i is the gravitational acceleration component along the i-th coordinate direction). S_3 denotes the coordinate system's rotation. Moreover, τ_{ij} denotes the viscous shear stress tensor, accordingly, τ_{ij}^R denotes the rotation form of τ_{ij}. Significantly, with the exception of S_i, other subscripts represent the corresponding coordinate direction.

The κ-ε turbulence model is used for the simulation, and the fluid volume is selected as $8L_{lander} \times 8W_{lander} \times 10H_{lander}$(length, width, height (LWH)); the dimension of the lander is 0.99 × 0.97 ×1.04 m. The wall conditions and wall roughness comprise an adiabatic wall and is rated at zero, respectively. The two-scale wall functions (2SWF) model [36] was used to describe boundary layers on a fine mesh; the number of cells across a boundary layer is 8, and the corresponding approach is called the "thick boundary layer". To select an appropriate basic mesh size and obtain high calculation accuracy with as few cells as possible, six different basic mesh sizes were used to calculate the hydrodynamic drag force during the lander's descent process, and the incoming flow velocity is set at 0.1 m/s and 1.0 m/s. The so-called mesh irrelevance verification is shown in Table 2.

Table 2. The mesh irrelevance verification.

Basic Mesh [1]/(m)	Cells	Drag/(N) in 0.1 m/s	Drag/(N) in 1.0 m/s
0.17	131,459	7.18	721.98
0.15	210,683	6.02	601.49
0.13	311,779	6.98	704.71
0.11	525,830	6.97	696.15
0.09	898,422	6.83	686.12
0.07	1,974,685	6.81	681.81

[1] The basic mesh is used for the x/y/z direction.

As observed in Table 2, when the basic mesh is less than 0.09 m, the drag force gradually stabilizes and undergoes a small change. Considering limited computing resources, a basic mesh of 0.09 m is finally selected for the next CFD simulation.

3.2. Calculation of Damping Force Coefficients

In the CFD simulation, the descent velocity of the benthic lander is changed by adjusting the incoming flow velocity. The incoming flow velocity is set as 0.1–1.0 m/s, and the interval is 0.1 m/s. In contrast, Figure 2a, a cut plot, provides a contoured view of the velocity parameter distribution in which the range is divided by colored intervals; thus, each interval has its own color. Obviously, the symmetrical velocity distribution of the left and right sides also means that the damping force (D(υ)) of the lander is essentially symmetrical. Therefore, when only hydrodynamic forces are considered, the lander is relatively stable in its attitude. The damping force is a quadratic function of velocity only. From the simulated forces, the least mean square method is used to determine the hydrodynamic parameters based on Equation (3), and the results are shown in Figure 2b. The corresponding quadratic function is shown in Equation (6). As with the lander, the damping force function of a single floatation sphere is shown in Equation (7). For the full CFD data, we refer the readers to Supplementary Table S1: CFD data.

$$D(\upsilon) = 686.5\upsilon^2 - 0.3\upsilon \tag{6}$$

$$D(\upsilon) = 119.4\upsilon^2 + 0.5\upsilon \tag{7}$$

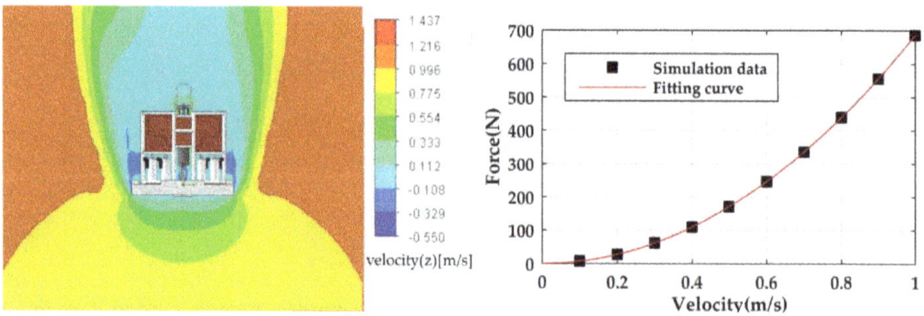

(a) (b)

Figure 2. (**a**) Cut plot when the incoming flow velocity is 1 m/s; (**b**) damping force quadratic function curve of the lander.

3.3. Calculation of Added Mass Coefficients

When an underwater vehicle moves in a fluid, the inertia of the fluid opposes the motion; that effect is equal to having a virtual mass added to the mass of the vehicle. In other words, when the underwater vehicle exhibits unsteady motion, it will produce added mass; we refer the reader to Equation (3). The velocities with sine functions [37] or ramp functions [38] are widely used in unsteady motions. In this study, the velocity with ramp functions was selected; we refer the reader to Equation (8), where "a" is the constant acceleration and the parameter is set as 3 groups: 0.100, 0.125 and 0.150 m/s^2.

$$\upsilon = a \cdot t \tag{8}$$

In the SOLIDWORKS Flow Simulation, the simulation project selects the time-dependent physical feature. The fluid's unsteady motion requires a certain time period (for example, in group 3, the time approx. 13 s) to reach relatively stable acceleration velocities, and the actual acceleration is different from the set value. The actual velocity development is shown in Figure 3a, where the black dashed line represents the actual acceleration (or slope) when the steady state is attained. As shown in Equation (3), the total drag force comprises two parts: the damping force calculated in Equation (6) and the difference force generated by the added mass. The results of the explained difference force are shown in Figure 3b (in group 3, the data from groups 1 and 2 are shown in Figures A3 and A4, respectively). The dashed line is the boundary at which the fluid reaches a relatively stable acceleration; obviously, the difference force is relatively constant at this boundary. Finally, bringing simulation data into Equation (3), we can obtain the added mass; we refer the reader to Table 3.

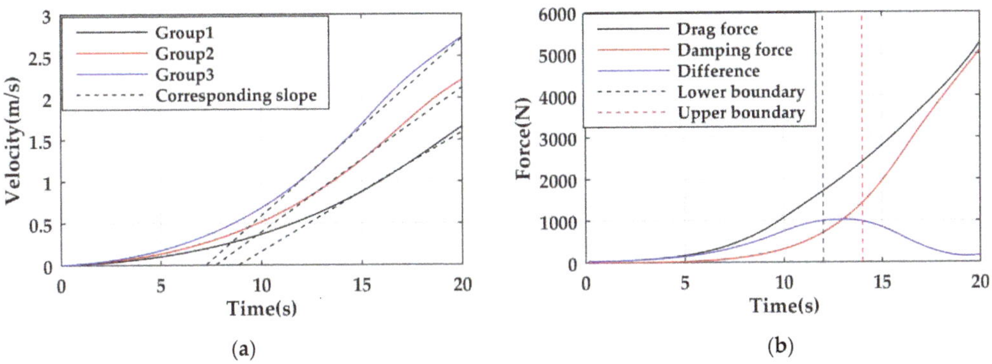

Figure 3. (**a**) The actual velocity development; (**b**) the difference force generated by the added mass in group 3.

Table 3. Summary of added mass calculation.

Group	Setting Acceleration (m/s^2)	Actual Acceleration (m/s^2)	Boundary(s)	M_A(kg) [1]
1	0.100	0.143	14–18	4008.4
2	0.125	0.173	12–16	4138.2
3	0.150	0.212	12–14	4163.9
	The mean value of M_A(kg) in 3 groups			4103.5

[1] Note: the M_{RB} is 588.6 kg.

4. MATLAB Simulink

To study the influence of the lander's configuration parameters on the variables of the descent process and based on the established dynamic model and calculated hydrodynamic coefficients, the MATLAB Simulink simulation environment was used to design the block diagram model (as we can see from Figure 4), and the corresponding file is shown in Supplementary Model S1: Simulink dynamic model. This simulation model has three input variables—the number of floatation spheres (N), the mass of the weight stack (m_{stack}) and the bottom area of the weight stack (A_{stack}) —and three output variables—the descent velocity (V_{des}), descent time (T_{des}) and subsidence depth (D_{sub}) with the seafloor. The variable explanation is as follows, and the nomenclature is shown in Table 4.

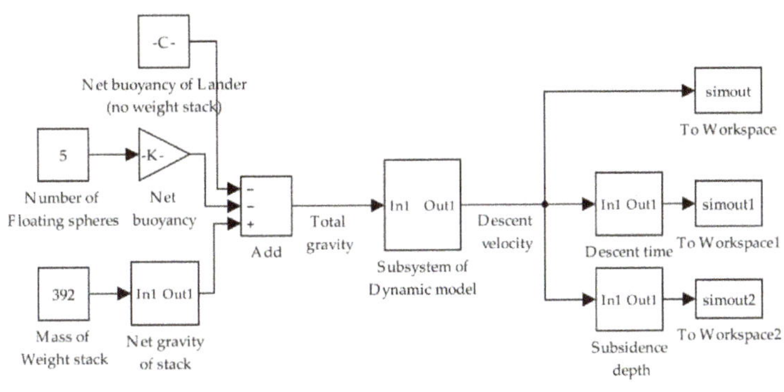

Figure 4. Simulink block diagram model.

Table 4. Nomenclature.

Symbol	Description	Unit	Symbol	Description	Unit
N	number of floatation spheres	-	T_{des}	descent time	H
m_{stack}	mass of the weight stack	kg	V_{des}	descent velocity	m/s
A_{stack}	bottom area of the weight stack	m^2	D_{sub}	subsidence depth	mm

- A_{stack} influences the hydrodynamic coefficients (as we can see from Figure A5): The corresponding function relationship is calculated by using SOLIDWORKS Flow Simulation. The results are shown in Equations (9) and (10);
- T_{des} takes the water depth of 5133 m during the sea trial as the background: The corresponding function is shown in Equation (11);
- D_{sub} was tested by a previous study [39,40] for the same benthic lander: The corresponding function is shown in Equation (12).

$$D_{quad.} = 14.29{A_{stack}}^2 - 23.17A_{stack} + 9.92 \tag{9}$$

$$D_{lin.} = 14.29{A_{stack}}^2 - 23.17A_{stack} + 9.92 \tag{10}$$

$$T_{des} = \frac{5133}{3600\upsilon} \tag{11}$$

$$D_{sub} = 30.7\upsilon + 2.65 \tag{12}$$

The dynamic model in Equation (3) is a nonlinear, first-order differential equation that may be integrated (in the Simulink diagram model, the Integrator block is used; we refer the readers to Figure A5) numerically to yield vehicle translational velocities when provided with the suitable initial conditions. The fixed-step size is 0.02 s by using the ode4 (Runge–Kutta) solver to solve the nonlinear differential equation.

Keeping the number of floatation spheres (N) constant at five in the descent process, the steady state motion was calculated by changing the mass of the weight stack (m_{stack}) and the bottom area of the weight stack (A_{stack}). The developmental rules of the subsidence depth (D_{sub}) and descent time (T_{des}) are shown in Figure 5. Obviously, the subsidence depth increases with m_{stack}; the smaller the bottom area, the greater m_{stack} affects the subsidence depth. This shows that when the bottom area is small, it is easier to change the subsidence depth of the lander by adjusting m_{stack}. In contrast, the descent time decreases with m_{stack}; therefore, m_{stack} shall be appropriately increased in the sea trial in pursuit of operational efficiency.

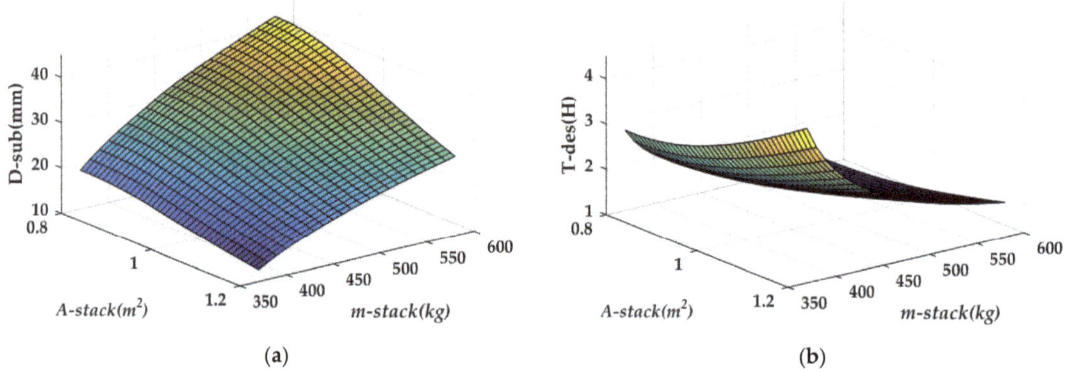

Figure 5. (**a**) Subsidence depth vs. the mass of the weight stack and bottom area of the weight stack; (**b**) descent time vs. mass of the weight stack and bottom area of the weight stack.

Keeping the bottom area of the weight stack (A_{stack}) constant at 0.96 m^2 in the descent process, the steady state motion was calculated by changing the mass of the weight stack (m_{stack}) and the number of floatation spheres (N). The results of the descent time (T_{des}) and subsidence depth (D_{sub}) are shown in Figure 6. As the number of floatation spheres increases, T_{des} increases but D_{sub} decreases. When m_{stack} is small, T_{des} and D_{sub} are more sensitive to changes in the number of floatation spheres. For example, when N is six, T_{des} increased dramatically and lost its practical significance; thus, this situation was deleted from this Figure. Therefore, m_{stack} should not be too small in value in order to reduce the influence of T_{des} and D_{sub}.

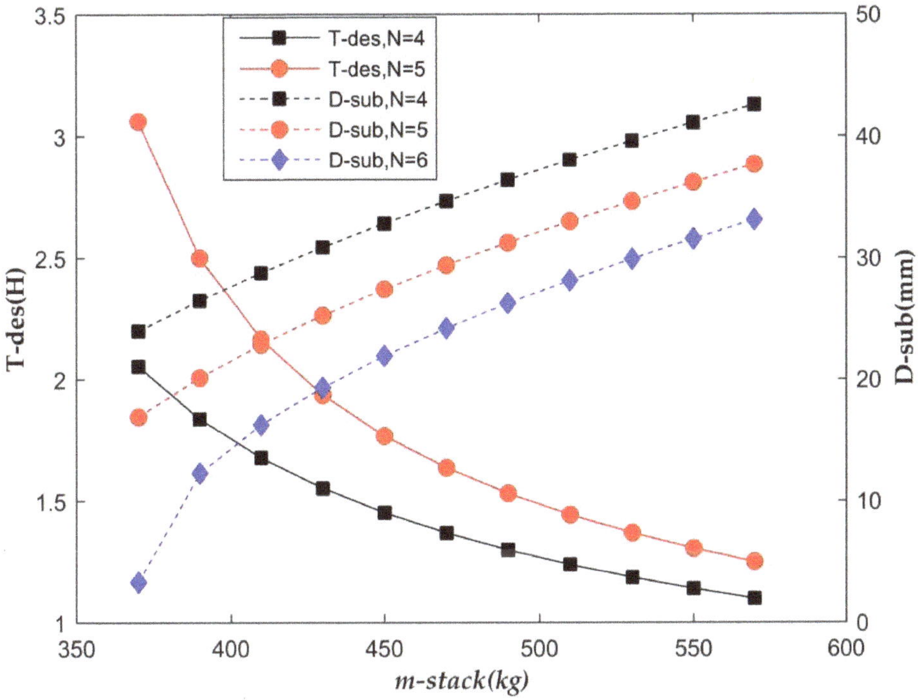

Figure 6. Subsidence depth, descent time vs. mass of the weight stack and number of floatation spheres.

5. Multi-Objective Optimization

Based on the analysis of MATLAB Simulink, we know the relationship between specific configuration parameters and descent time (T_{des}) and subsidence depth (D_{sub}) of the lander, but making a choice is still difficult because there are various configuration schemes. Second, in practical applications, the descent time should be as small as possible to increase the operation efficiency in sea trials, and the subsidence depth should be as small as possible to avoid retrieval failure. As we can see from Figure 5, the descent time and subsidence depth are two conflicting objectives; thus, it is a multi-objective optimization problem (MOP). In this case, the objective of optimization is to minimize the descent time and subsidence depth simultaneously, which can be formulated [41] as follows:

$$\begin{cases} \min\limits_{X} F(X) = (f_1(X), f_2(X))^T \\ \text{s.t. } g_i(X) \le 0, i = 1,2,3 \end{cases} \tag{13}$$

where $X = (x_1, x_2, x_3)^T$ is the decision vector; decision variable $x_1 \in [4,5,6]$ is the number of floatation spheres (N), which is configured by previous sea trial experience; $x_2 \in (370 - 580)$ is the mass of the weight stack (m_{stack}), which is limited by the boundary of the bottom area of the weight stack (A_{stack}); and $x_3 \in (0.8-1.2)$ is the bottom area of the weight stack (A_{stack}), which is limited by the number of scientific device carried by the lander. F(X) is the objective vector, objective value $f_1(X)$ is the descent time (T_{des}) defined in Equation (11), and $f_2(X)$ is the subsidence depth (D_{sub}) defined in Equation (12). $g_i(X)$ denotes inequality constraints, where $g_1(X)$ ensures that the total net gravity is positive, $g_2(X)$ ensures that the descent time is less than 3 h, and $g_3(X)$ ensures that the subsidence depth is less than 45 mm. The details are described as follows, and the corresponding MATLAB code is shown in **Algorithms B1**.

$$g_1(X) = 26N - \frac{1869.7}{2869.7}x_2 + 85 \leq 0 \tag{14}$$

$$g_2(X) = f_1(X) - 3 \leq 0 \tag{15}$$

$$g_3(X) = f_2(X) - 45 \leq 0 \tag{16}$$

Based on the above MOP, the platform for evolutionary multi-objective optimization (PlatEMO) [42] was used to solve this problem, and the corresponding multi-objective evolutionary algorithms (MOEAs) comprise NSGA-II; [43]. The population size is set to 100, and the evaluation number is 10,000. The Pareto-optimal front (objective value) is obtained as shown in Figure 7a, and the corresponding Pareto-optimal set (decision variable) is shown in Figure 7b. The black, red and blue points distinguish the different numbers of floatation spheres. Finally, parts of the configuration from the Pareto-optimal front and set are shown in Table 5 (Supplementary Table S2: multi-objective optimization), which provides an accurate configuration reference depending on the specific T_{des} and D_{sub}.

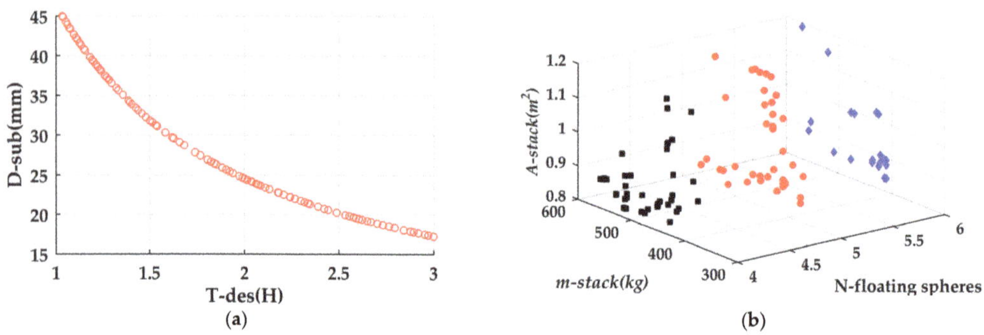

Figure 7. (**a**) The Pareto-optimal front of optimization results; (**b**) the distribution of the Pareto-optimal set.

Table 5. Summary of optimization results.

Decision Variables			Objective Values		Decision Variables			Objective Values	
N	m_{stack} (kg)	A_{stack} (m^2)	T_{des} (H)	D_{sub} (mm)	N	m_{stack} (kg)	A_{stack} (m^2)	T_{des} (H)	D_{sub} (mm)
4	557.7	0.89	1.03	45.0	4	511.8	0.92	1.18	39.8
4	510.0	0.83	1.10	42.3	4	508.1	0.87	1.13	41.3
5	392.0	0.96	1.76	27.5	5	441.9	1.20	2.63	19.3
5	468.3	1.19	2.35	21.3	5	444.5	1.10	2.24	22.2
6	410.4	0.89	3.00	17.2	6	553.6	0.92	1.44	33.1
6	436.5	0.88	2.32	21.5	6	414.0	0.89	2.88	17.9

The goal of MOEAs is to provide the last population with good convergence and diversity when the evaluation ends; convergence and diversity are also two important performance indicators for evaluating the optimization results. Diversity is usually measured by the diversity metric (DM) [44], and larger values are better. Convergence is usually measured by the inverted generational distance (IGD) [45], and smaller values are better.

Finally, the results of the DM and IGD indicators in this study are shown in Figure 8. Obviously, the IGD tends to be stable when the number of evaluations is greater than 500; the DM has some oscillation, but the average value is approximately 0.7, both of which meet the requirements of engineering practice.

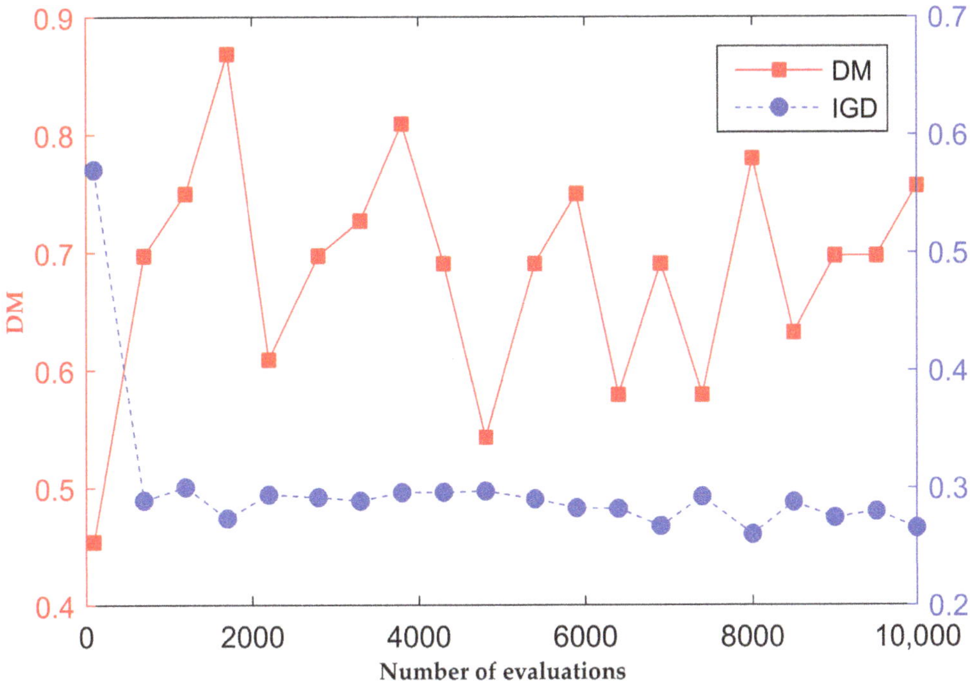

Figure 8. **The** DM and IGD indicator of optimization results.

6. Sea Trials

The benthic lander sea trial was conducted in the South China Sea in July 2018 [39,40], and the newest sea trial was conducted in the western Pacific Ocean (location: 134.84°E, 16.96°N) in January 2021 (Figure 9). The corresponding video is shown in Supplementary Video S1: sea trial. The newest sea trial will be discussed in this study, and the basic configuration is N = 5, m_{stack} = 392 kg, and A_{stack} = 0.96 m^2, one of the optimization results from Table 5. After the lifting device is released from the benthic lander, gravity provided by the weight stack drives the lander to freely land on the bottom of the seafloor. The original document that shows the range function of the acoustic release that obtains the relationship between the descent depth and descent time (Table 6 and Figure 10a) can be found in Supplementary Document S1: Lander deployment log sheet, and the following details are noteworthy:

- The depth information reflected by the range data of the acoustic release is relatively accurate and reliable. First, the total mass of the lander system (603 kg) is large, and the sea current has little influence on it. Second, the successful retrieval of the lander at the original location in the later sea trials proved that the horizontal drifts during the descent process can be ignored;
- Removal of instability during the initial launch and final landing, and the steady descent (6–70 min) velocity fluctuates between 0.68 and 0.89 m/s, which meets the expected results.

As observed in Figure 10b, the Simulink simulation velocity is basically consistent with the mean descent velocity in the sea trial; the correctness of the Simulink simulation results and CFD hydrodynamic calculation was verified; the mean descent velocity was calculated in the steady descent (6–70 min) process. As the added mass is considered in Section 2.1, the acceleration process of the lander is relatively long; at approx. 50 s, its velocity reaches the maximum and remains relatively stable. The descent time (T_{des}) calculated based on the average descent velocity of the sea trial is 1.85 h. Table 5 shows that, under the sea trial configuration, the descent time obtained by using multi-objective optimizations is 1.76 h, and the relative error between the two is 5.1%. Thus, the correctness of multi-objective optimizations was verified.

(a) (b)

Figure 9. (**a**) The lander on the deck; (**b**) the lander to be released.

Table 6. The depth information comes from acoustic release.

Descent Time (min)	Descent Depth (m)	Corresponding Velocity (m/s)	Descent Time (min)	Descent Depth (m)	Corresponding Velocity (m/s)
0	0	-	28	1536	0.89
2	88	0.73	41	2083	0.70
4	343	2.13	58	2908	0.81
6	434	0.76	70	3396	0.68
11	694	0.87	95	5071	1.12
17	951	0.71	105	5133	0.10

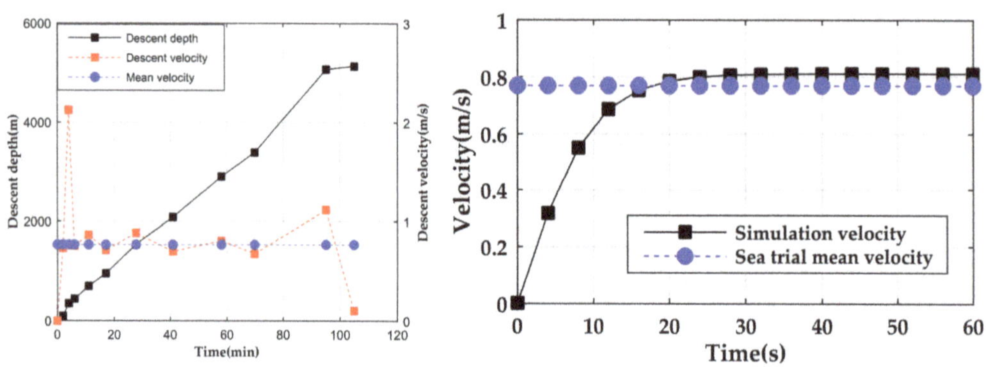

(a) (b)

Figure 10. (**a**) Descent information from the sea trial; (**b**) Simulink simulation data vs. sea trial data.

7. Discussion

In this study, based on the analysis of MATLAB Simulink and CFD methods, the multi-objective optimization algorithm was used to achieve a rational configuration scheme for a benthic lander that realizes reasonable control of descent velocity. This scheme makes up the shortcomings of previous studies, for example, using crossed wings (Mortensen et al. [31]) or integrating a hydrofoil (Jun [32] and Gang et al. [33]) into the benthic lander. On the other hand, the main reasons for the higher Simulink simulation velocity (see Figure 10b) can be concluded as follows. First, the hydrodynamic drag coefficients of the lander and the floatation spheres are considered, but the coupling effects between the two are neglected. Second, the CFD damping force calculation is based on the static steady descent velocity, but the actual descent velocity has some oscillations (Figure 10a). Finally, as described in Section 6, the horizontal drifts of the lander during the descent process can be ignored, but the drifts exist, which leads to the actual descent depth being smaller than the range data from the acoustic release. However, we can conclude that the overall trend is basically correct.

In specific applications, we need to decide the descent time (T_{des}) tolerated by the sea trials and the subsidence depth (D_{sub}) tolerated by the lander according to Figure 7a, and the two values were used to choose a rational configuration from Table 5 (the full table see Supplementary Table S2: multi-objective optimization). Significantly, the proper number of floatation spheres (N) should be carefully selected because N has a large influence on T_{des} and D_{sub} (Figure 6). The population size and number of evaluations should be appropriately increased when using multi-objective optimization, which can improve the choice of configuration scheme and enhance the convergence/diversity of the population, respectively.

The results were calculated by a multi-objective optimization algorithm, and only configurations with five floatation spheres were verified in the sea trial. In future work, the number of floatation spheres at four and six still require further verification, which can comprehensively verify the correctness of the proposed scheme. The applicability of CFD simulation also needs to be tested more directly, especially since the calculation of the added mass coefficients usually has a relatively large error. In future studies, a spherical example should be introduced to test the relevant settings of SOLIDWORKS Flow Simulation in order to reduce simulation errors.

Supplementary Materials: The following supporting information can be downloaded at: https://www.mdpi.com/article/10.3390/jmse11010224/s1, Document S1: Lander deployment log sheet; Model S1: Simulink dynamic model; Table S1: CFD data; Table S2: multi-objective optimization; Video S1: sea trial.

Author Contributions: Conceptualization, D.Z. and Z.S.; methodology, D.Z. and Q.Z.; software, Q.Z.; validation, D.Z.; formal analysis, D.Z. and Q.Z.; investigation, Q.Z. and C.D.; data curation, D.Z. and C.D.; writing—original draft preparation, Q.Z. and C.D.; writing—review and editing, Q.Z., C.D., Z.S. and D.Z.; visualization, Q.Z.; supervision, D.Z. and Z.S.; project administration, D.Z. and C.D.; funding acquisition, Z.S. and D.Z. All authors have read and agreed to the published version of the manuscript.

Funding: This research was funded by the High-Tech Research and Development Program of China, grant number 2012AA092102.

Institutional Review Board Statement: Not applicable.

Informed Consent Statement: Not applicable.

Data Availability Statement: Not applicable.

Acknowledgments: We thank Gao Wei for his help with the CFD simulation and MATLAB Simulink modeling. We thank Tian Ye for his help with the multi-objective optimization. We extend our gratitude to Jianhua Xu and Dongmei Zhang for their help at sea trials. We also thank Zhao Qi for his help with the water tank experiment.

Conflicts of Interest: The authors declare no conflict of interest.

Appendix A

Figure A1. Water tank experiment.

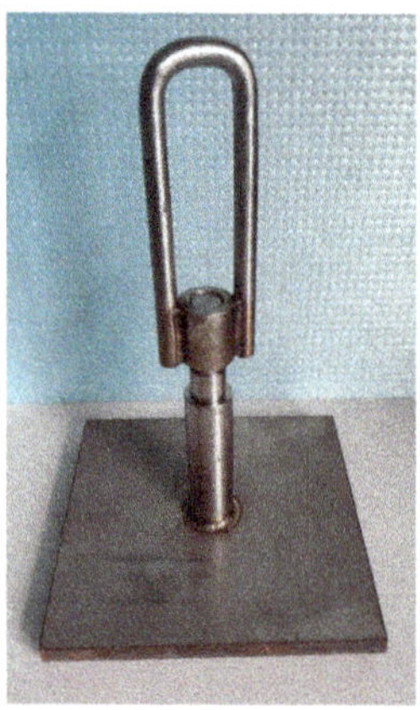

Figure A2. Bearing bracket.

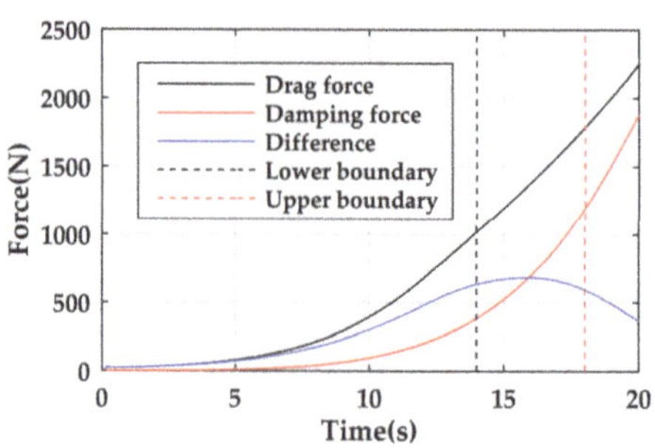

Figure A3. Difference force in group 1.

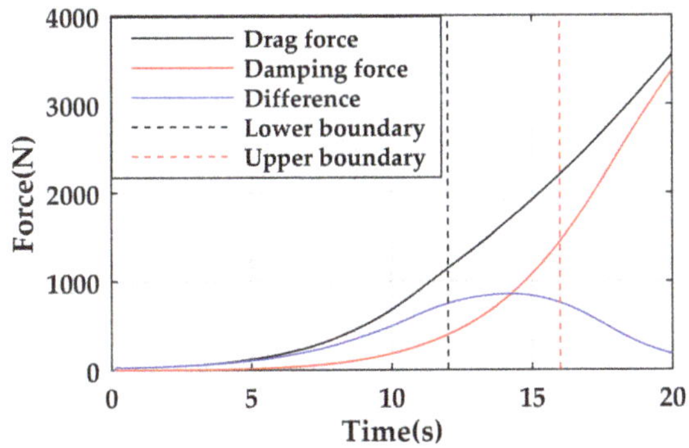

Figure A4. Difference force in group 2.

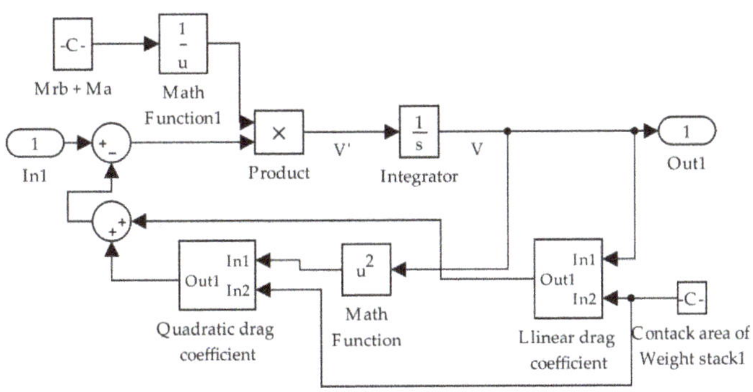

Figure A5. Subsystem of the dynamic model.

Appendix B

Algorithms B1% Usage: 1.put Lander_optimization.m file to PlatEMO-1.6(can be another version)\PlatEMO\Problems\CF folder

```
% 2. corresponding command in matlab: main('-algorithm',@NSGAII,'-problem', @Lander_optimization,'-N',100,'-M',2,'-D',3,'-evaluation',10000,'-save',1);
% Reference: [46]
classdef Lander_optimization < PROBLEM
% <problem> <CF>
% Constrained benchmark MOP
    methods
    %% Initialization
        function obj = Lander_optimization()
          obj.Global.M = 2; %number of objects
          if isempty(obj.Global.D)
                obj.Global.D = 3; %number of decision variables
            end
            obj.Global.lower = [4370,0.8]; %lower boundary of decision variables
            obj.Global.upper = [6580,1.2]; %upper boundary of decision variables
            obj.Global.encoding = 'real';
        end
        %% Calculate objective values
        function PopObj = CalObj(obj,X)
            m_frame = 211; %mass of frame(without weight stack)(kg)
            G = 9.8; %gravitational acceleration(m/s^2)
            Bfq = 26; %buoyancy of signal floatation sphere(kg)
```

```
                Rho_ballast = 2869.69; %the density of the ballast(kg/m^3)
                Rho_water = 1000; %the density of the water(kg/m^3)
                k1 = 686.5; %quadratic hydrodynamic coefficient of lander
                k2 = 119.4; %quadratic hydrodynamic coefficient of signal floatation sphere
                %round function is to towards nearest integer for the numbers of floata-
tion sphere
                k = k2*round(X(:,1)) + k1.*(14.29.*X(:,3).^2 − 23.17.*X(:,3) + 9.92);
                %85 is the net gravity of lander without the weight stack
                g = (−Bfq.*round(X(:,1)) + X(:,2)./Rho_ballast.*(Rho_ballast-Rho_water)
−85).*G./(m_frame+X(:,2));
                v = sqrt((m_frame+X(:,2)) .*g./k); %descent velocity
                PopObj(:,1) = 5133./(3600.*v); %descent time
                PopObj(:,2) = 30.7.*v + 2.65; %subsidence depth
            end
            %% Calculate constraint violations
            function PopCon = CalCon(obj,X)
                PopObj = obj.CalObj(X);
                %g_1(X) is to make sure the total net gravity is positive
                PopCon(:,1) = 26.*round(X(:,1)) − X(:,2)./2869.69.*(2869.69−1000) + 85;
                %g_2(X) is to make sure the descent time less than 3 h
                PopCon(:,2) = PopObj(:,1) − 3;
                %g_3(X) is to make sure the subsidence depth less than 45 mm
                PopCon(:,3) = PopObj(:,2) − 45;
            end
            %% Sample reference points on Pareto front
            function P = PF(obj,N)
                P(:,1) = (0.33:1/20:3)'; %the values of f1(x)
                P(:,2) = (30.7*5)./(9.*P(:,1)) + 2.65; %the values of f2(x)
            end
        end
    end
```

References

1. Marshall, N.B. Glimpses into Deep-Sea Biology. In *The Arctic Ocean: The Hydrographic Environment and the Fate of Pollutants*; Rey, L., Ed.; Palgrave Macmillan: London, UK, 1982; pp. 263–271.
2. McLellan, B.C. Sustainability Assessment of Deep Ocean Resources. *Procedia Environ. Sci.* **2015**, 28, 502–508. [CrossRef]
3. Harrould-Kolieb, E.R.; Herr, D. Ocean Acidification and Climate Change: Synergies and Challenges of Addressing Both under the UNFCCC. *Clim. Policy* **2012**, 12, 378–389. [CrossRef]
4. Cooley, S.R.; Bello, B.; Bodansky, D.; Mansell, A.; Merkl, A.; Purvis, N.; Ruffo, S.; Taraska, G.; Zivian, A.; Leonard, G.H. Overlooked Ocean Strategies to Address Climate Change. *Glob. Environ. Chang.* **2019**, 59, 101968. [CrossRef]
5. Nunoura, T.; Takaki, Y.; Hirai, M.; Shimamura, S.; Makabe, A.; Koide, O.; Kikuchi, T.; Miyazaki, J.; Koba, K.; Yoshida, N.; et al. Hadal Biosphere: Insight into the Microbial Ecosystem in the Deepest Ocean on Earth. *Proc. Natl. Acad. Sci. USA* **2015**, 112, E1230–E1236. [CrossRef]
6. Bowen, A.D.; Yoerger, D.R.; Taylor, C.; McCabe, R.; Howland, J.; Gomez-Ibanez, D.; Kinsey, J.C.; Heintz, M.; McDonald, G.; Peters, D.B.; et al. The Nereus Hybrid Underwater Robotic Vehicle for Global Ocean Science Operations to 11,000 m Depth. In Proceedings of the OCEANS 2008, Quebec, QC, Canada, 15–18 September 2008; pp. 1–10. [CrossRef]
7. Wynn, R.B.; Huvenne, V.A.I.; Le Bas, T.P.; Murton, B.J.; Connelly, D.P.; Bett, B.J.; Ruhl, H.A.; Morris, K.J.; Peakall, J.; Parsons, D.R.; et al. Autonomous Underwater Vehicles (AUVs): Their Past, Present and Future Contributions to the Advancement of Marine Geoscience. *Mar. Geol.* **2014**, 352, 451–468. [CrossRef]
8. Smith, K.L., Jr.; Clifford, C.H.; Eliason, A.H.; Walden, B.; Rowe, G.T.; Teal, J.M. A Free Vehicle for Measuring Benthic Community Metabolism1. *Limnol. Oceanogr.* **1976**, 21, 164–170. [CrossRef]
9. Duineveld, G.C.A.; Lavaleye, M.S.S.; Berghuis, E.M. Particle Flux and Food Supply to a Seamount Cold-Water Coral Community (Galicia Bank, NW Spain). *Mar. Ecol. Prog. Ser.* **2004**, 277, 13–23. [CrossRef]
10. Lavaleye, M.; Duineveld, G.; Lundälv, T.; White, M.; Guihen, D.; Kiriakoulakis, K.; Wolff, G. Cold-Water Corals on the Tisler Reef: Preliminary Observations on the Dynamic Reef Environment. *Oceanography* **2009**, 22, 76–84. [CrossRef]

11. Zhao, G.; Yu, X.; Li, X. Benvir: A in Situ Deep-Sea Observation for Benthic Environmental Monitoring. *High Technol. Lett.* **2015**, *25*, 54–60.
12. Linke, P.; Sommer, S.; Rovelli, L.; McGinnis, D.F. Physical Limitations of Dissolved Methane Fluxes: The Role of Bottom-Boundary Layer Processes. *Mar. Geol.* **2010**, *272*, 209–222. [CrossRef]
13. Jamieson, A.J.; Fujii, T.; Solan, M.; Priede, I.G. HADEEP: Free-Falling Landers to the Deepest Places on Earth. *Mar. Technol. Soc. J.* **2009**, *43*, 151–160. [CrossRef]
14. Chen, J.; Zhang, Q.; Zhang, A.; Tang, Y. 7000M Lander Design for Hadal Research. In Proceedings of the 2014 Oceans—St. John's, St. John's, NL, Canada, 14–19 September 2014; pp. 1–4.
15. Peoples, L.M.; Norenberg, M.; Price, D.; McGoldrick, M.; Novotny, M.; Bochdansky, A.; Bartlett, D.H. A Full-Ocean-Depth Rated Modular Lander and Pressure-Retaining Sampler Capable of Collecting Hadal-Endemic Microbes under in Situ Conditions. *Deep. Sea Res. Part I Oceanogr. Res. Pap.* **2019**, *143*, 50–57. [CrossRef]
16. Wei, Z.-F.; Li, W.-L.; Li, J.; Chen, J.; Xin, Y.-Z.; He, L.-S.; Wang, Y. Multiple in Situ Nucleic Acid Collections (MISNAC) from Deep-Sea Waters. *Front. Mar. Sci.* **2020**, *7*. [CrossRef]
17. Dong, C.; Ma, T.; Liu, R.; Lai, Q.; Shao, Z. *Hydrocarboniclastica marina* gen. nov., sp. nov., a Marine Hydrocarbonoclastic Bacterium Isolated from an in Situ Enriched Hydrocarbon-Degrading Consortium in Sea Sediment. *Int. J. Syst. Evol. Microbiol.* **2019**, *69*, 2250–2257. [CrossRef] [PubMed]
18. Barclay, D.R.; Simonet, F.; Buckingham, M.J. Deep Sound: A Free-Falling Sensor Platform for Depth-Profiling Ambient Noise in the Deep Ocean. *Mar. Technol. Soc. J.* **2009**, *43*, 144–150. [CrossRef]
19. Yonggang, J.; Chaoqi, Z.; Liping, L.; Dong, W. Marine Geohazards: Review and Future Perspective. *Acta Geol. Sin.—Engl. Ed.* **2016**, *90*, 1455–1470. [CrossRef]
20. Montagner, J.-P.; Karczewski, J.-F.; Romanowicz, B.; Bouaricha, S.; Lognonné, P.; Roult, G.; Stutzmann, E.; Thirot, J.-L.; Brion, J.; Dole, B.; et al. The French Pilot Experiment OFM-SISMOBS: First Scientific Results on Noise Level and Event Detection. *Phys. Earth Planet. Inter.* **1994**, *84*, 321–336. [CrossRef]
21. Sommer, S.; Pfannkuche, O.; Linke, P.; Luff, R.; Greinert, J.; Drews, M.; Gubsch, S.; Pieper, M.; Poser, M.; Viergutz, T. Efficiency of the Benthic Filter: Biological Control of the Emission of Dissolved Methane from Sediments Containing Shallow Gas Hydrates at Hydrate Ridge. *Glob. Biogeochem. Cycles* **2006**, *20*. [CrossRef]
22. Sommer, S.; Linke, P.; Pfannkuche, O.; Schleicher, T.; Deimling, J.S.V.; Reitz, A.; Haeckel, M.; Flögel, S.; Hensen, C. Seabed Methane Emissions and the Habitat of Frenulate Tubeworms on the Captain Arutyunov Mud Volcano (Gulf of Cadiz). *Mar. Ecol. Prog. Ser.* **2009**, *382*, 69–86. [CrossRef]
23. Linke, P.; Suess, E.; Torres, M.; Martens, V.; Rugh, W.D.; Ziebis, W.; Kulm, L.D. In Situ Measurement of Fluid Flow from Cold Seeps at Active Continental Margins. *Deep. Sea Res. Part I Oceanogr. Res. Pap.* **1994**, *41*, 721–739. [CrossRef]
24. Smith, K.L.; Baldwin, R.J.; Karl, D.M.; Boetius, A. Benthic Community Responses to Pulses in Pelagic Food Supply: North Pacific Subtropical Gyre. *Deep. Sea Res. Part I Oceanogr. Res. Pap.* **2002**, *49*, 971–990. [CrossRef]
25. Ekeroth, N.; Kononets, M.; Walve, J.; Blomqvist, S.; Hall, P.O.J. Effects of Oxygen on Recycling of Biogenic Elements from Sediments of a Stratified Coastal Baltic Sea Basin. *J. Mar. Syst.* **2016**, *154*, 206–219. [CrossRef]
26. Khripounoff, A.; Caprais, J.-C.; Crassous, P.; Etoubleau, J. Geochemical and Biological Recovery of the Disturbed Seafloor in Polymetallic Nodule Fields of the Clipperton-Clarion Fracture Zone (CCFZ) at 5000-m Depth. *Limnol. Oceanogr.* **2006**, *51*, 2033–2041. [CrossRef]
27. Kononets, M.; Tengberg, A.; Nilsson, M.; Ekeroth, N.; Hylén, A.; Robertson, E.K.; van de Velde, S.; Bonaglia, S.; Rüt-ting, T.; Blomqvist, S.; et al. In Situ Incubations with the Gothenburg Benthic Chamber Landers: Applications and Quality Control. *J. Mar. Syst.* **2021**, *214*, 103475. [CrossRef]
28. Tréhu, A.M.; de Moor, A.; Madrid, J.M.; Sáez, M.; Chadwell, C.D.; Ortega-Culaciati, F.; Ruiz, J.; Ruiz, S.; Tryon, M.D. Post-Seismic Response of the Outer Accretionary Prism after the 2010 Maule Earthquake, Chile. *Geosphere* **2019**, *16*, 13–32. [CrossRef]
29. Person, R.; Aoustin, Y.; Blandin, J.; Marvaldi, J.; Rolin, J.F. From Bottom Landers to Observatory Networks. *Ann. Geophys.* **2006**, *49*. [CrossRef]
30. Spagnoli, F.; Penna, P.; Giuliani, G.; Masini, L.; Martinotti, V. The AMERIGO Lander and the Automatic Benthic Chamber (CBA): Two New Instruments to Measure Benthic Fluxes of Dissolved Chemical Species. *Sensors* **2019**, *19*, 2632. [CrossRef]
31. Mortensen, A.C.; Lange, R.E. Design Considerations of Wing Stabilized Free-Fall Vehicles. *Deep. Sea Res. Oceanogr. Abstr.* **1976**, *23*, 1231–1240. [CrossRef]
32. Jun, C. Research on Techniques of Hadal Lander and Applications on Biology. Ph.D. Thesis, University of Chinese Academy of Sciences, Beijing, China, 2018.
33. Gang, X.; Yanjun, L.; Yifan, X. Hydrodynamic Characteristics Research and Structure Optimization of Hadal Lander with Hydrofoil. *J. Mech. Eng.* **2022**, *58*, 1–15.
34. Yuanyuan, S.; Donghui, Z. Design of a Deep-Sea Microbe Enrichment Device. *J. Hangzhou Dianzi Univ.* **2017**, *37*, 57–60.
35. Thor, I.F. *Guidance and Control of Ocean Vehicles, Chapter 4*; John Wiley and Sons Ltd.: New York, NY, USA, 1994.
36. SOLIDWORKS Flow Simulation. *Flow Simulation 2017 Technical Reference*; Dassault System: Vélizy-Villacoublay, France, 2017.
37. Kim, H.; Akimoto, H.; Islam, H. Estimation of the Hydrodynamic Derivatives by RaNS Simulation of Planar Motion Mechanism Test. *Ocean. Eng.* **2015**, *108*, 129–139. [CrossRef]

38. Lack, S.; Rentzow, E.; Jeinsch, T. Experimental Parameter Identification for an Open-Frame ROV: Comparison of Towing Tank Tests and Open Water Self-Propelled Tests. *IFAC-PapersOnLine* **2019**, *52*, 271–276. [CrossRef]
39. Yu, W. Dynamic Analysis of Sediment Sediment Bottom Sitting Process of Deep Sea Lander. Master's Thesis, Hangzhou Dianzi University, Hangzhou, China, 2020.
40. Yu, Z.; Zhang, C.; Chen, J.; Ren, Z. Dynamic Analysis of Bottom Subsidence of Benthic Lander. *J. Mar. Sci. Eng.* **2022**, *10*, 824. [CrossRef]
41. Release PlatEMO v1.6 (2018/9/9), BIMK/PlatEMO, GitHub. Available online: https://github.com/BIMK/PlatEMO/releases/tag/v1.6 (accessed on 20 November 2022).
42. Tian, Y.; Cheng, R.; Zhang, X.; Jin, Y. PlatEMO: A MATLAB Platform for Evolutionary Multi-Objective Optimization [Educational Forum]. *IEEE Comput. Intell. Mag.* **2017**, *12*, 73–87. [CrossRef]
43. Deb, K.; Pratap, A.; Agarwal, S.; Meyarivan, T. A Fast and Elitist Multiobjective Genetic Algorithm: NSGA-II. *IEEE Trans. Evol. Comput.* **2002**, *6*, 182–197. [CrossRef]
44. Deb, K.; Jain, S. Running Performance Metrics for Evolutionary Multi-Objective Optimizations. In Proceedings of the Fourth Asia-Pacific Conference on Simulated Evolution and Learning (SEAL'02), Orchid Country Club, Singapore, 18–22 November 2002; pp. 1–18.
45. Coello, C.A.C.; Cortés, N.C. Solving Multiobjective Optimization Problems Using an Artificial Immune System. *Genet. Program. Evolvable Mach.* **2005**, *6*, 163–190. [CrossRef]
46. Zhang, Q.; Zhou, A.; Zhao, S.; Suganthan, P.; Liu, W.; Tiwari, S. Multiob-jective optimization test instances for the CEC 2009 special session and competition University of Essex, Colchester, UK and Nanyang technological University, Singapore, special session on performance assessment of multi-objective optimization algorithms. *Tech. Rep.* **2008**, *264*, 1–30. Available online: https://www.researchgate.net/publication/265432807 (accessed on 20 November 2022).

Journal of
Marine Science and Engineering

MDPI

Article

An Underwater Wet-Mateable Electrical Connector with Dual-Bladder Pressure-Balanced Oil-Filled (PBOF) Technology

Wentao Song [1,2,3], Cuibo Yang [4], Weicheng Cui [2,3,*], Changhui Song [2,3,5], Ping Yang [2,3], Jin Hong [4], Yi Lei [4], Qimeng Liu [1,2,3] and Zhenhua Wang [1,2,3]

1 Zhejiang University–Westlake University Joint PhD Program, Zhejiang University, Hangzhou 310058, China
2 Key Laboratory of Coastal Environment and Resources of Zhejiang Province, School of Engineering, Westlake University, Hangzhou 310024, China
3 Institute of Advanced Technology, Westlake Institute for Advanced Study, Hangzhou 310024, China
4 Shanghai Kingboom Transmission Technology Co., Ltd., Shanghai 201806, China
5 Zhejiang Nektron Intelligent Technology Co., Ltd., Hangzhou 310024, China
* Correspondence: weichengcui@westlake.edu.cn

Abstract: Underwater wet-mateable connectors have been widely used to reduce the cost and the time of installation, maintenance and reconfiguration in many fields, such as the oil and gas (O&G) industry, offshore renewable energy (ORE), and undersea observatories. In the past few years, the authors' group has made some efforts in developing wet-mateable connectors. This paper presents a methodology for designing and testing a wet-mateable electrical connector. First, an innovative wet-mateable electrical connector with dual-bladder pressure-balanced oil-filled (PBOF) technology is proposed. Second, the generalized equations of differential pressure are derived. Then, a procedure of thermal-electric-structure (TES) coupling simulation is proposed, and a series of finite element analysis (FEA) involving coupled multi-field problems is conducted, including thermal-electric coupling analysis, static structural analysis, and dynamic analysis. Finally, a prototype of the proposed connector is developed successfully, and its electrical performance is verified by the online test in a hydrostatic pressure environment with an ocean depth of 3000 m, which has reached the leading level in China. This paper is the first discloser on wet-mateable connectors in the aspects of design, theory, simulation and testing, which might be helpful to many ocean scientists in developing countries who are technically blocked or could not afford the high cost.

Keywords: wet-mateable; electrical connector; pressure-balanced oil-filled (PBOF); coupled multi-field; prototype test

Citation: Song, W.; Yang, C.; Cui, W.; Song, C.; Yang, P.; Hong, J.; Lei, Y.; Liu, Q.; Wang, Z. An Underwater Wet-Mateable Electrical Connector with Dual-Bladder Pressure-Balanced Oil-Filled (PBOF) Technology. *J. Mar. Sci. Eng.* **2023**, *11*, 156. https://doi.org/10.3390/jmse11010156

Academic Editor: Spyros Hirdaris

Received: 27 November 2022
Revised: 13 December 2022
Accepted: 5 January 2023
Published: 9 January 2023

1. Introduction

Energy demand drove the industrial revolution in the 1800s, and it may drive another new round of industrial revolution in the 21st century [1]. The ocean has abundant resources (e.g., minerals, renewable energy, genes, space, etc.), which are essential for human beings to achieve sustainable development [2]. However, we have just explored less than 20% of the ocean, and only 5% in detail [3,4]. Although submersibles are the main working force for the research fleet, various deep-sea equipment deployed in the oil and gas (O&G) industry, offshore renewable energy (ORE), and undersea observatories are also fundamental to the effective exploration and utilization of the ocean world and marine resources.

Underwater wet-mateable connectors can be mated and de-mated under the sea, they have been extensively used in the above fields to reduce the cost and the time of installation, maintenance and reconfiguration [5]. The O&G industry is the biggest consumer of wet-mateable connectors. With the increase in the depth of exploitation, wet-mateable electrical connectors that can be used in a harsh environment with high ocean pressure and submarine seismic risks are the main demand. As is experienced in the O&G exploration and production, most of the needed wet-mateable connectors with high voltage rates (less than 12 kV) should be operated at depths of up to 3000 m [6]. ORE farms also need a large number of wet-mateable electrical connectors. With the development of technologies of offshore wind energy, tidal stream, and ocean current energy, the requirement for wet-mateable connectors with high power is also increasing [7,8]. In this field, the voltage rate of the required wet-mateable connectors ranges from 1 kV to 33 kV, while their depth rate is only up to 100 m [6,8]. The application of wet-mateable connectors in ORE farms has the advantage of allowing the transformer to be lifted out of the water with no need to lift the long and heavy cable [9]. Another important application field of wet-mateable connectors is undersea observatories, for example, the Astronomy with a Neutrino Telescope and Abyss environmental RESearch (ANTARES) project [10] and the next-generation neutrino telescopes KM3NeT [11]. Figure 1 presents a schematic of the complete link between the string power module/string control module (SPM/SCM) and the junction box in ANTARES. Whereas the mounting of the jumpers is performed on shore, the completion of the wet-mate connection can be realized by a remotely operated vehicle (ROV) [10]. So, wet-mateable connectors can provide an effective solution to the expansion and networking of undersea observatories [3].

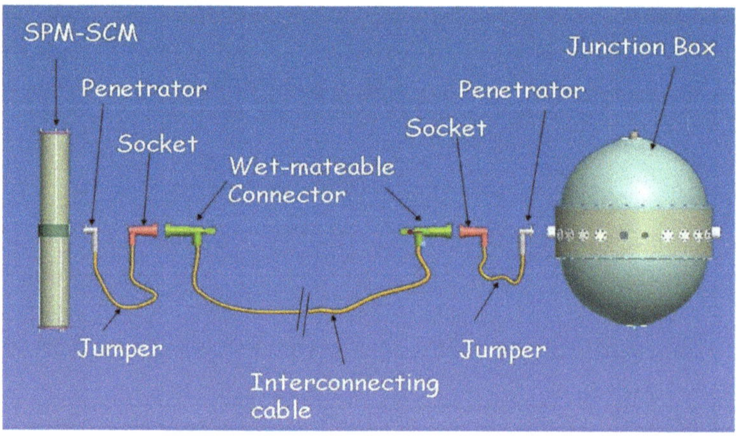

Figure 1. A schematic of the complete link between the string power module/string control module (SPM/SCM) and the junction box in ANTARES [10].

However, only a few developed countries can produce wet-mateable connectors worldwide. The key technologies are mostly confidential and protected by patents, so there are few published papers on this topic, which makes wet-mateable connectors become one of the bottleneck techniques in China [3,5]. Conceivably, the market price of wet-mateable connectors is very expensive. For example, the cost of a wet-mateable hybrid connector rated at 11 kV is expected to be in the range of £250 k/unit [8], and even low-spec products may cost as much as £20 k/unit [6]. Today, the top manufacturers of wet-mateable connectors include SEACON, MacArtney Group, Teledyne ODI, Siemens Energy, and so on. Each vendor has its own unique technology [3,7,8].

In the past few years, the authors' group has made some efforts in developing wet-mateable connectors. In 2021, Song and Cui [3] gave a very detailed overview of wet-mateable connectors and proposed the concept of functional units to present a better understanding of the key technologies involving electrical/optical connection, pressure-balanced oil-filled (PBOF), penetrable self-sealing, automatic interlocking/docking (AID) and so on. Readers can refer to the literature [3] for details. Among these key technologies, PBOF is the most essential for wet-mateable connectors, which can eliminate the differential pressure of connectors, making the wet-mate possible [3,5,7,12]. In 2022, Song et al. [5] presented a very detailed investigation of PBOF technology and developed the core component used in wet-mateable connectors: a rubber-made resilient bladder [5]. Compared to commercial off-the-shelf (COTS) products, this resilient bladder has a simple structure and lower cost, and it is more suitable for mass production. The results of the tests verified that the depth rate of the new bladder is greater than 6500 m. In the next step, the authors' group aims to develop a wet-mateable electrical connector. The simulation technique involving a coupled multi-field problem is a proven way to save the cost and the time of research and development (R&D) [13]. The finite element method (FEM) can be easily adapted to constitutive equations of different physical fields [14]. So, finite element analysis (FEA) will play a very important role in this work.

In the future, the next-generation submersible will be more inspired by fish in nature. Compared with traditional submersibles, bio-inspired fish robots have the advantages of high efficiency, high maneuverability, low noise, and so on [15,16]. Meanwhile, the next-generation submersible will have more functions, such as underwater charging. If submersibles land on an underwater docking station, the power transfer can be achieved by using wet-mateable connectors [17,18]. Foreseeably, wet-mateable connectors must have a bright application prospect.

In this paper, the authors proposed an underwater wet-mateable electrical connector with dual-bladder PBOF technology. First, the generalized differential pressure equations of dual-bladder PBOF technology are derived. Second, a procedure of coupled multi-field simulation for wet-mateable electrical connectors is proposed. Then, a series of FEA is conducted, including thermal-electric coupling analysis, static structural analysis, and dynamic analysis. Finally, a prototype of the proposed connector is developed successfully, and its electrical performance is verified by a series of tests. The technical level of this prototype has reached the leading position in China. This paper might be the first discloser on wet-mateable connectors in the aspects of design, theory, simulation, and testing, which might be helpful to many ocean scientists in developing countries who are technically blocked or could not afford the high cost.

2. Design and Theory

2.1. Overall Design

The main specifications of the proposed wet-mateable electrical connector are listed in Table 1. Considering the current demands in the O&G industry and undersea observatories in China, the operating depth of wet-mateable connectors will not exceed 3000 m in the next five years. Other specifications mainly refer to some first-class COTS products, such as SECAON HydraElectric, Nautilus^{TM} WM1.7-30, and Siemens DigiTRON.

Table 1. Main specifications of the proposed wet-mateable electrical connector.

Number of Contacts	Depth Rating	Contact Resistance	Insulation Resistance	Operating Current	Operating Voltage
4	3000 m	$\leq$5 mΩ	$\geq$10 GΩ	$\leq$30 A	$\leq$1.5 kV DC

Figure 2 presents a schematic of the proposed connector which consists of two parts: the plug unit and the receptacle unit. The dual-bladder PBOF technology is introduced to improve reliability, and its core components are two kinds of resilient bladder (parts 12 and 13). The shuttle pin design is used in electric connections to achieve penetrable self-sealing, and its core component is a moveable stopper (part 11) and the contact pin with an insulating layer (part 3). The proposed connector also has the AID technology which can facilitate ROVs to operate under the sea, and its core component is a latching member (part 4). Compared to available products, the proposed connector is a smaller-sized one, whose maximum radial dimension is less than 100 mm and the axial dimension in the mated state is less than 210 mm. The key technologies of the proposed connector are patented by China National Intellectual Property Administration [19].

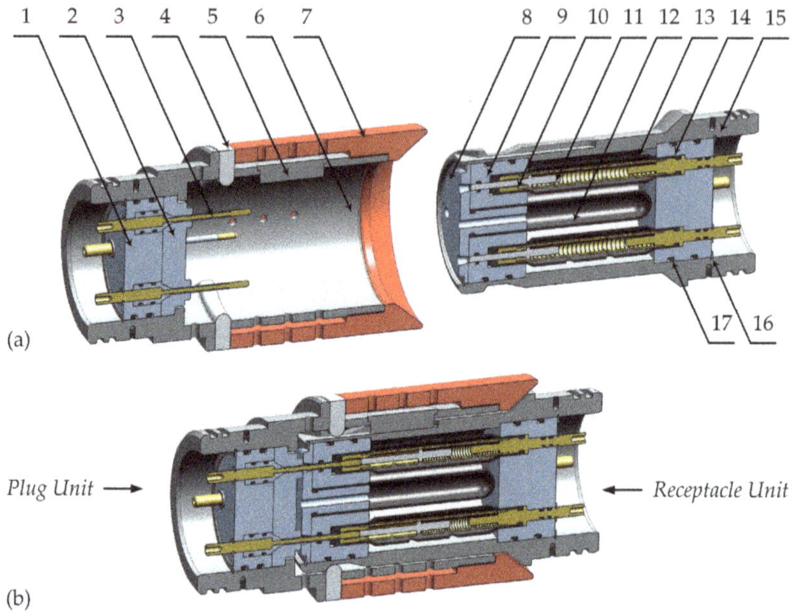

Figure 2. A schematic illustration of (**a**) the de-mated state and (**b**) the fully mated state of the proposed wet-mateable electrical connector: 1-plug pedestal; 2-plug orifice plate; 3-contact pin with an insulating layer; 4-latching member; 5-aligning key; 6-plug shell; 7-dust cap; 8-receptacle orifice plate; 9-receptacle front pedestal; 10- electric connection; 11-moveable stopper; 12-primary resilient bladder; 13-secondary resilient bladder; 14-receptacle rear pedestal; 15-receptacle shell; 16-retaining ring; 17-O-ring.

2.2. PBOF Design

PBOF technology can provide a small mating/de-mating force and sealing pressure with better reliability. Thus, the general O-ring sealing design can be used in the proposed connector. Figure 3 presents a schematic of the dual-bladder PBOF technology used in the proposed connector. The primary oil chamber is formed outside of the primary resilient bladder, while the independent secondary oil chambers are formed inside each secondary bladder. The oil chambers are all filled with the dielectric oil by a custom hypodermic tube of a syringe which can be inserted through each passageway of the moveable stopper. In the ocean, the primary bladder separates seawater and filled-oil. The ocean pressure can be transmitted to the oil in the primary oil chamber first and then to that in the secondary oil chamber, so an internal pressure is produced to balance the ocean pressure.

In particular, to fill the primary oil chamber, one secondary bladder must be punched through with holes to connect the primary oil chamber and its secondary oil chamber. So, this electrical channel protected by only one bladder can be used for unimportant circuits or ground terminals. The frequently used filled oil is the XIAMETER^{TM} PMX-200 silicon fluid (previously known as DC 200) produced by Dow Corning, which has good water-repellency, lubrication, and dielectric property [3]. Fluorosilicone rubber is preferred to make such bladders that shall not react with silicone oil and seawater.

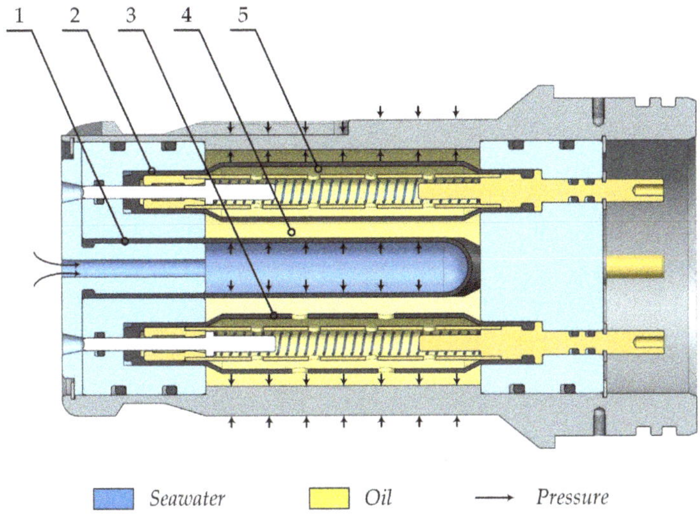

Figure 3. A schematic illustration of the dual-bladder PBOF technology used in the proposed wet-mateable electrical connector: 1-primary resilient bladder; 2-secondary resilient bladder; 3-secondary resilient bladder with through-holes; 4-primary oil chamber; 5-secondary oil chamber.

2.3. PBOF Theory

Due to the compressibility of oil, a differential pressure between the ambient pressure and the oil pressure will arise, which is a very important indicator of PBOF technology. The authors' group has studied this problem systematically and the results are presented in a previous publication [5]. On this basis, this paper further gives the generalized differential pressure equations of the dual-bladder PBOF technology. For the proposed connector, the differential pressure of the primary resilient bladder can be determined by the following equations.

$$C_{11} \cdot \Delta P_1^2 + C_{21} \cdot \Delta P_1 + C_{31} = 0 \tag{1}$$

$$\begin{cases} C_{11} = \frac{(2-\mu_1)^2 R_1^4}{2E_1^2\delta_1^2}\left(\frac{L_1}{2} - 2k_{11} + k_{21}\right) \\ C_{21} = \frac{2(2-\mu_1)R_1^3}{E_1\delta_1}\left(\frac{L_1}{2} - k_{11}\right) + \frac{R_0^2 L_0 - (R_1-\delta_1)^2 L_1}{E_B} \\ C_{31} = -\frac{R_0^2 L_0 - (R_1-\delta_1)^2 L_1}{E_B} \cdot P \end{cases} \tag{2}$$

$$\begin{cases} k_{11} = \frac{1}{\beta_1}\left(1 - e^{-\frac{\beta_1 L_1}{2}} \cos\frac{\beta_1 L_1}{2}\right) \\ k_{21} = \frac{1}{4\beta_1}\left[3 - e^{-\beta_1 L_1}(2 + \sin\beta_1 L_1 + \cos\beta_1 L_1)\right] \\ \beta_1 = \frac{\sqrt[4]{3(1-\mu_1^2)}}{\sqrt{R_1\delta_1}} \end{cases} \tag{3}$$

where

P is the ocean pressure, in the unit of MPa;
ΔP_1 is the differential pressure of the primary resilient bladder, in the unit of MPa;
E_1 is the apparent elastic modulus of the primary resilient bladder, in the unit of MPa;
E_B is the bulk modulus of the filled-oil, in the unit of MPa;
L_0 is the length of the primary oil chamber, in the unit of mm;
L_1 is the length of the primary resilient bladder, in the unit of mm;
R_0 is the internal radius of the primary oil chamber, in the unit of mm;
R_1 is the external radius of the primary resilient bladder, in the unit of mm;

δ_1 is the thickness of the primary resilient bladder, in the unit of mm;
μ_1 is the Poisson's ratio of the primary resilient bladder.

Similarly, the differential pressure equation of the secondary resilient bladder can be obtained by the following equations.

$$C_{12} \cdot \Delta P_2^2 + C_{22} \cdot \Delta P_2 + C_{32} = 0 \tag{4}$$

$$\begin{cases} C_{12} = \frac{(2-\mu_2)^2 R_2^4}{2E_2^2\delta_2^2}\left(\frac{L_2}{2} - 2k_{12} + k_{22}\right) \\ C_{22} = \frac{2(2-\mu_2)R_2^3}{E_2\delta_2}\left(\frac{L_2}{2} - k_{12}\right) + \frac{R_0^2 L_0 - (R_2-\delta_2)^2 L_2}{E_B} \\ C_{32} = -\frac{R_0^2 L_0 - (R_2-\delta_2)^2 L_2}{E_B} \cdot P \end{cases} \tag{5}$$

$$\begin{cases} k_{12} = \frac{1}{\beta_2}\left(1 - e^{-\frac{\beta_2 L_2}{2}} \cos\frac{\beta_2 L_2}{2}\right) \\ k_{22} = \frac{1}{4\beta_2}\left[3 - e^{-\beta_2 L_2}(2 + \sin\beta_2 L_2 + \cos\beta_2 L_2)\right] \\ \beta_2 = \frac{\sqrt[4]{3(1-\mu_2^2)}}{\sqrt{R_2\delta_2}} \end{cases} \tag{6}$$

where

ΔP_2 is the differential pressure of the secondary resilient bladder, in the unit of MPa;
E_2 is the apparent elastic modulus of the secondary resilient bladder, in the unit of MPa;
L_2 is the length of the secondary resilient bladder, in the unit of mm;
R_2 is the external radius of the secondary resilient bladder, in the unit of mm;
δ_2 is the thickness of the secondary resilient bladder, in the unit of mm;
μ_2 is the Poisson's ratio of the secondary resilient bladder.

Therefore, the total differential pressure of the proposed connector is as follows:

$$\Delta P = \Delta P_1 + \Delta P_2 \tag{7}$$

For the proposed connector, we have R_0 = 28 mm, L_0 = 67 mm, and E_B = 1034 MPa. For two resilient bladders, we have $E_1 = E_2$ = 22 MPa, $R_1 = R_2$ = 7 mm, $\delta_1 = \delta_2$ = 1 mm, $\mu_1 = \mu_2$ = 0.5, L_1 = 65 mm, and L_2 = 67 mm. Then, the differential pressures of the proposed connector can be determined. Figure 4 presents a three-dimensional (3D) view of two bladders, as well as the theoretical results of their differential pressures. The results are given as follows:

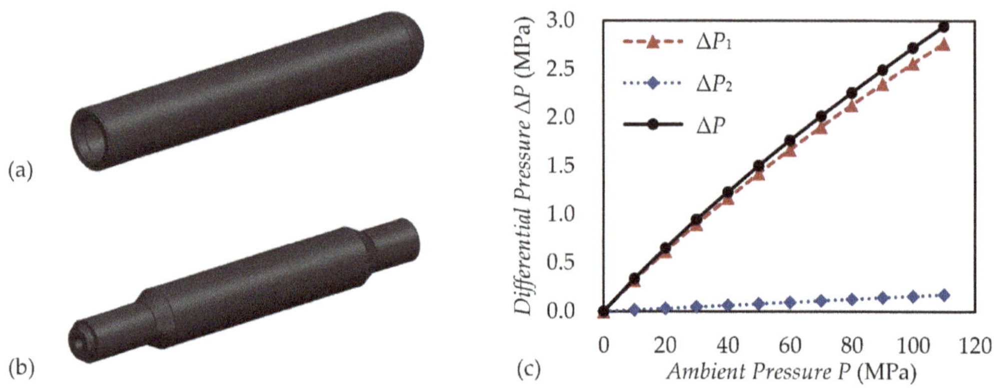

Figure 4. A 3D view of (**a**) the primary resilient bladder and (**b**) the secondary resilient bladder, and (**c**) the change curves of the differential pressures in the proposed wet-mateable electrical connector.

(1) The total differential pressure is mainly determined by the primary bladder, which is almost less than 3.5% of the ocean pressure.

(2) The differential pressure of the secondary bladder is much smaller than that of the primary bladder. This reminds us that the differential pressure provided by the bladder under the action of internal pressure is much larger than that provided by such a bladder with a similar dimension but under the action of external pressure.

(3) Although the dual-bladder PBOF technology can improve reliability, it may also increase the total differential pressure. Nevertheless, due to the pressure-balanced mode of the secondary bladder, there is only a slight impact on the performance.

2.4. Shuttle Pin Design

The shuttle pin design can provide the sealing performance in the dynamic process of mating and de-mating, which belongs to a penetrable seal-sealing [3,20]. As is depicted in Figure 5, when the contact pin is inserted into the electric socket, it will push the moveable stopper inside and make full contact with the crown spring. Thus, the current will flow through the pin to the socket, and then to the conductive sleeve and the conductive shaft in turn. When the pin is de-mated, the stopper will be followed by the pin closely under the action of the spring force. Therefore, the sealing performance can always be guaranteed by this penetrable self-sealing design.

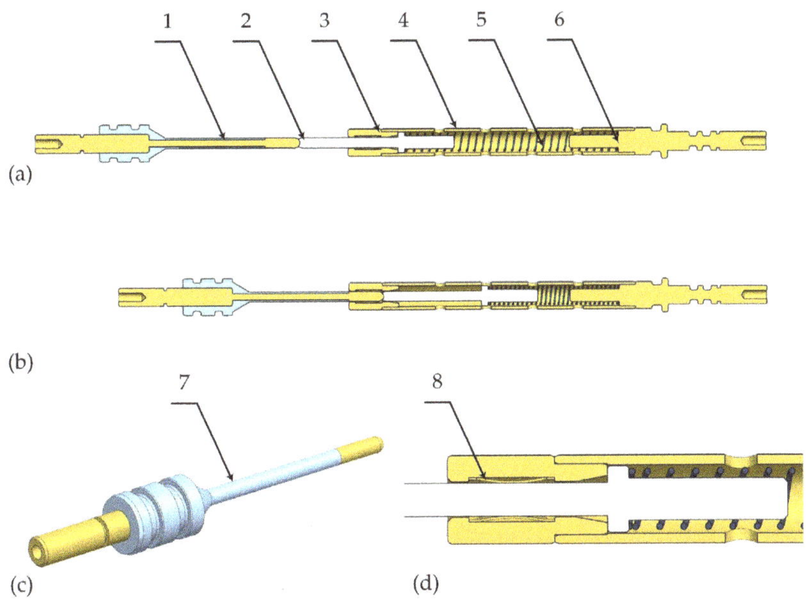

Figure 5. A schematic illustration of the shuttle pin design in (**a**) the disconnecting state and (**b**) the connecting state, and the detail views of (**c**) the contact pin and (**d**) the electric socket: 1-contact pin; 2-moveable stopper; 3-electric socket; 4-conductive sleeve; 5-spring; 6-conductive shaft; 7-insulating layer; 8-crown spring.

The shuttle pin design involves the following essential components and subtle design skills:

(1) The moveable stopper must be an insulating part, which is made of polyetheretherketone (PEEK). The diameter of the stopper is only 3 mm, which almost reaches the limit of commercial off-the-shelf (COTS) products. The stopper's small size brings great challenges to the manufacturing and the penetrable self-sealing. Therefore, it presents one of the most difficult challenges in developing wet-mateable electrical connectors.

(2) The contact pin, as shown in Figure 5c, has an insulating layer made of PEEK to prevent seawater from contacting the conductor after the intrusion, which is almost as expensive as the resilient bladder due to the difficulty of processing.

(3) The maximum deformation of the spring should satisfy the hypodermic tube of the syringe and can reach the position of oil-filling. Additionally, the authors recommend using the piano-wire spring which has good performance in strength and fatigue resistance.

(4) According to the requirement of the maximum operating current, the standard parts can be selected for the crown spring, which is usually made of beryllium copper. The other conductive components can be made of copper alloys, such as brass or lead brass. Moreover, gold plating can be used on the electrical contact surface to decrease contact resistance.

(5) An easily occurred and fatal failure mode of the shuttle pin design is that the moveable stopper cannot eject to the original position after de-mating. To avoid this, the end of the stopper and the end of the fitted hole in the electric socket are both designed as cone types, which are shown in Figure 5d. This skill can also increase allowable fit tolerances to decrease the cost of manufacturing.

(6) It is worth noting that PEEK plastic is a very important dielectric material used in wet-mateable electrical connectors, which has good insulating properties in harsh environments under the sea. Also, PEEK is a self-lubricating material with higher tensile and yield strength and excellent fatigue resistance [3].

3. Finite Element Analysis

3.1. Coupled Multi-Field Simulation

Due to the Joule heating effect, wet-mateable electrical connectors will inevitably generate heat when they transfer electricity. However, the excessive temperature will reduce the elasticity of electric contact and soften the insulating material, resulting in the failure of connection and insulation. Moreover, thermal stress may also cause a fatal effect on the reliability of the structure. For example, if high stress produces a large deformation of the moveable stopper, the function of the penetrable self-sealing will be totally lost. Therefore, the coupled multi-field simulation is indispensable for the development of wet-mateable electrical connectors.

Here, the basic equations of thermal-electric coupling are given as follows. Equations (8) and (9) are derived from the heat flow equation and the charge continuity equation respectively, and Equation (10) is the constitutive equation for the dielectric medium [14,21].

$$\rho C\frac{\partial T}{\partial t} + \nabla\cdot([\Pi]\cdot\boldsymbol{J}) - \nabla\cdot([\lambda]\cdot\nabla T) = \dot{q} \tag{8}$$

$$\nabla\cdot\left([\varepsilon]\cdot\nabla\frac{\partial\varphi}{\partial t}\right) + \nabla\cdot([\sigma]\cdot[\alpha]\cdot\nabla T) + \nabla\cdot([\sigma]\cdot\nabla\varphi) = 0 \tag{9}$$

$$\boldsymbol{D} = [\varepsilon]\cdot\boldsymbol{E} \tag{10}$$

where

C is the specific heat capacity, in the unit of J/(kg·°C);
T is the absolute temperature, in the unit of °C;
q is the heat generation rate, in the unit of W/m^3;
ρ is the density, in the unit of kg/m^3;
$\boldsymbol{D}$ is the electric flux density vector, in the unit of C/m^2;
$\boldsymbol{E}$ is the electric field intensity vector, in the unit of V/m;
$\boldsymbol{J}$ is the electric current density vector, in the unit of A/m^2;
$[\Pi]$ is the Peltier coefficient matrix, $[\Pi] = T\cdot[\alpha]$, in the unit of V;
$[\alpha]$ is the Seebeck coefficient matrix, in the unit of V/°C;
$[\varepsilon]$ is the dielectric permittivity matrix, in the unit of F/m;
$[\lambda]$ is the thermal conductivity matrix, in the unit of W/m·°C;
$[\sigma]$ is the electrical conductivity matrix, in the unit of S/m.

By applying the Galerkin FEM procedure to the above-derived coupled equations, the thermoelectric finite element equations can be obtained as follows [14,21]:

$$\begin{bmatrix} C^{TT} & 0 \\ 0 & C^{\varphi\varphi} \end{bmatrix} \begin{pmatrix} \dot{T}_e \\ \dot{\varphi}_e \end{pmatrix} + \begin{bmatrix} K^{TT} & 0 \\ K^{\varphi T} & K^{\varphi\varphi} \end{bmatrix} \begin{pmatrix} T_e \\ \varphi_e \end{pmatrix} = \begin{pmatrix} Q \\ I \end{pmatrix} \tag{11}$$

where

C^{TT} is the thermal damping matrix;
$C^{\varphi\varphi}$ is the dielectric damping matrix;
K^{TT} is the thermal stiffness matrix;
$K^{\varphi T}$ is the Seebeck stiffness matrix;
$K^{\varphi\varphi}$ is the electric stiffness matrix;
I is the electric current load vector;
Q is the combined heat load vector;
T_e is the nodal temperature vector;
φ_e is the nodal electric potential vector.

As is depicted in Figure 6, the authors proposed a procedure of thermal-electric-structure (TES) coupling simulation for wet-mateable electrical connectors, which can be achieved by FEM based on ANSYS software. The interface of ANSYS Workbench can easily realize coupled-field analysis. First, an FEA model involving thermal-electric coupling is proposed to verify the thermoelectricity of the above-designed connector. Second, the obtained temperature distribution should be transferred to the static structural FEA model. Then, the thermal stress distribution can be determined to verify the reliability of the structure. Furthermore, the pre-stress in the structure cannot be ignored in dynamic analysis. Thus, thermal stress should be transferred to the FEA model of modal analysis. Only in this step can accurate dynamic analyses be done. The simulation results of the proposed connector will be presented and discussed next. All the material specifications used in the FEA models are listed in Table 2, which are from ANSYS GRANTA Materials Database.

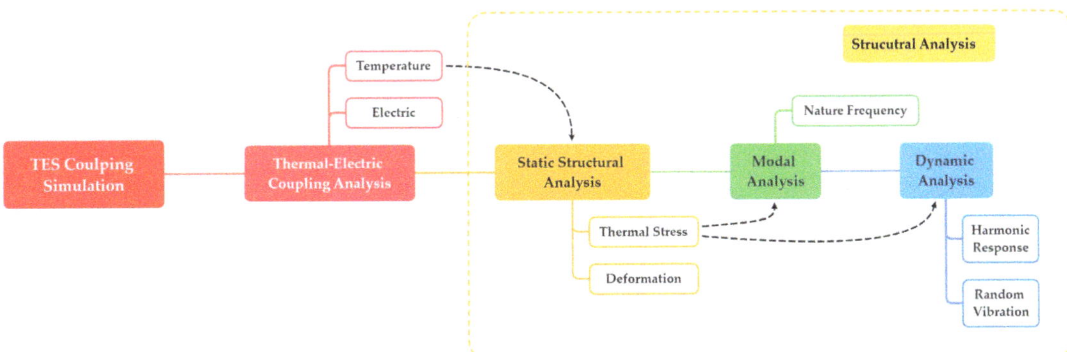

Figure 6. A schematic illustration of the TES coupling simulation for wet-mateable electrical connectors.

Table 2. Material specifications of the proposed wet-mateable electrical connector.

Property \ Material	Copper Alloy	PEEK Plastic	Fluorosilicone Rubber *	Stainless Steel
Density (kg/m^3)	8940	1310	1120	7900
Isotropic Thermal Conductivity (W/m·°C)	394	0.25	0.245	15
Isotropic Resistivity (S/m)	1.72×10^{-8}	9.95×10^{13}	1.22×10^{12}	7.07×10^{-7}
Coefficient of Thermal Expansion (1/°C)	1.70×10^{-5}	5.48×10^{-5}	2.74×10^{-4}	1.70×10^{-5}
Young's Modulus (GPa)	131	3.85	-	198
Poisson's Ratio	0.345	0.4		0.27
Yield Strength (MPa)	247	90.9	8.97	243
Tensile Strength (MPa)	320	96.3	8.97	546

* The hyper-elastic of the fluorosilicone rubber is described by Mooney-Rivlin model with the constants $C_{10} = 2.59 \times 10^5$ Pa and $C_{01} = 6.5 \times 10^4$ Pa.

3.2. Thermal-Electric Coupling Analysis

A thermal-electric coupling FEA model is proposed based on ANSYS Mechanical, and the corresponding boundary conditions include the following aspects:

(1) The current I is applied to the end-face of the contact pin (here I = 40 A). The voltage U of the end-face of the conductive shaft is set to be zero (U = 0).

(2) Due to the existence of contact resistance $R_{contact}$, a heat flow Q_{HF} is applied on the surface of the electric contact (here, $Q_{HF} = I^2 \cdot R_{contact} = 3.2$ W).

(3) According to different conditions of the environment, the convective film coefficient $h_{convective}$ is set on the corresponding heat dissipation surface.

To begin with, Figure 7 presents the temperature distribution of the electric contact with PBOF technology, as well as corresponding boundary conditions. The results are shown as follows:

(1) The maximum temperature of the electric contact in Figure 7a is about 74.61 °C, while it is decreased by 2.85% to about 72.48 °C in Figure 7b. The maximum temperature of the crown spring in Figure 7a is about 69.29 °C, while it is decreased by 3.09% to about 67.15 °C in Figure 7b.

(2) Due to the higher convective film coefficient of oil, PBOF technology can improve the heat dissipation capability of the electric contact. Importantly, because the use of the secondary resilient bladder increases the area of dissipation, the dual-bladder PBOF technology can further reduce thermal power by about 3%.

Figure 8 further gives the temperature distribution of the whole connector in terrestrial and underwater environments respectively, as well as the corresponding boundary conditions. Figure 9 presents the electric field distribution of some insulating components in an underwater environment, as well as the corresponding maximum intensity E_{max}. The results are shown as follows:

(1) The maximum temperature of the proposed connector in a terrestrial environment is about 53.92 °C, while it is decreased by about 35.17% to 34.98 °C in the underwater environment. The maximum temperature in both cases appears on the contact pin.

(2) The ambient temperature under the sea is much lower than the terrestrial temperature. Also, the convective film coefficient of seawater is much larger than that of air. This allows better heat dissipation of wet-mateable electrical connectors in the operating condition, which can further be improved by the dual-bladder PBOF technology. As a result, the overcurrent capacity of the proposed connector will exceed the rated current of the crown spring.

(3) The receptacle front pedestal has the strongest electric field with an intensity of 21.8 kV/mm, which almost reaches the breakdown field of PEEK. The electric field intensities of other insulating components are much smaller. The analysis of thermoelectric performance further confirms that the maximum operating current of the proposed connector can reach 40 A.

To this extent, the thermal-electric coupling analysis further reveals a characteristic of wet-mateable electrical connectors, that is, the overcurrent capacity can be enhanced by the underwater environment and PBOF technology. Therefore, we revised the maximum operating current in Table 1 as follows: 30 A in dry conditions and 40 A under the sea. Moreover, the glass fibre-reinforced PEEK can be used to provide better resistance to dielectric breakdown.

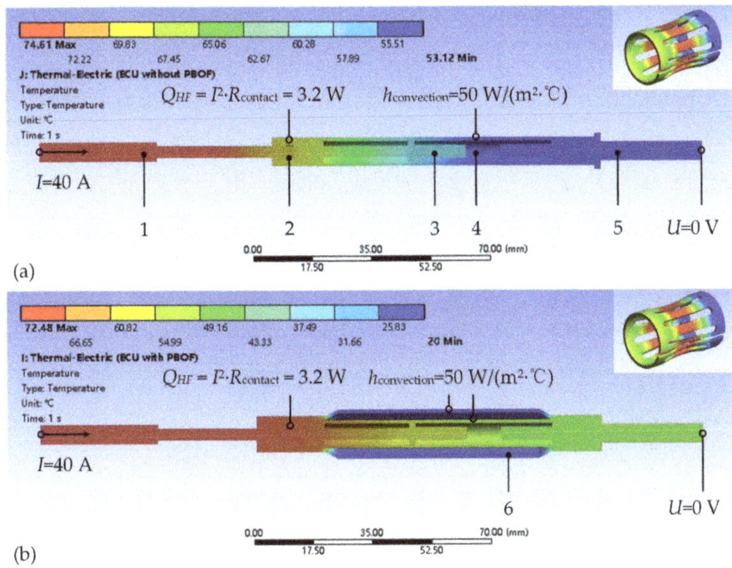

Figure 7. The temperature distribution of the electric contact with (**a**) PBOF technology and (**b**) dual-bladder PBOF technology: 1-contact pin; 2-electric socket; 3-moveable stopper; 4-conductive sleeve; 5-conductive shaft; 6-secondary resilient bladder.

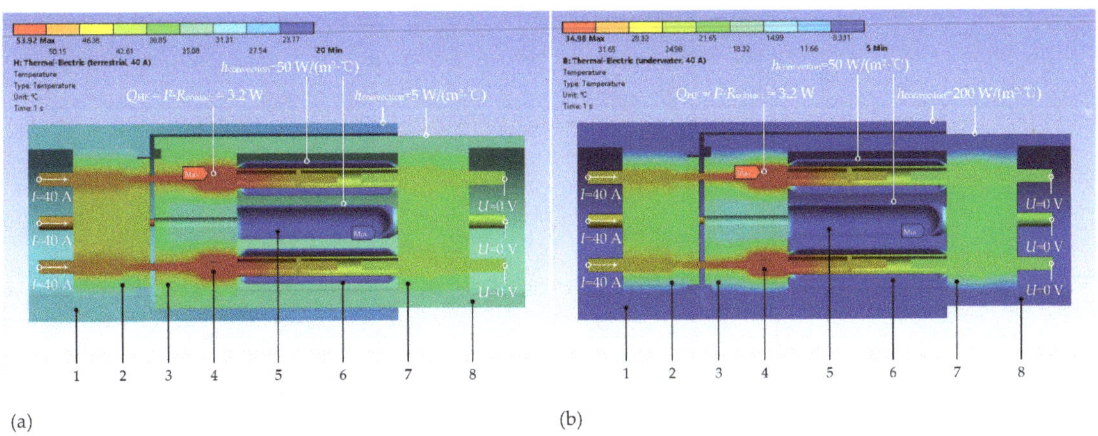

Figure 8. The temperature distribution of the proposed wet-mateable electrical connector in (**a**) terrestrial environment (the ambient temperature is 20 °C) and (**b**) underwater environment (the ambient temperature is 5 °C): 1-plug shell; 2-plug pedestal; 3-receptacle front pedestal; 4-electric contact; 5-primary resilient bladder; 6-secondary resilient bladder; 7-receptacle rear pedestal; 8-receptacle shell.

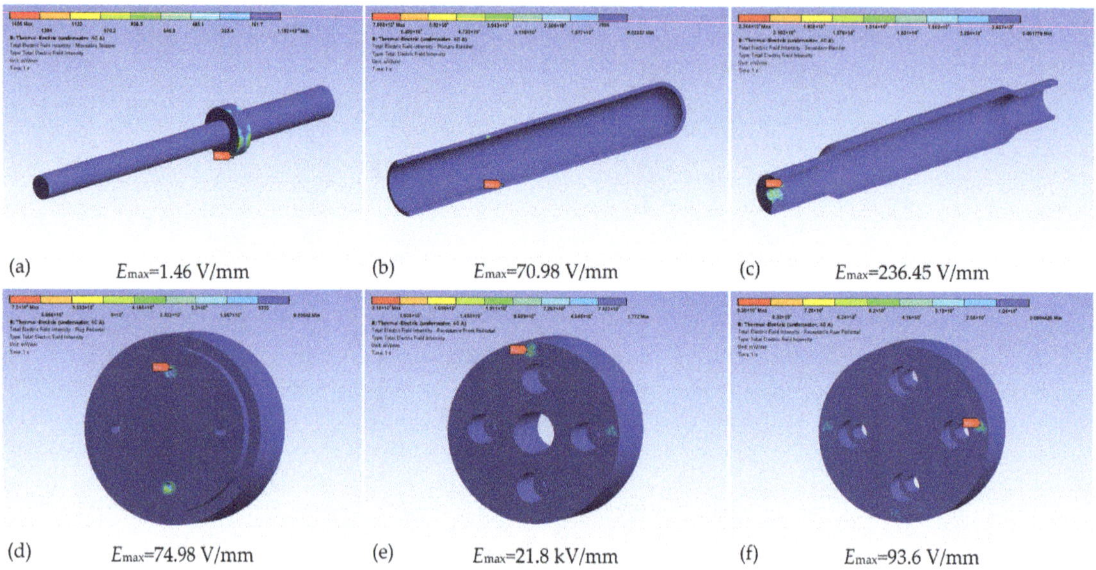

Figure 9. The voltage distribution of the main insulating components in an underwater environment (the ambient temperature is 5 °C): (**a**) moveable stopper, (**b**) primary resilient bladder, (**c**) secondary resilient bladder, (**d**) plug pedestal, (**e**) receptacle front pedestal, and (**f**) receptacle rear pedestal.

3.3. Static Structural Analysis

A static structural FEA model based on ANSYS Mechanical is proposed. The obtained temperature distribution from the thermal-electric analysis is coupled with this FEA model to calculate the thermal stress. Figure 10 presents the stress distribution of some core components, as well as the corresponding maximum stress σ_{max}. The results are shown as follows:

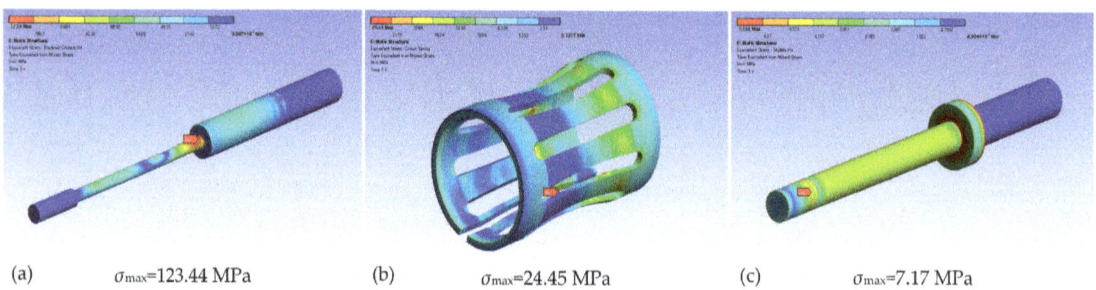

Figure 10. The thermal stress distribution of some core components of the proposed wet-mateable electrical connector: (**a**) contact pin, (**b**) crown spring, and (**c**) moveable stopper.

(1) The components having large deformation are the resilient bladders, and the maximum deformation is about 0.97 mm which has no influence on the reliability of the structure. The deformation of the moveable stopper is just about 0.02 mm at maximum, which has no effect on the function of the shuttle pin design.

(2) The components having large thermal stress are the contact pins, and the maximum equivalent stress is about 123.44 MPa which is less than the yield stress of copper alloy. The equivalent stress of the crown spring is about 24.45 Mpa in maximum, which has no effect on its elasticity. The thermal stress of the moveable stopper is much smaller.

In this subsection, a static structural FEA model of the AID technology is also proposed to verify the performance and provide guidance for the selection of the latching member's material. Figure 11 presents the mating process in the simulation, and Figure 12 gives the change curves of the mating/de-mating force and the equivalent stress of the latching member. The results are given as follows:

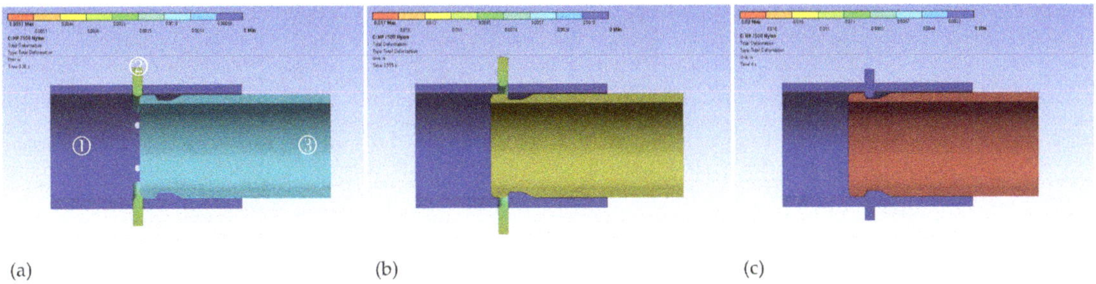

Figure 11. The mating process of the latching mechanism in the simulation: (**a**) the receptacle (part 3) starts engaging with the plug (part 1), (**b**) the receptacle is made the maximum engagement with the latching member (part 2), and (**c**) the plug and the receptacle are fully mated and interlocked by the latching member.

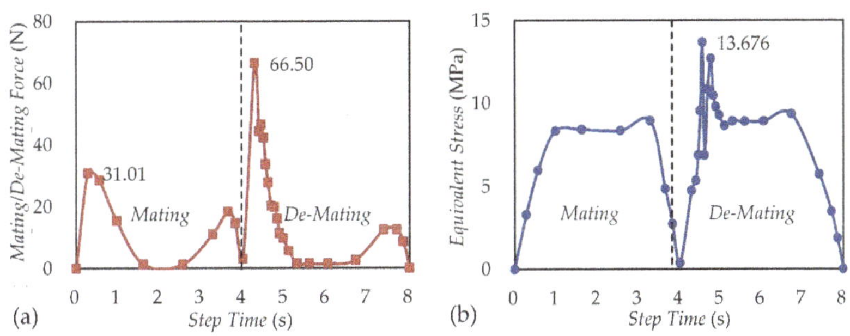

Figure 12. The change curves of (**a**) the mating/de-mating force and (**b**) the equivalent stress of the latching member.

(1) The latching member must have the appropriate elastic constants. If the Young's modulus of the material is too small, it cannot provide enough locking force, while if Young's modulus is too large, the mating force will be increased. According to the simulation results, the material with Young's modulus of about 70 MPa is the ideal material for the latching member.

(2) The de-mating force of the proposed AID technology is about 66.50 N, while the mating force is about 31.01 N. The de-mating force should be appropriately larger than the mating force, which can achieve easy docking and reliable interlocking of the wet-mateable connectors.

3.4. Dynamic Analysis

A pre-stress FEA model of modal analysis based on ANSYS Mechanical is proposed. The obtained thermal stress distribution from the TES coupling analysis is transferred to this FEA model. The results are given as follows:

(1) The natural frequencies of the proposed connector are around 24.2 Hz, 75.4 Hz, 96.2 Hz and 113.6 Hz, which should be avoided in the operating condition in engineering practice.

(2) The inherent vibration is concentrated on the primary resilient bladder, which indicates that the primary bladder has a good effect on vibration attenuation. Some typical mode shapes of the primary bladder are shown in Figure 13.

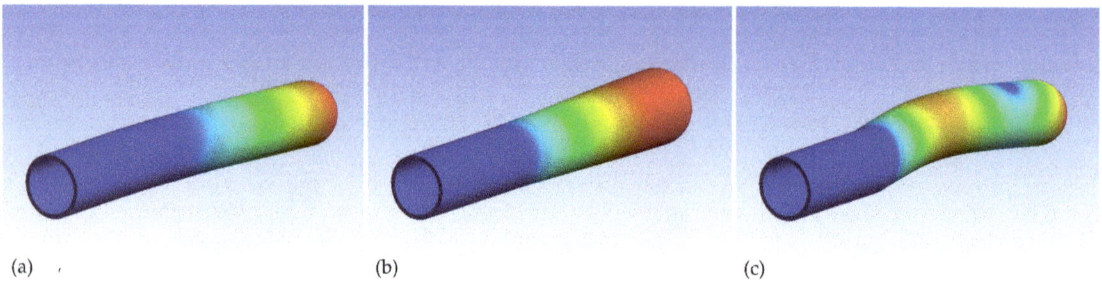

Figure 13. The mode shapes of the primary resilient bladder: (**a**) 1st mode, (**b**) 11th mode, and (**c**) 12th mode.

Moreover, a pre-stress FEA model of harmonic response based on ANSYS Mechanical is also proposed. The obtained thermal stress distribution from the TES coupling analysis is transferred to this FEA model, too. The frequency spacing ranges from 0 Hz to 70 Hz. Figure 14 gives the mode shape and the frequency response of the secondary resilient bladder. The results show that the harmonic response is concentrated on the secondary resilient bladder. Therefore, we can realize another advantage of the proposed dual-bladder PBOF technology; that is, the primary bladder has the function of inherent vibration attenuation, while the secondary bladder has the function of harmonic vibration attenuation.

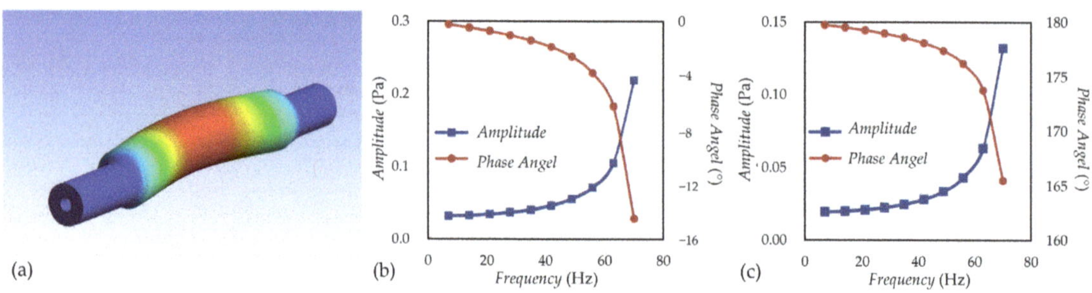

Figure 14. The harmonic response analysis of the secondary resilient bladder: (**a**) mode shape, (**b**) frequency response in the axial direction, and (**c**) frequency response in the radial direction.

4. Prototype and Test

4.1. Prototyping

According to the previously described design and analysis, a prototype of the proposed wet-mateable electrical connector has been manufactured. Figure 15 presents some important parts of prototyping. Because the primary resilient bladder has been developed successfully [5], four secondary resilient bladders are made of the same material by injection moulding, shown in Figure 15a. The insulating layer of the contact pin is made of PEEK also by injection moulding, shown in Figure 15b. The latching member is made of HP 7500 nylon by 3D-printing, shown in Figure 15c. Figure 15d presents four assemblies of electric contact. Figure 15e shows some custom syringes for oil-filling, in which a notch is formed in the end-face of hypodermic tubes by wire-electrode cutting. This is a cost-saving way compared with the way of manufacturing with computer numerical control (CNC). Importantly, before filling the oil, the air in the oil chambers should be extracted by the syringe, and then, the oil with the determined volume can be injected, which is shown in Figure 15e. Finally, we complete the assembly of the prototype, which is shown in Figure 16. For the convenience of the pressure test, four electrical channels of the plug are connected in pairs (Channel **I** to **IV**, Channel **II** to **III**), and two flanges are installed to the plug and the receptacle.

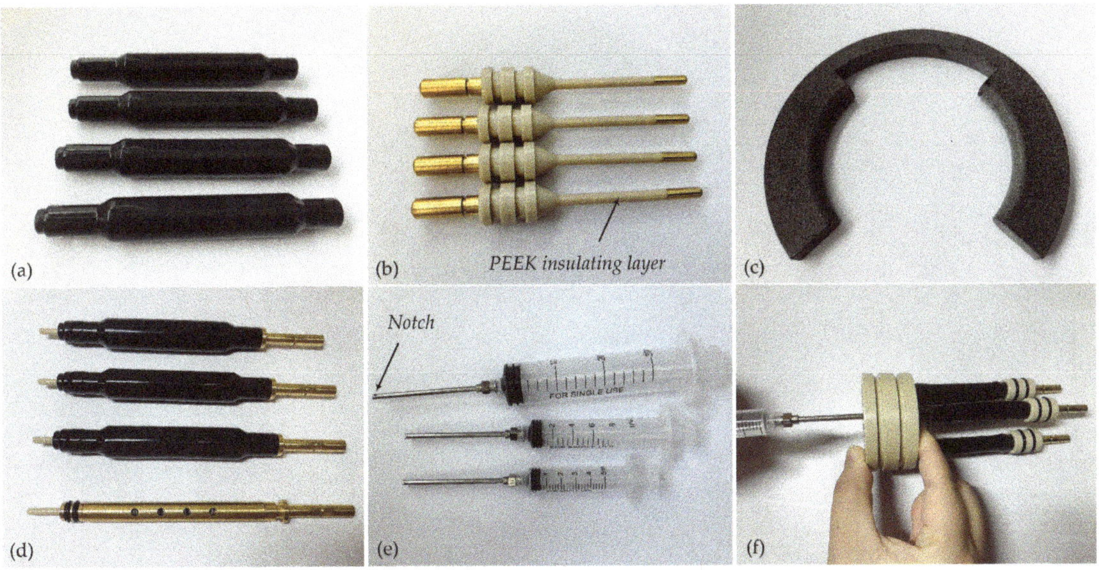

Figure 15. The core components in the prototyping of the proposed wet-mateable electrical connector: (**a**) secondary resilient bladders, (**b**) contact pins with PEEK insulating layers, (**c**) latching member, (**d**) assemblies of electric contact, (**e**) oil-filling syringe, (**f**) process of oil-filling.

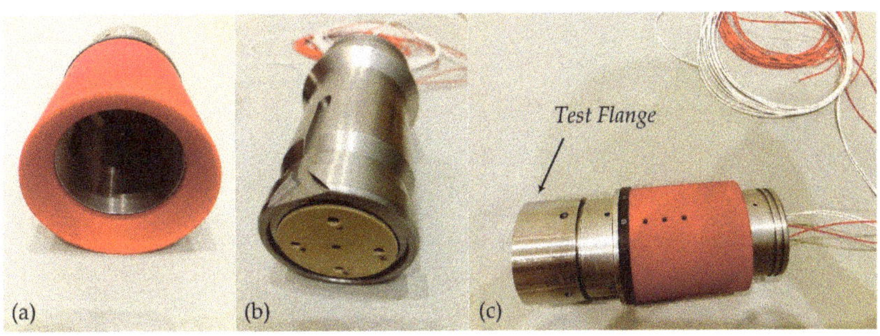

Figure 16. A prototype of the proposed wet-mateable electrical connectors: (**a**) the plug unit, (**b**) the receptacle unit, and (**c**) the fully-mated state of the prototype.

4.2. Electrical and Pressure Testing

The electrical performance of the prototype is qualified in accordance with the specification SEAFOM TSD-02, which is a key standard covering the requirement for designing and testing a wet-mateable connector [22]. First, the electrical performance of the prototype has been fully tested in a dry environment. The contact resistance is measured by the UNI-T® UT620C digital micro-ohm-meter with 4-wire Kelvin type measurement, and the insulation resistance is measured by the VICTOR® VC60E$^+$ insulation tester under the voltage of 2.5 kV/DC. The results (see Table 3) show that the contact resistance is lower than 2 mΩ and the insulation resistance is above 10 GΩ, which are satisfied with the design requirements. Also, the prototype has passed the proof voltage test of 1.0 kV/DC with the Tonghui® TH9320 hipot tester. The photos of the above tests are shown in Figure 17.

Table 3. Electrical test results of the prototype in a dry environment *.

Test Type	Contact Resistance	Insulation Resistance	Proof Voltage
Channel **I**	1.750 mΩ	-	-
Channel **II**	1.819 mΩ	-	-
Channel **III**	1.785 mΩ	-	-
Channel **IV**	1.764 mΩ	-	-
Channel **I–II**	-	12.78 GΩ	1.0 kV/DC
Channel **III–IV**	-	12.99 GΩ	1.0 kV/DC

* All the tests are performed in accordance with SEAFOM TSD-02.

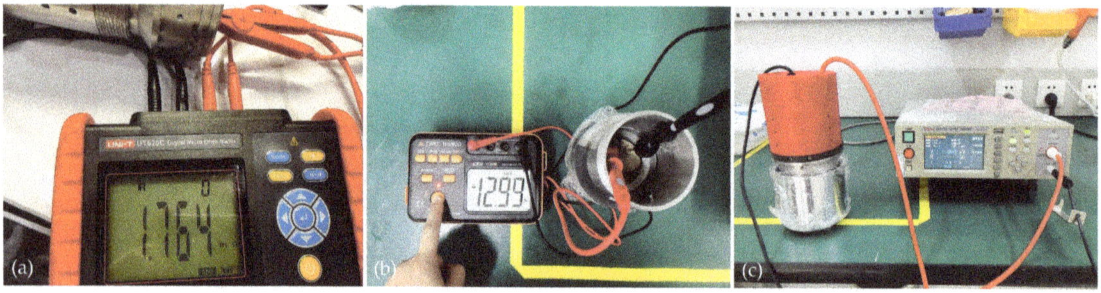

Figure 17. Electrical testing of the prototype: (**a**) contact resistance test, (**b**) insulation resistance test, and (**c**) proof voltage test.

Then, the online test of electrical performance was conducted in a hydrostatic pressure environment, and the schematic is illustrated in Figure 18. Here, a pressure chamber in the Hadal Science and Technology Research Centre (HSTRC) of Shanghai Ocean University is used. The fully-mated prototype is installed inside the pressure chamber where four test wires are penetrated through it. The electrical performance of the prototype has been verified in the pressure environment with an ocean depth of 3000 m. Figure 19 gives the electrical performance curves of the prototype in a hydrostatic pressure environment, which shows that:

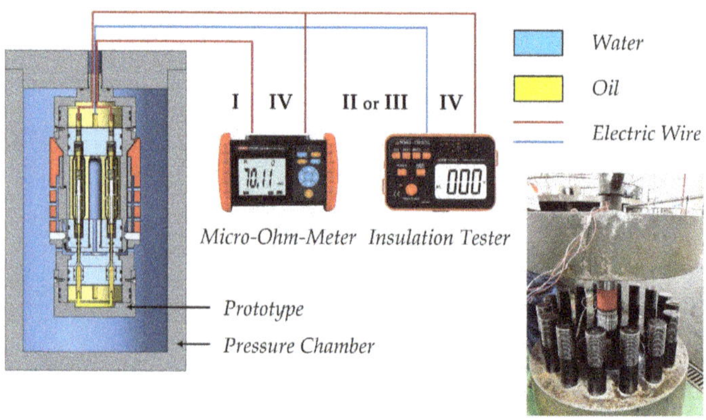

Figure 18. A schematic illustration of the prototype's online test of electrical performance in pressure environment.

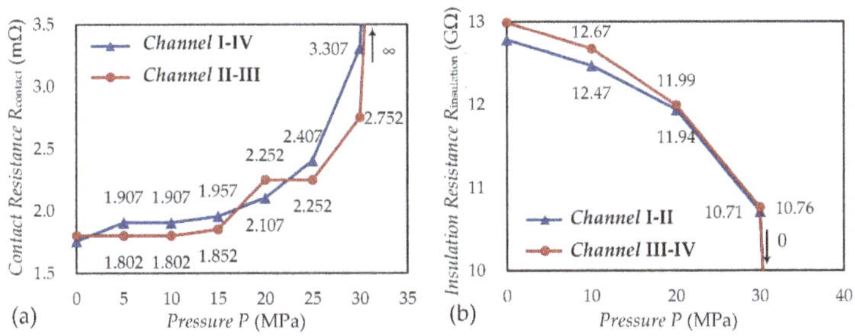

Figure 19. The electrical performance curves of the prototype in a hydrostatic pressure environment: (**a**) contact resistance and (**b**) insulation resistance.

(1) When the test pressure is less than 30 MPa, the electrical performance of the prototype can meet the design requirements. The contact resistance is about 3.3 mΩ in maximum while the insulation resistance is about 10.71 GΩ at minimum.

(2) As the test pressure increases, the contact resistance is increased while the insulation resistance is decreased. It reminds us that leakage plays a decisive role in the electrical performance of wet-mateable electrical connectors.

(3) When the test pressure exceeds 30 MPa, the differential pressure of the prototype will be higher than the sealing pressure of the O-rings used in the shuttle pin design. Thus, the failure of the penetrable self-sealing will cause a large amount of water to intrude into the electric contact, which results in infinite contact resistance and zero insulation resistance.

4.3. Failure Analysis

The above tests also revealed the failure modes of wet-mateable electrical connectors, which are discussed as follows:

(1) The main failure mode of wet-mateable connectors is sealing failure [23]. For the proposed connector, because the dual-bladder PBOF technology is introduced, only the common O-ring is used in the shuttle pin design. Additionally, its small size limits the sealing pressure of the penetrable self-sealing.

(2) Another important failure mode is the PBOF failure. Some ultimate pressure tests were conducted on the prototype. Under the condition with a very large pressure rate (about 10 MPa/min), the destruction of the primary resilient bladder was observed. As is shown in Figure 20, a very clean cut appears on the primary bladder. The PBOF failure can lead to the rapid loss of the sealing performance, which must be avoided.

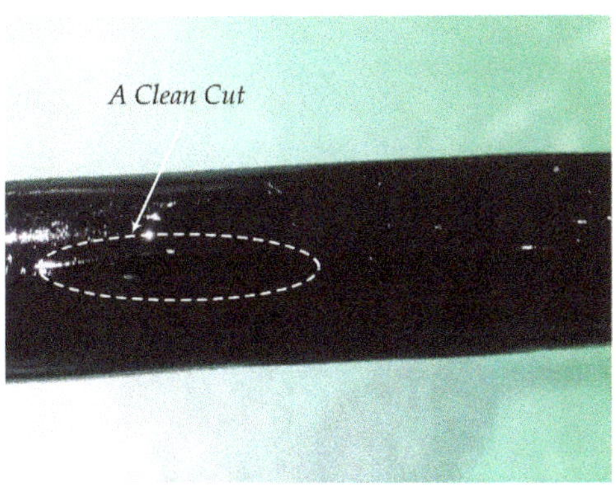

Figure 20. A destructed primary resilient bladder in the ultimate pressure test.

According to the failure analysis, we can realize how to improve the depth rate of the prototype in the future, and two aspects should be focussed on as follows:

(1) Replace the rod seals in the shuttle pin design to improve the sealing performance of the penetrable seal-sealing. For example, O-rings made of fluorosilicone rubber with high hardness or other rod seals with high sealing pressure can be used.

(2) Woven fabrics can be added to the material of resilient bladders to enhance the strength, which can improve the reliability of PBOF technology.

5. Discussion and Conclusions

This paper presents a methodology for designing and testing a wet-mateable electrical connector, which might be the first discloser on this topic in the aspects of design, theory, simulation and testing. The innovations of the proposed connector can be summarized as follows:

(1) The dual-bladder PBOF design used in the proposed connector is mainly based on the low-cost primary resilient bladder, which is different from some COTS products' designs (e.g., SECAON HydraElectric, Nautilus^{TM} WM1.7-30 and Siemens DigiTRON). The bladders used in COTS products are often very large, which are usually very difficult to manufacture by injection moulding.

(2) The AID design used in the proposed connector is also different from what is used in Nautilus^{TM} WM1.7-30 and Siemens DigiTRON. Also, the latching member can be made directly by 3D printing, which is cost-effective.

(3) In fact, the cost of developing a wet-mateable connector is very high. We have almost reached the limit of cost control in the R&D process. From another perspective, with lower cost and a shorter development cycle, we have made some major breakthroughs in this bottleneck field in China, and this benefited from the previous review research and the use of innovative simulation techniques.

It should be noted that the wet-mating online test of electrical performance in a high-pressure environment has not been carried out in this paper. To realize the wet-mating operation in a high-pressure environment, a specially-customed pressure chamber should be required. Unfortunately, such equipment is unavailable to us at present, which also involves a lot of key technologies. Limited by our financial conditions, the depth rate of the prototype and the test condition can only reach the present level, which can be improved in the future.

The main conclusions can be summarized as follows:

(1) An innovative wet-mateable connector has been proposed, which involves the dual-bladder PBOF technology, the penetrable self-sealing design, and the AID technology.

(2) The generalized differential pressure equations of the dual-bladder PBOF technology have been derived. The results show that the differential pressure provided by the bladder under the action of internal pressure is much larger than that provided by such a bladder with a similar dimension but under the action of external pressure.

(3) A procedure of thermal-electric-structure (TES) coupling simulation for wet-mateable electrical connectors has been proposed, and a series of FEA involving coupled multi-field problems have been conducted, including thermal-electric coupling analysis, static structural analysis, and dynamic analysis. The results show that the underwater environment allows better heat dissipation of wet-mateable electrical connectors, which can further be improved by the dual-bladder PBOF technology. Also, the primary resilient bladder has the function of inherent vibration attenuation while the secondary resilient bladder has the function of harmonic vibration attenuation.

(4) A prototype of the proposed connector has been developed successfully, and many process problems of core components have been solved. The electrical performance of the prototype has been verified by the online test in the hydrostatic pressure environment with an ocean depth of 3000 m. The technical level has reached the leading position in China.

(5) The failure modes of wet-mateable connectors have been revealed, mainly including the sealing failure and the PBOF failure. The rod seals with high sealing pressure should

be used in the shuttle pin design. The woven fabric-reinforced fluorosilicone rubber can be used to make resilient bladders to improve the reliability of PBOF technology. Additionally, the glass fibre-reinforced PEEK can be used to provide better resistance to dielectric breakdown.

Author Contributions: Conceptualization, W.S., C.Y., W.C., C.S., J.H. and Y.L.; methodology, W.S., W.C., C.S., P.Y., J.H., Y.L., Q.L. and Z.W.; software, W.S. and Z.W.; validation, W.C.; formal analysis, W.S.; investigation, W.S., C.Y., J.H. and Y.L.; resources, W.S., C.Y., W.C., C.S., P.Y., Y.L. and Q.L.; data curation, W.S. and W.C.; writing—original draft preparation, W.S.; writing—review and editing, W.S. and W.C.; visualization, W.S.; supervision, C.Y. and W.C.; project administration, C.Y. and W.C.; funding acquisition, C.Y., W.C., C.S. and P.Y. All authors have read and agreed to the published version of the manuscript.

Funding: This work was supported by Zhejiang Key R&D Program of China (Grant No. 2021C03157), the Startup funding of New-Joined PI of Westlake University (Grant No. 041030150118), and the Scientific Research Funding Project of Westlake University (Grant No. 2021WUFP017).

Institutional Review Board Statement: Not applicable.

Informed Consent Statement: A Not applicable.

Data Availability Statement: Not applicable.

Acknowledgments: We also would like to thank the colleagues of Hadal Science and Technology Research Centre (HSTRC) of Shanghai Ocean University.

Conflicts of Interest: The authors declare no conflict of interest.

References

1. Borthwick, A.G.L. Marine renewable energy seascape. *Engineering* **2016**, *2*, 69–78. [CrossRef]
2. Song, W.T.; Cui, W.C. Review of deep-ocean high-pressure simulation systems. *Mar. Technol. Soc. J.* **2020**, *54*, 68–84. [CrossRef]
3. Song, W.T.; Cui, W.C. An overview of underwater connectors. *J. Mar. Sci. Eng.* **2021**, *9*, 813. [CrossRef]
4. Menandro, P.S.; Bastos, A.C. Seabed mapping: A brief history from meaningful words. *Geosciences* **2020**, *10*, 273. [CrossRef]
5. Song, W.T.; Yang, C.B.; Cui, W.C.; Lei, Y.; Hong, J.; Wang, Z.H.; Hu, Z.Y. Study of pressure-balanced oil-filled (PBOF) technology. *Ocean Eng.* **2022**, *260*, 111757. [CrossRef]
6. Collin, A.J.; Nambiar, A.J.; Bould, D.; Whitby, B.; Moonem, M.A.; Schenkman, B.; Atcitty, S.; Chainho, P.; Kiprakis, A.E. Electrical components for marine renewable energy arrays: A techno-economic review. *Energies* **2017**, *10*, 1973. [CrossRef]
7. Rémouit, F.; Ruiz-Minguela, P.; Engström, J. Review of electrical connectors for underwater applications. *IEEE J. Ocean. Eng.* **2018**, *43*, 1037–1047. [CrossRef]
8. Wood Group Kenny. *Wet Mate Connector Market Study*; Technical Report 2500014-01-D-3-001; ORE Catapult: Glasgow, UK, 2014; Available online: https://ore.catapult.org.uk/wp-content/uploads/2018/01/Wet-mate-connector-study.pdf (accessed on 13 December 2022).
9. Weiss, P.; Beurthey, S.; Chardard, Y.; Dhedin, J.F.; Andre, T.; Rabushka, K.; Tourcher, C.; Gauch, F.; Micoli, C. Novel wet-mate connectors for high voltage and power transmissions of ocean renewable energy systems. In Proceedings of the 4th International Conference on Ocean Energy 2012, Dublin, Ireland, 17–19 October 2012; ICOE: Lisbon, Portugal, 2022. Available online: https://www.icoe-conference.com/publication/novel_wet_mate_connectors_for_high_voltage_and_power_transmissions_of_ocean_renewable_energy_systems/ (accessed on 13 December 2022).
10. ANTARES Collaboration. ANTARES: The first undersea neutrino telescope. *Nucl. Instrum. Methods A* **2011**, *656*, 11–38. [CrossRef]
11. KM3NeT Collaboration. Deep-sea deployment of the KM3NeT neutrino telescope detection units by self-unrolling. *J. Instrum.* **2020**, *15*, P11027. [CrossRef]
12. Han, Q.; Chen, H.Y.; Yang, W.C.; Zhang, Y.; Yang, J.K.; Chen, Y.F. Analysis of reciprocating O-ring seal in the pressure-balanced oil-filled wet-mate electrical connectors for underwater applications. *Lubr. Sci.* **2019**, *31*, 335–345. [CrossRef]
13. Song, W.T.; Cui, W.C. Design of a practical metal-made cold isostatic pressing (CIP) chamber using finite element analysis. *Materials* **2022**, *15*, 3621. [CrossRef]
14. Antonova, E.E.; Looman, D.C. Finite elements for thermoelectric device analysis in ANSYS. In Proceedings of the ICT 2005 24th International Conference on Thermoelectrics, Clemson, CA, USA, 19–23 June 2005; IEEE: New York, NY, USA, 2015; pp. 215–218. [CrossRef]
15. Sun, B.A.; Li, W.K.; Wang, Z.H.; Zhu, Y.P.; He, Q.; Guan, X.Y.; Dai, G.M.; Yuan, D.H.; Li, A.; Cui, W.C.; et al. Recent progress in modelling and control of bio-inspired fish robots. *J. Mar. Sci. Eng.* **2022**, *10*, 773. [CrossRef]
16. Liu, Q.M.; Chen, H.; Wang, Z.H.; He, Q.; Chen, L.K.; Li, W.K.; Li, R.P.; Cui, W.C. A manta ray robot with soft material based flapping wing. *J. Mar. Sci. Eng.* **2022**, *10*, 962. [CrossRef]

17. Lin, M.L.; Lin, R.; Yang, C.J.; Li, D.J.; Zhang, Z.Y.; Zhao, Y.C.; Ding, W.J. Docking to an underwater suspended charging station: Systematic design and experimental test. *Ocean Eng.* **2022**, *249*, 110766. [CrossRef]
18. Fan, F.H.; Ishibashi, S. Underwater applications of light emitting diodes. In Proceedings of the 2015 IEEE Underwater Technology (UT), Chennai, India, 23–25 February 2015; IEEE: New York, NY, USA, 2015; p. 15143567. [CrossRef]
19. Song, W.T.; Cui, W.C. Underwater Wet Plugging Electric Connector Capable of Realizing Pressure Equilibrium. China Patent CN 112600029 B, 29 October 2021.
20. Cairns, J.L. Spark-Proof Hostile Environment Connector. U.S. Patent 5,194,012, 16 March 1993.
21. Thongsri, J. Transient thermal-electric simulation and experiment of heat transfer in welding tip for reflow soldering process. *Math. Probl. Eng.* **2018**, *2018*, 4539054. [CrossRef]
22. SEAFOM. Functional Design and Test Specification for Subsea Electrical and Optical Connectors and Jumpers; SEAFOM TSD-02; Epsom, UK, 2016. 2016. Available online: https://seafom.com/?mdocs-file=78 (accessed on 13 December 2022).
23. Lamare, P.; Vernin, P. Underwater mateable electro-optical connectors: The feedback from ANTARES. *Nucl. Instrum. Meth. A* **2009**, *602*, 252–254. [CrossRef]

Journal of
Marine Science and Engineering

MDPI

Article

Semi-Analytical Methods for the Joint Strength and Sealing Performance in the Failure Process of the Subsea Pipeline Compression Connector

Zhenyu Li [1], Gang Wang [2,*], Xiangyu Wang [3], Shaoming Yao [1], Feihong Yun [1], Peng Jia [1] and Liquan Wang [1]

1 College of Mechanical and Electrical Engineering, Harbin Engineering University, Harbin 150001, China; jade10623@sina.com (Z.L.); yao.sm@hotmail.com (S.Y.); yunfeihong@hrbeu.edu.cn (F.Y.); 13633605161@139.com (P.J.); wangliquan@hrbeu.edu.cn (L.W.)
2 College of Shipbuilding Engineering, Harbin Engineering University, Harbin 150001, China
3 Yantai Research Institute of Harbin Engineering University, Yantai 264006, China; wangxiangyu325@126.com
* Correspondence: wanggang@hrbeu.edu.cn

Abstract: Radial seals are sensitive to axial overload failure and may cause leaks. This paper presents two semi-analytical methods for the joint strength and sealing performance of the subsea pipeline compression connector under axial overload failure. The method for the joint strength consists of two parts: One is the analytical model for the joint strength of the connection and seal under axial tension and compression conditions. The models are based on membrane theory, considering the hardening and bending effects. The other is a two-dimensional, axisymmetric finite element model for the joint strength of the radial metal seal. The semi-analytical method for the overload sealing performance is derived using a finite element model and the Reynolds equation of the laminar flow. The effects of critical parameters on the joint strength and the overload sealing performance are analyzed. The experiments are carried out with specimens and prototypes to evaluate the evolution of the sealing interface and the joint strength. The results show that both the internally turned sealing surface and the deflection of the pipe can improve the joint strength. In addition, the compression-type connector can remain sealed under the maximum axial overload. The proposed methods allow the prediction and identification of the overload joint strength and the sealing condition of the compression-type connector and provide a better understanding of the radial metal seal under the axial overload condition.

Keywords: semi-analytical method; compression connector; joint strength; sealing performance; failure; membrane theory; radial seal

Citation: Li, Z.; Wang, G.; Wang, X.; Yao, S.; Yun, F.; Jia, P.; Wang, L. Semi-Analytical Methods for the Joint Strength and Sealing Performance in the Failure Process of the Subsea Pipeline Compression Connector. *J. Mar. Sci. Eng.* **2023**, *11*, 1417. https://doi.org/10.3390/jmse11071417

Academic Editors: Bruno Brunone and José António Correia

Received: 20 June 2023
Revised: 13 July 2023
Accepted: 13 July 2023
Published: 14 July 2023

1. Introduction

The axial force overload is a significant factor contributing to the pipeline failure in subsea connections [1]. For non-welded connectors [2], when subjected to the overload, connectors with axial seals may fail due to the separation of sealing surfaces, while connectors with radial seals may fail due to the axial sliding of sealing surfaces. This study is performed to analyze the joint strength and sealing performance during the axial overload of the subsea compression connector with the radial metal seal.

The key problem of the radial metal seal is the leakage, and the complexity of the metal seal derives from its surface topography [3]. Due to the existence of stochastic gaps at the sealing interface, the two sealing surfaces cannot be fully contacted on the microscopic scale, and the risk of leakage increases if the gaps are connected across the seal [4]. In the studies of the leakage ratio on randomly rough surfaces, the topography formed by turning shows unique sealing properties [5]. This issue can be traced back to the groundbreaking studies on the surface waviness theory [6] and its further developments on the surface roughness [7] and sealing design factors [8]. In the aspect of the gasket, through experimentation, Nakamura and Funabashi [9] studied the leakage on a spiral-like topography; then, Polycarpou and Etsion [10] developed a simplified semi-empirical model to analyze the leakage and emphasized the significance of micro-profile parameters in determining the sealing performance of the metal gasket. Arghavani et al. [11] compared the deformation and pressure distribution of the sealing surfaces of flanged gaskets manufactured by different machining methods, and the surface topographies were found to be the decisive factors affecting the leakage ratio. In terms of experiments, a polymer film method was introduced, and the spiral contact marks and leakage flow paths were proposed to observe the true contact condition of the topography on the gasket [12], and the real contact areas on the sealing surfaces were measured [13]. To provide further clarification on the sealing capabilities of the turning topography, Geoffroy and Prat [3] developed a model for the leakage of turning metal gaskets and identified the transition from radial to spiral leakage. One of the advantages of their progress is to establish a more realistic geometric model of local surfaces, which is considered a substitute for the geometrical representation of ideal surfaces. In addition, Robbe-Valloire and Prat [4] further studied the radial and circumferential leakage paths by specifying the typical contact patterns of metal static seals. Ledoux et al. [14] simulated the leakage ratios utilizing a modal discrete decomposition method and concluded that the leakage ratios can be greatly reduced if the turned-like waviness is controlled. Bourniquel et al. [15] studied the leakage prediction using a similar method, with the idea that the roughness may not have a significant effect on the overall leakage ratio.

The turning topography related to the studies of the asperity [16] and the probabilistic models [17] of rough surfaces. By creating a two-dimensional (2D) stochastic model, Lorenz and Persson [18] proved that the leakage ratio sensitively depended on the skewness in the height probability distribution. Then, by developing a three-dimensional (3D) finite element (FEA) model, Zhang et al. [19] proposed a novel approach to calculate the leakage ratio for specific surface topographies. Besides, multiscale and fractal methods are utilized as effective means for morphological analysis. Marie et al. [20] created a functional prototype and implemented a multiscale computational approach for the leakage ratio of the sealing structure. Liao et al. [21] proposed a technique for decomposing 3D surfaces into multiscale surfaces by the wavelets to predict surface functions and identify machining errors. Deltombe et al. [22] used a multiscale analysis to evaluate the performance of the seal, and micro-roughness played a major role in the leakage. Yan and Fan [23] used a multiscale model to investigate the impact of the fluid pressure on the sealing performance of pipe connections. In other respects, So and Liu [24] proposed an elastic–plastic model to consider the anisotropy of the rough contact surfaces. Shao et al. [25] predicted the processing surface by high-definition metrology. Tang et al. [26] developed a hydraulic–thermal FEA model for the sealing surface of connectors and showed that temperature was significant to the thermal deformation of both metal and non-metal sealing surfaces. Ernens et al. [27] conducted experiments for a variety of thread compounds and applied pin/box surface coatings to describe the sealing mechanism of the metal seals.

Scholars have also conducted important research on compression connections. Based on the elasticity superposition theorem, Wei et al. [28] derived the formula for the sealing pressure of a compression connector and determined the sealing condition using the FEA method. Wang et al. [29] developed an optimization method for the compression connector using a static metal sealing mechanism and created a FEA model with the zero-order method to optimize its structure. Weddeling et al. [30] developed an analytical model for calculating the required charging energy for a radial connection with a mandrel placed inside the pipe. In another study, Weddeling et al. [31] formulated a transferable load formula incorporating various groove shapes, using membrane theory. With the experimental data and FEA methods, Henriksen et al. [32] analyzed the forces and deformations involved in the process of connecting pipe flanges and found that the increased radial stiffness is the most reliable indicator of a clamped connection. Agrawal et al. [33] presented a novel approach for compression connectors and conducted both tension and compression tests, indicating that the compression connector exhibited a considerable strength with welded joints. Onyegiri et al. [34] studied compression connectors for sandwich pipes by incorporating 2D axisymmetric and 3D FEA models. Quispe et al. [35] presented a FEA model for a threaded compression connection with square teeth and a metal static seal. The model was specifically designed for connecting subsea sandwich pipes. Yan and Fan [23] used a multiscale model to investigate the impact of the fluid pressure on the sealing performance of a pipe connection. Liu et al. [36] enhanced the anchoring structure by implementing a nonlinear FEA technique. Li et al. [37] proposed a semi-analytical method to predict the radial sealing performance of the compression connector, and the leakage condition was obtained as a function of the minimum radial deflection. Wu et al. [38] presented a method for calculating the axial load capacity of the steel pipe-to-sleeve grouted connections.

For the compression connection, there is still no analytical method to determine the joint strength and the overload sealing performance. In practice, the FEA approach remains the exclusive choice in many instances. A more comprehensive understanding of the capacity for an enduring overload and sealing performance of compression connections is needed. In addition, most studies on metal seals focus on the axial sealing mechanisms. There is limited research available on radial seals.

This paper presents two semi-analytical methods for analyzing the joint strength and the overload sealing performance of the subsea compression-type connector. Specimen and prototype tests are carried out, and the results are analyzed and compared.

2. Structure and Mechanism of the Compression Connector

The compression pipeline connector is a mechanical connection (ASTM F1387 [39] and ISO 8434-1 [40]), which is used to connect two pipes, particularly in the underwater environment. This type has been developed by many well-known companies, including the Phastite® connector from Parker Hannifin (Columbus, OH, USA), the Tube-Mac® connector from PYPLOK® Corp. (Stoney Creek, Ontario, Canada), the compression connector from HAELOK® (Schlieren, Switzerland), etc.

The studied connector and its assembly tools, as well as the assembly process, are shown in Figure 1. It is designed to connect the single-wall subsea pipeline between 2 and 12 inches in diameter. The connector was directly attached to the pipe by the mechanical deformation and created a metal static seal at the same time [27]. The connector comprises a base and two press rings, as shown in Figure 1a,c, which are both made of metal and symmetrically positioned around the pipe, as shown in Figure 1c,d. The base features one or more sealing ridges. The dedicated assembly tools shown in Figure 1b were used for the automation of the assembly process.

(a) (b)

Figure 1. *Cont.*

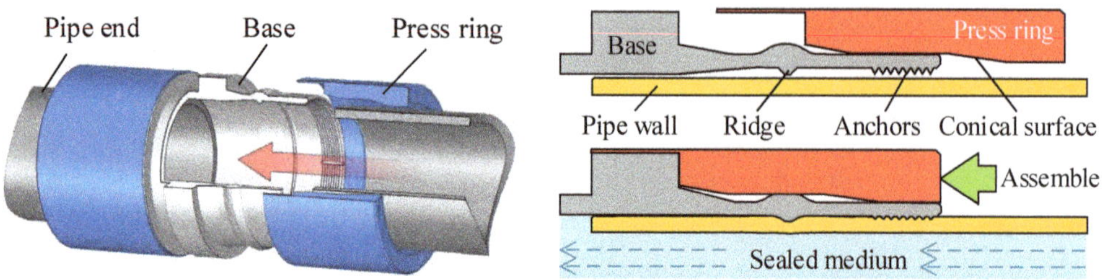

(c) (d)

Figure 1. (**a**) Compression connector components, (**b**) assembly tools, (**c**) connector parts, and (**d**) connecting process.

3. Semi-Analytical Model of the Joint Strength

After assembly, the radial deflection of the connected pipe not only creates sealing pressure on the sealing ridge, but also provides partial joint strength in the axial direction. In our previous study [37], analytical models for the radial deflection of the base and pipe were developed. With these models, the joint strength of the deflected pipe can be derived. The relevant deflection formulas are introduced below.

3.1. Analytical Model of the Pipe Deflection

In this study, the following basic assumptions were made:

- The pipe is a cylindrical shell with an even thickness, and the stress distribution is uniform across the pipe thickness.
- During the axial loading, the small clearance between the pipe and the base is ignored.
- All surfaces are in the ideal condition without any heat treatment, coating, and defects.
- The material of the pipe is isotropic and bilinear hardening.
- The deformation is quasi-static, with strain rate > 10^{-2} s^{-1} (ASTM E8/E8M [41]).

The pipe surface is assumed to be subjected to a concentrated radial sealing pressure, P, which is uniformly distributed in the circumferential direction, and the radial deflection of the connected pipe can be expressed as [37]:

$$\left.\begin{aligned} w(x) &= -\frac{P}{8K_2^3 D} e^{-K_2 x}(\cos K_2 x + \sin K_2 x) \\ K_2 &= \left(\frac{E\delta}{4R_a^2 D}\right)^{\frac{1}{4}},\ D = \frac{E\delta^3}{12(1-v^2)} \end{aligned}\right\}. \quad (1)$$

The model set the origin at the center of the sealing ridge, with the positive x-axis pointing to the right side of the pipe axis. P is the sealing pressure, $w(x)$ is the deflection, with a positive direction pointing outside of the pipe, E is the Young's modulus, D is the flexural rigidity, δ is the pipe thickness, v is the Poisson's ratio, and R_a is the radius of the mid-surface of the pipe wall before deflection.

Assuming the half-cone angle of the deflection as an unsigned angle, $\psi(x)$, it can be expressed as:

$$\psi(x) = \frac{P}{4K_2^2 D} e^{-K_2 x} \sin K_2 x. \quad (2)$$

Accordingly, the deflection, $w(x)$, and half-cone angle, $\psi(x)$, are plotted in Figure 2. The following conclusions were drawn from Figure 2:

- The $w(x)$ reached the maximum at $x = 0$, and it was basically inversely proportional to x.
- The influence length of the $w(x)$ could be approximated as $x \in \left(0, \frac{7\pi}{4K_2}\right)$.
- In the range: $x \in \left(0, \frac{\pi}{K_2}\right]$, the angle varied with x.

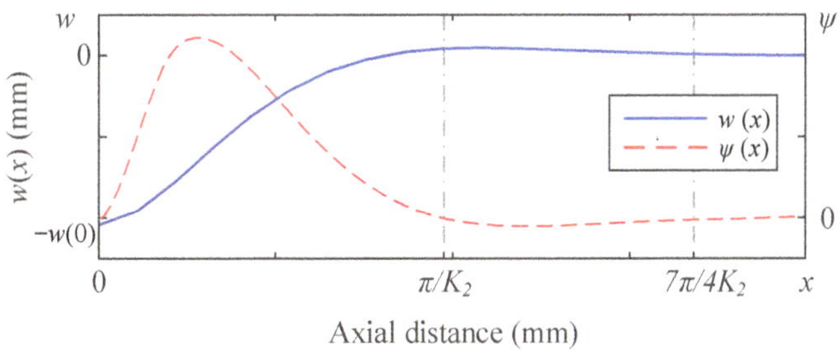

Figure 2. Deflection, $w(x)$, and half-cone angle, $\psi(x)$, of the connected pipe.

When subjected to axial overload, the deflected section will deform and provide additional resistance. To simplify the formulation, an average half-cone angle $\alpha = \overline{\psi(x)}$.

3.2. Joint Strength of the Connected Pipe

The joint strength of the deflected pipe can be derived from the above deflection parameters. Referring to studies on the connector that is supported by the inner mandrel [30,31], the membrane theory was used in the formulation, and material hardening and bending stress were also considered. Since the stress condition of the deflected pipe was different between tension and compression, they were separately derived.

3.2.1. Tension Strength Model

When the pipe was under the axial tension load, F_L, the stress state was as illustrated in Figure 3. The origin was fixed at the projection of the ridge on the axis of the pipe. The positive x-axis was designated as the direction opposite to F_L.

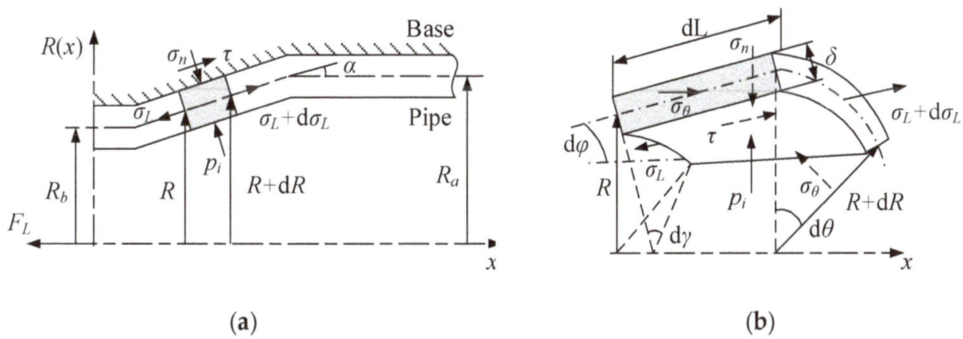

Figure 3. (**a**) Diagram of the deflected pipe under axial tension, and (**b**) a thin-walled cylindrical cell from (**a**).

In Figure 3a, R_b is the radius after deflection, $R(x)$ is the radius of the deflection section, and p_i is the internal pressure. A thin-walled cylindrical cell was taken from the deflected section of the pipe and shown in Figure 3b, where σ_L is the stress parallel to the pipe surface, σ_n is normal to the pipe surface, and σ_θ is the circumferential stress.

The differential equilibrium equation normal to the pipe surface is:

$$\sigma_L R\mathrm{d}\theta\mathrm{d}L - 2\sigma_\theta \sin\frac{\theta}{2}\delta\mathrm{d}L\cos\alpha = 0. \quad (3)$$

The differential equilibrium equation parallel to the pipe surface is:

$$(\sigma_L + \mathrm{d}\sigma_L)(R+\mathrm{d}R)(\delta+\mathrm{d}\delta)\mathrm{d}\theta - \sigma_L R\mathrm{d}\theta\delta + 2\sin\frac{\mathrm{d}\theta}{2}\sin\alpha\sigma_\theta\delta\mathrm{d}L + \mu\sigma_n R\mathrm{d}\theta\mathrm{d}L = 0, \quad (4)$$

where μ is the friction coefficient.

We combined Equations (3) and (4) and ignored the second-order infinitesimal, which yielded:

$$R\frac{d\sigma_L}{dR} + \sigma_L + \sigma_\theta(1+\mu\cot\alpha) = 0. \quad (5)$$

For the plastic condition of the hardening, an isotropic bilinear hardening model [42] was adopted, and the constitutive relationship of the hardening took the form of:

$$\sigma^T = \sigma_s + K_1 \cdot \varepsilon^T, \quad (6)$$

where σ^T is the true stress, σ_s is the yield strength, ε^T is the accumulated plastic strain, and K_1 is the hardening modulus.

According to the Tresca yield criterion [43], the plastic state is:

$$\sigma_L + \sigma_\theta = K_3\sigma^T, \quad (7)$$

where the coefficient $K_3 \in \left[1, \frac{\sqrt{3}}{2}\right]$ [44]. Substituting Equations (6) and (7) into Equation (5) yielded:

$$R\frac{d\sigma_L}{dR} - a_1\sigma_L - a_2R + a_3 = 0, \quad (8)$$

where:

$$\left.\begin{array}{l} a_1 = \mu\cot\alpha \\ a_2 = \frac{K_1K_3}{R_i}(1+a_1) \\ a_3 = K_3(\sigma_s+K_1)(1+a_1) \end{array}\right\}. \quad (9)$$

The antiderivative of Equation (8) is:

$$(a_1-1)\sigma_L = CR^{a_1} - a_2R + (a_1-1)\frac{a_3}{a_1}, \quad (10)$$

where C is a constant. By setting the boundary conditions as $R = R_a$ and $\sigma_L = -p_i$, the final expression for axial stress with hardening under the tension condition was derived as:

$$\sigma_L = \frac{K_1K_3(\mu\cot\alpha+1)}{R_a(\mu\cot\alpha-1)}\left[R_a\left(\frac{R}{R_a}\right)^{\mu\cot\alpha} - R\right] + K_3(\sigma_s+K_1)\left(\frac{1}{\mu\cot\alpha}+1\right)\left[1-\left(\frac{R}{R_a}\right)^{\mu\cot\alpha}\right] - p_i\left(\frac{R}{R_a}\right)^{\mu\cot\alpha}. \quad (11)$$

The bending moments at both ends of the curved section in Figure 3a also affected σ_L. To describe the effect, for example, we took a bending section of the pipe at $R = R_a$, as shown in Figure 4.

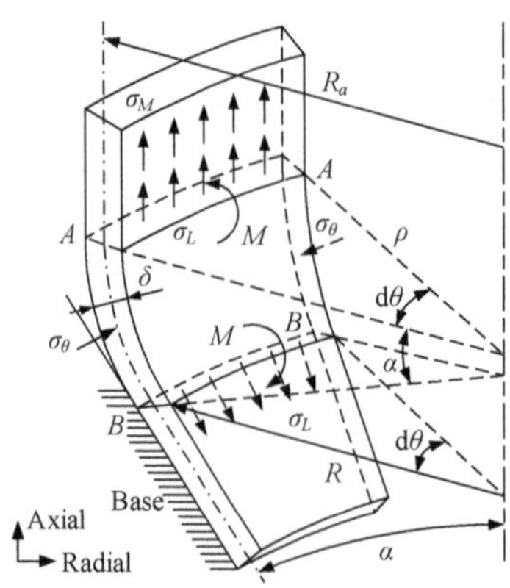

Figure 4. A bending section of the deflected pipe.

Within the bending area (from cross-section A-A to B-B), the curvature varied as the stress increased. Assuming the presence of a secondary stress, σ_M, caused by the curvature, and that the curvature change was entirely caused by the work carried out by σ_M, the equilibrium equation of work for the cross-section A-A is given by:

$$\sigma_M \delta R_a \mathrm{d}\theta \rho \alpha = M\alpha, \tag{12}$$

where ρ is the radius of curvature, and M is the bending moment. The equilibrium equation of forces for the cross-section B-B is:

$$\sigma_L \delta R_a \mathrm{d}\theta \rho (1 - \cos\alpha) = M, \tag{13}$$

while the bending moment can be expressed as [45]:

$$M = \frac{\delta^2 R_a}{4} \mathrm{d}\theta \sigma_s. \tag{14}$$

Substituting Equation (14) into Equations (12) and (13) yielded:

$$\sigma_M = \sigma_L (1 - \cos\alpha). \tag{15}$$

Hence, the axial stress considering both hardening and bending is:

$$\sigma_{LM} = (3 - 2\cos\alpha)\sigma_L. \tag{16}$$

Then, the strength under tension overload could be determined as:

$$F_L = 2\pi\delta R \sigma_{LM}. \tag{17}$$

The maximum tension strength along the deflected segment is:

$$F_{Lmax} = 2\pi\delta R_b \sigma_{LM}. \tag{18}$$

The analytical model is validated by the FEA model in Section 4. The maximum axial strength of the FEA reached 5.08×10^4 N, the corresponding theoretical value was 5.30×10^4 N, and the relative error was 4.17%. Compared to the results without hardening, it was reduced by 4.37%. The error was caused by the simplification of the models, the average half-cone angle, etc.

3.2.2. Compression Strength Model

The axial compression load changed the stress in the pipe. We defined the axial pressure as σ_P, and according to Figure 3a,b, the differential equilibrium equations both normal and parallel to the pipe surface can be expressed as:

$$\left.\begin{array}{l} \sigma_n R d\varphi \frac{\mathrm{d}R}{\sin\alpha} + 2\sigma_\theta \delta \frac{\mathrm{d}R}{\sin\alpha} \sin\frac{\mathrm{d}\gamma}{2} = 0 \\ (\sigma_P + \mathrm{d}\sigma_P)(R + \mathrm{d}R)\delta d\theta - \sigma_P \delta R d\theta - 2\sigma_\theta \delta \frac{\mathrm{d}R}{\sin\alpha} \sin\frac{\mathrm{d}\varphi}{2} + \sigma_n R d\varphi \frac{\mathrm{d}R}{\sin\alpha} = 0 \end{array}\right\}. \tag{19}$$

Since $\mathrm{d}\varphi = \sin\alpha \mathrm{d}\theta$ and $\mathrm{d}\gamma = \cos\alpha \mathrm{d}\theta$, Equation (19) can be written as:

$$R\frac{\mathrm{d}\sigma_P}{\mathrm{d}R} + \sigma_P - \sigma_\theta(1 + \mu\cot\alpha) = 0. \tag{20}$$

Via the Tresca yield criterion with the material hardening:

$$\sigma_\theta - \sigma_n = K_3\left[\sigma_s + K_1\left(1 - \frac{R}{R_a}\right)\right]. \tag{21}$$

Ignoring the small quantity σ_n, and substituting Equation (21) into Equation (20), yielded:

$$R\frac{d\sigma_P}{dR} + \sigma_P - K_3\left[\sigma_s + K_1\left(1 - \frac{R}{R_a}\right)\right](1 + \mu\cot\alpha) = 0. \quad (22)$$

Integrating the equation with the boundary conditions: $R = R_b$ and $\sigma_P = p_i$, the axial stress under compression including the hardening is:

$$\sigma_P = K_3(1 + \mu\cot\alpha)\left(1 - \frac{R_b}{R}\right)\left[\sigma_s + K_1\left(1 - \frac{R + R_b}{2R_a}\right)\right] - p_i\frac{R_b}{R}. \quad (23)$$

Similar to Equation (16), the stress including the bending moment and material hardening is:

$$\sigma_{PM} = (3 - 2\cos\alpha)\sigma_P. \quad (24)$$

The load capacity of the pipe is:

$$F_P = 2\pi R\delta\sigma_{PM}. \quad (25)$$

The maximum compression strength is:

$$F_{Pmax} = 2\pi R_a\delta\sigma_{PM}. \quad (26)$$

The results are validated by the FEA in Section 4. During the overload process, the maximum axial strength of the FEA was 5.46×10^4 N, while the theoretical value was 5.76×10^4 N. The relative error was 5.23%.

3.3. Joint Strength of the Seal

In engineering applications, the turning topography is usually left on the sealing surface to achieve a better bonding strength [46]. The topography is an anisotropic helix pattern, similar to a thread [37]. In the assembly, the topography is embedded in the pipe surface. To determine the maximum shear strength of the sealing interface, as shown in Figure 5, the calculation method for the thread strength [47] was used.

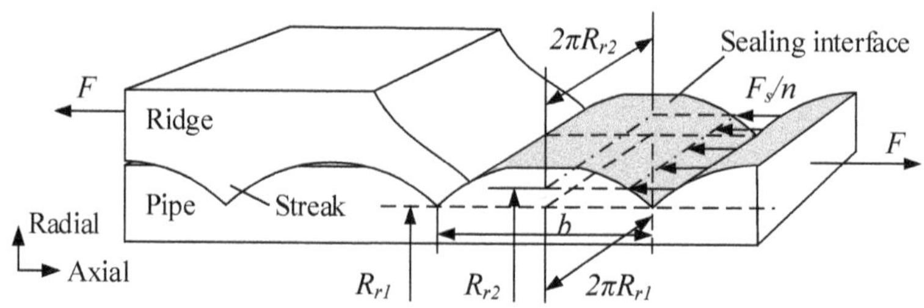

Figure 5. Model of the engaged sealing interface.

Taking the dent on the pipe side as the study object, the width of the embedded streaks was much more than the height; thus, by disregarding the bending moment effect, the maximum of axial shearing force, F_s, is given by:

$$F_{smax} = 2\pi R_{r1} b n \tau_p. \quad (27)$$

where b is the turning feed rate, R_{r1} is the radial radius of the bottom of the dent, n is the number of streaks, and τ_p is the allowable shear stress of the material. The average axial overload on each streak, F_s/n, is acting on the pitch radius, R_{r2}.

The results are validated by the FEA in Section 4. The maximum axial strength of the FEA was 2.49×10^5 N, compared to the analytic value of 2.47×10^4 N. The relative error was only 0.96%. However, this method can only estimate the maximum strength before overload. Due to the nonlinear and prominent deformation of the topography, the F_s changed during overload, so F_s is further studied by the FEA method in Section 4.

Combining the models obtained in the previous sections and using the superposition theorem [28], a semi-analytical model for total axial joint strength, F, can be derived as follows:

$$F = F_{L,P} + F_s + F_A. \quad (28)$$

where F_L or F_p need to be determined according to the force condition, F_s is obtained by FEA, and F_A is the strength of the anchor section according to its configuration, which was considered constant in this paper. This model can be used to obtain both the maximum axial strength and the strength variation during overload. It can also be used to predict the overload forces in sealing experiments, as in Section 6.

4. FEA Modeling

To validate the proposed models and obtain nonlinear evolution of the sealing interface under overload, FEA models were created. The numerical analysis was performed using ABAQUS®/Standard [48].

4.1. FEA Model for the Pipe

A simplified 2D axisymmetric model was created to investigate the axial overload behavior in the deflected section of the pipe, as shown in Figure 6. The model uses the CAX4R element, which is a four-node, bilinear, axisymmetric, quadrilateral element with reduced integration and hourglass control. The details of the ridge and sealing surfaces were ignored, and the base was considered rigid with a high yield strength.

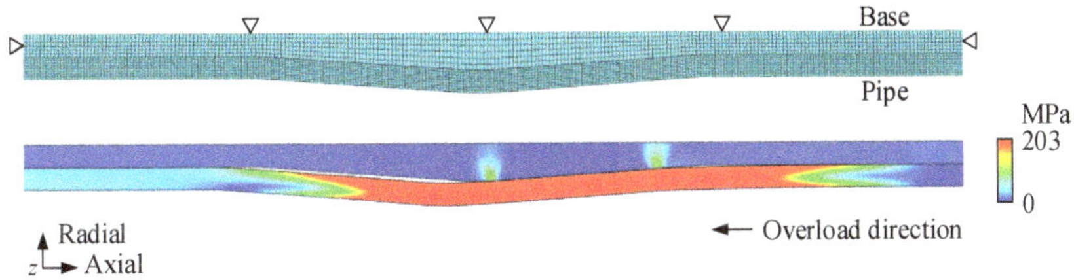

Figure 6. FEA of the axial strength components, F_L and F_p.

4.2. FEA Model for Sealing Surface

To determine the evolution of the sealing interface during overload, a 2D axisymmetric FEA model featuring the turning topography was created, as shown in Figure 7. The new model retains the same element as the previous one and incorporates topography parameters from the validated FEA model in the previous study [37]. The mesh was optimized, as shown in Figure 7b–d. The mesh sensitivity was examined and verified.

The material and geometric parameters used in the theoretical analysis are listed in Table 1.

Table 1. Geometrical parameters and material properties.

Name	Unit	Symbol	Value
Type	inch	—	6
Sealing width	mm	—	3
Radius	mm	R_a	82.5
Thickness	mm	δ	5
Embedded depth	mm	—	2
Average half-cone angle	rad	α	0.034
Turning speed	m/min	—	100
Turning feed	mm/r	—	0.5
Depth of cut	mm	—	0.2

Table 1. *Cont.*

Name	Unit	Symbol	Value
Yield strength	MPa	σ_s	235
Shear strength	MPa	τ_p	188
Young's modulus	MPa	E	2.10×10^5
Friction coefficient	—	μ	0.2
Poisson coefficient	—	v	0.3

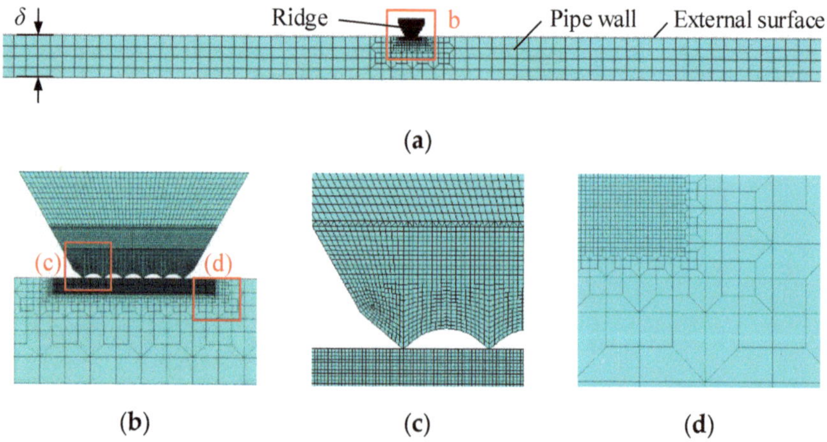

Figure 7. (**a**) The FEA model for analyzing joint strength and leakage, (**b**) the local view of the sealing section, (**c**) the local view of the streaks, and (**d**) the view of the refinement.

The sealing topography used in the leakage model is described in Section 5 to derive the theoretical sealing performance. The results of both models are compared with experiments in Section 6.

5. Semi-Analytical Model of the Leakage Ratio

A helical leakage channel was formed when there was relative sliding of the sealing surfaces. The cross-sectional area of this channel varied with the relative displacement, x_r, between the sealing surfaces. Figure 8 shows a typical profile of a leakage channel, which is enclosed between curves 1 and 2. To facilitate the comparison, the Abbott–Firestone curve [49] was used. The leakage channel can be mapped to the positive, semi-axial curve 4 via curve 3. The area enclosed by the coordinate axes and curve 4 remained constant with the channel area.

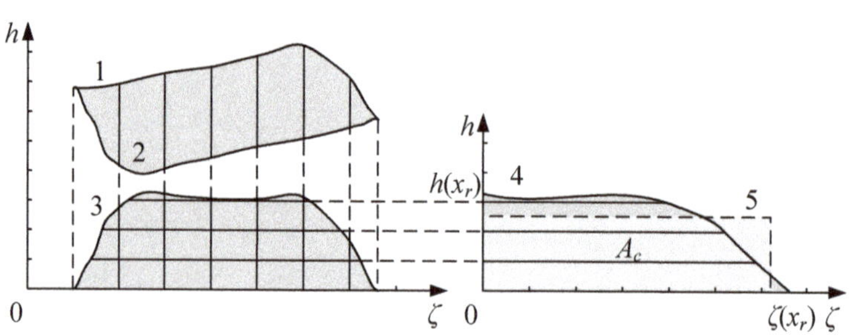

Figure 8. A typical profile of a leakage channel and its corresponding semi-axis curves.

The cross-section area, $A_c(x_r)$, can be expressed by the radial height, $h(x_r)$, and the axial width, $\zeta(x_r)$, as:

$$A_c = \int_0^{\zeta} h \, \mathrm{d}x. \tag{29}$$

To determine the leakage ratio, the rectangular area (curve 5) can be used as an approximation of the cross-sectional shape of curve 4. The equivalent height of the rectangular leakage channel is:

$$\bar{h}(x_r) = \frac{A_c(x_r)}{\zeta(x_r)}. \tag{30}$$

The length of the leakage channel is:

$$L_c = 2\pi R_{r1}(n-1). \tag{31}$$

The governing equation for the laminar flow can be expressed by the Reynolds equations [50] as:

$$\begin{aligned} &\nabla \cdot (\vec{q}_v) = 0.; \vec{q}_v = -\tfrac{h^3}{12\eta}\Delta p; \\ &p = p_i, \ \ at\ x = x_i; \\ &p = p_o, \ \ at\ x = x_o, \end{aligned} \tag{32}$$

where $\vec{q}_v$ is the volume flow ratio per unit width, η is the viscosity of the fluid, and p is the fluid pressure. The circumferential leakage ratio, Q_c, as a function of x_r, can be derived as:

$$Q_c(x_r) = \frac{(p_i^2 - p_o^2)\zeta(x_r)\bar{h}^3(x_r)}{48\eta p_o(n-1)\pi R_{r1}}. \tag{33}$$

6. Results and Discussion

6.1. Parameter Analysis

In this section, the influence of the parameters on the joint strength was analyzed using the proposed strength models, and the overload sealing performance was also analyzed.

A critical sealing factor for radial seals is the maximum pipe deflection, which is also a critical factor for the axial strength. This study used a necking coefficient, $(w/R)_{x=0}$, to examine the correlation between the deflection and the strengths, as shown in Figure 9a. The change rates for the strengths and w/R are presented in the semi-logarithmic coordinates in Figure 9b.

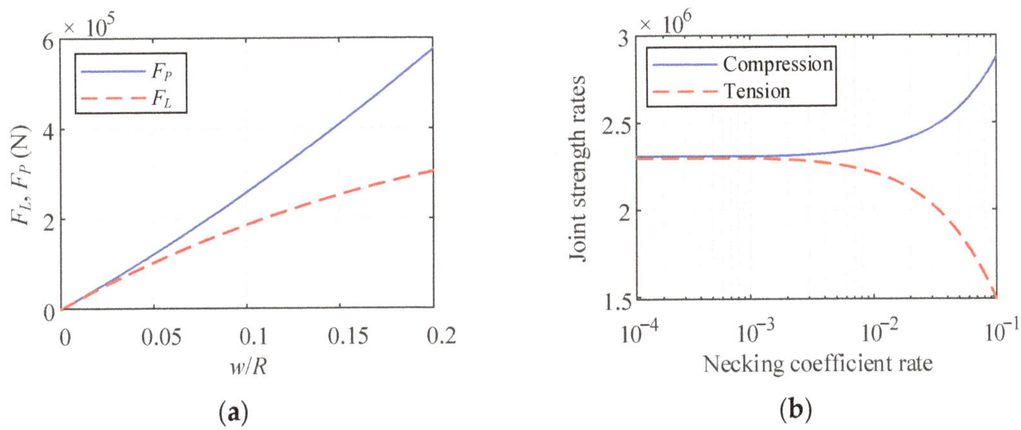

Figure 9. (**a**) Tension and compression joint strengths, F_L and F_p, as a function of the necking coefficient, w/R, and (**b**) change rates for the strengths and the necking coefficient, both at $x = 0$.

The following conclusions were drawn from Figure 9:

- Compression strength was proportional to the w/R.
- The correlation between strength and w/R was approximately linear for compression, and gradually decreased for tension.
- Trends in the rates of tension and compression strength along with w/R were opposite, and there was a threshold for the w/R rate, within which the strength increase rate was independent of the w/R rate.
- Compared to the crimped connector [31], a similar trend of F_L was observed.

Thickness, δ, is also an important parameter that can affect the joint strength. The strengths as a function of the ratio of δ/R are illustrated in Figure 10a. The change rates of the strengths and δ/R are presented in the semi-logarithmic coordinates in Figure 10b.

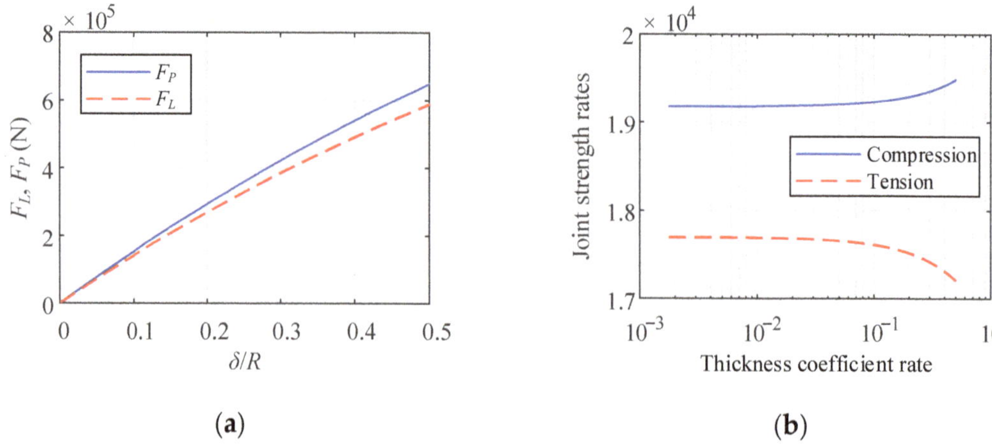

Figure 10. (**a**) Tension and compression joint strengths, F_L and F_p, as a function of the thickness coefficient, δ/R, and (**b**) change rates for the strengths and the thickness coefficient.

The following conclusions were drawn from Figure 10:

- The joint strengths increased in proportion to δ/R, with an approximate quadratic relationship.
- The trends of strengths with δ/R were similar to those with w/R. Within the thin-wall threshold [51], $\delta/R < 0.05$, the change rates of strengths tended to be constant; when the $\delta/R \geq 0.05$, the change rates of strengths were inversely proportional to the rate of δ/R.
- Comparing Figure 10 to Figure 9, the impacts of δ/R and w/R on the strengths were different in magnitude.
- Compared to the crimped connector [31], a similar trend of δ/R was observed within $\delta/R < 0.4$, and since then, the trend changed due to the differences in the structure.

The influence of the average half-cone angle, α, is also an important control variable during design. The strengths as a function of α are shown in the Figure 11a, while the change rates of the strengths and α are presented in semi-logarithmic coordinates in Figure 11b.

The following conclusions were drawn from Figure 11:

- Trends in both strengths followed a similar pattern as α increased, and α had a significant impact on the strengths within $\alpha < 1°$; when $\alpha > 2°$, α had a negligible impact on strength.
- When $\alpha < 1°$, there were obvious differences in the influence on strengths, and α affected the compression strength more than the tension.

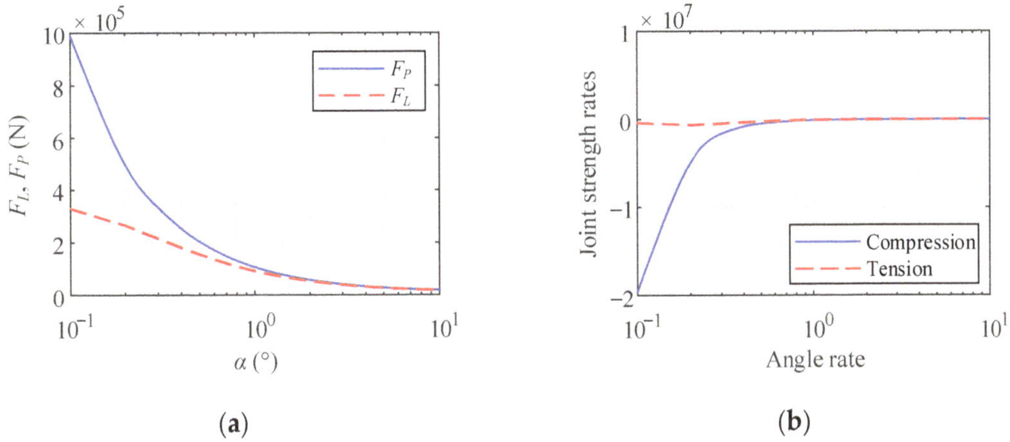

Figure 11. (**a**) Tension and compression joint strengths, F_L and F_p, as a function of the average half-cone angle, α, and (**b**) change rates for the strengths and α.

Apart from these parameters, the principles governing the relationship in each formula are self-evident; here, we focus on the sealing performance under overload conditions. During the axial overload, there was a continuous evolution between the sealing surfaces. Figure 12a,b show the evolution of the sealing interface during the overload by Von Mises stress and axial displacement fields, respectively. The result shows that there was a gradual separation of the interface, leading to a slide between surfaces.

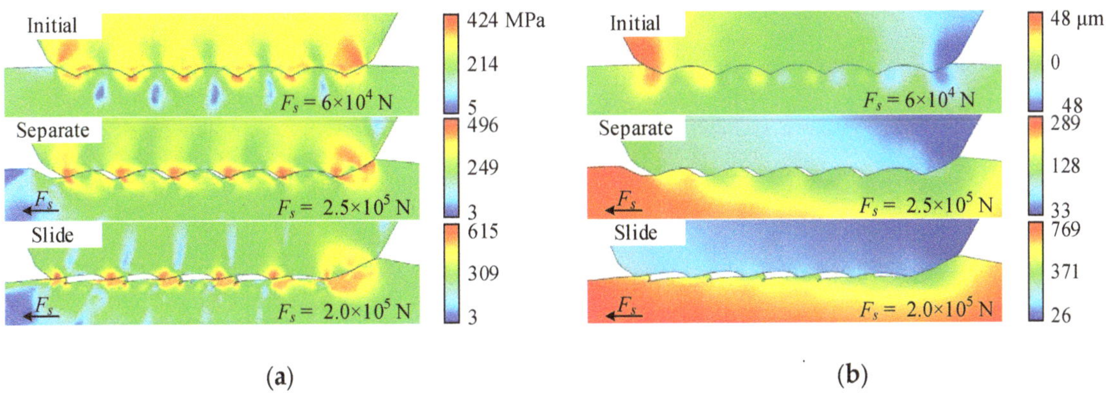

Figure 12. The evolution of the sealing interface during axial overload: (**a**) Von Mises stress field and (**b**) axial displacement field.

To demonstrate the evolution, six channel profiles at representative moments are presented, taken from the middle channel on the interface. Binary progression profiles are illustrated in Figure 13, with the white portion denoting the channel. The order of change is from labels a to f. The data below are the average of all channels along the seal.

According to Section 5, the channel profile in Figure 13 can be converted into the corresponding semi-axis curves, as shown in Figure 14. The height of the leakage channel, $h(x_r)$, increased then decreased, while the width of the leakage channel, $\zeta(x_r)$, gradually increased.

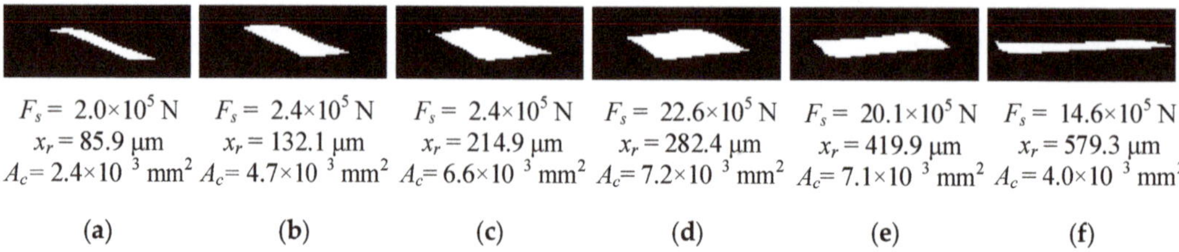

Figure 13. Binary progression profiles of the cross-section of the leakage channel. The white portion denotes the channel, and the order of change is from labels (**a**) to (**f**).

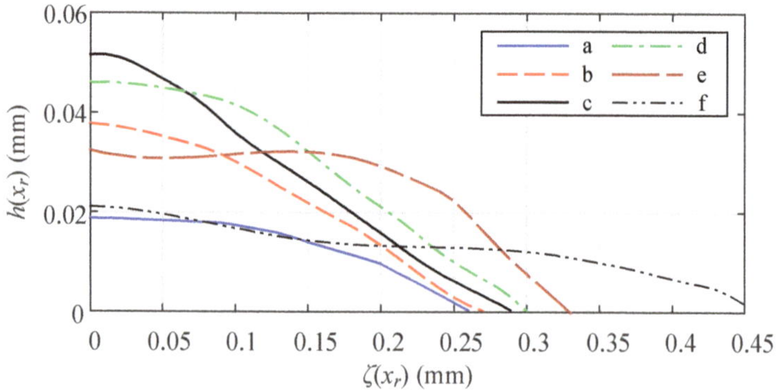

Figure 14. Semi-axis curves of the typical cross-sections.

Using the above procedure, the leakage ratio, Q_c, under overload conditions was determined by sampling from the FEA results.

Next, the relative displacement under overload conditions was analyzed. Figure 15 shows the joint strength, F_s, the axial displacement of the midpoint of the ridge sealing surface, x_R, and the change rate of the relative displacement, x_r', as a function of the axial displacement of the midpoint of the pipe sealing area, x_P.

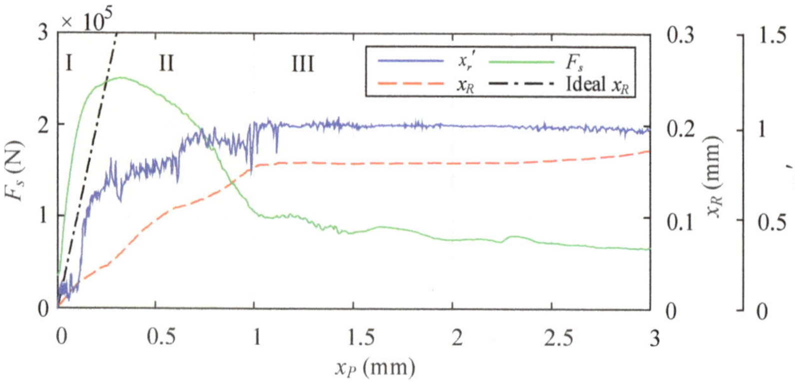

Figure 15. Joint strength, F_s, displacement, x_R, and change rate of the relative displacement, x_r', of the seal under the overload condition.

As shown in Figure 15, the development of the sealing interface can be classified into three stages: Follow (I), Separate (II), and Slide (III). The following conclusions were drawn:

- In the Follow stage (I), the ridge was pulled by the pipe and the two contact surfaces moved simultaneously. In this stage, F_{smax} was reached.

- In the Separate stage (II), x_r' was increasing but still less than one, with some plateau periods, indicating that the ridge was still following with the pipe. In this stage, the helical leakage channel occurred.
- In the Slide stage (III), x_r' was close to one, indicating that the ridge was completely sliding relative to the pipe. The surfaces in this stage would be severely deformed, ploughed [52], and worn.
- Besides, compared to the crimped connector [31] and the die-less connector [45], a similar trend of F_s was observed within $x_p < 1$; hence, it can be further analyzed by the force–strain curve [45] to identify the plastic deformation of the whole prototype and the beginning of the relative movement between the joint partners.

Based on the above analysis, the theoretical leakage ratio can be evaluated. The typical leakage ratios of air ($\eta = 1.77 \times 10^{-5}$ Pa·s), water ($\eta = 1.01 \times 10^{-3}$ Pa·s), and oil ($\eta = 1.50$ Pa·s), with inlet pressure, $p_i = 1$ MPa, and outlet pressure, $p_o = 101$ kPa, were calculated and presented in the semi-logarithmic coordinates in Figure 16. The corresponding F_s and x_r' are also displayed.

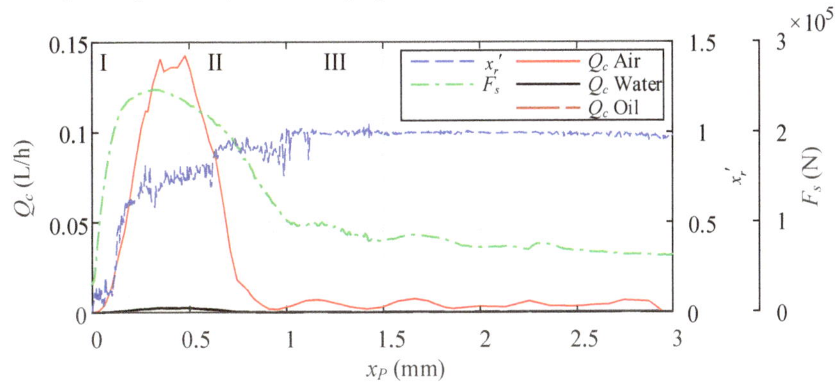

Figure 16. The leakage ratio, Q_c, and the corresponding F_s and x_r' of the seal during overload.

The following conclusions were drawn from Figure 16:

- The highest leakage ratio occurred in the Separate stage, while the Follow and Slide stages had relatively low leakage ratios.
- The maximum Q_c occurred after F_{smax}, indicating that the compression-type connector has an excellent overload ability.
- In the Slide stage, Q_c experienced minor fluctuations, in accordance with topography engagement. Although L_c was shortened, the change in the leakage ratio was low because $\bar{h}$ was reduced.

In addition, the FEA results showed that there was a material accumulation on the colliding side where the sealing edge meets the ridge, potentially impacting the leak.

6.2. Experimental Tests

Two types of overload tests were conducted: one was an axial shear test that uses specimens to observe the topography evolution under the axial shear load, and the other was a tension and compression test that uses full-scale prototypes to qualitatively evaluate the overload sealing performance.

6.2.1. Axial Shear Test

The evolution of the sealing interface under the shear load was assessed using a common approximation method for bolted joints, which suggests flattening the sealing surface from the cylinder into a plane [53]. A schematic diagram of the test setup is shown in Figure 17a. The test process was to apply an equivalent vertical sealing pressure, P', to achieve the sealing state, followed by an equivalent horizontal shear force, F_s', to induce the shear load at the interface and produce the relative displacement, x_r. To apply loads and displacement, an YZJ-50 direct-shear device with a maximum capacity of 5×10^5 N was used.

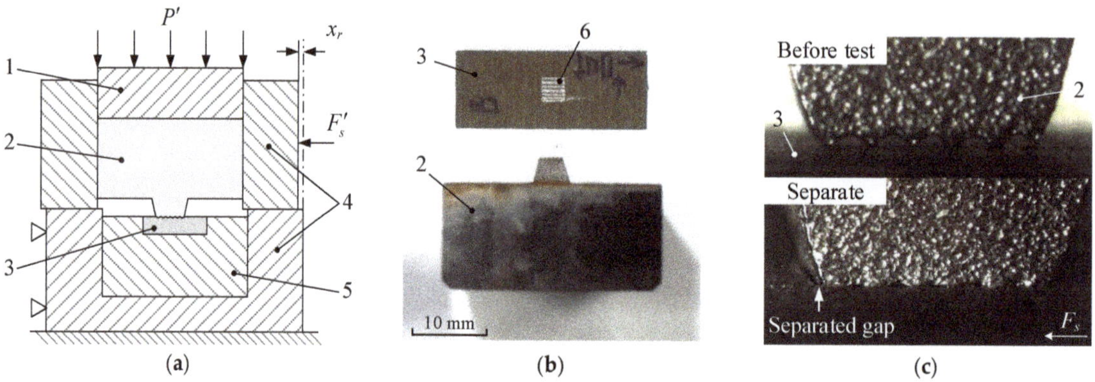

Figure 17. (**a**) The schematic diagram of the test setup, (**b**) the specimens, and (**c**) the local view of the sealing interface. The structure consists of: (1) vertical compression part, (2) ridge specimen, (3) pipe specimen, (4) container, (5) specimen bed, and (6) dent area.

Specimens are shown in Figure 17b, and the local view of the seal is shown in Figure 17c. The turning surface was simplified to an even-square area with straight streaks. To enhance the identification of deformation, the streaks were made more obvious than the actual turning surface. The thickness of the pipe specimen was designed to minimize the impact on the surface deformation. All specimens were designed for the easy replacement.

To analyze the deformation of the topography, an optical stereomicroscope was used, the Stereo Discovery V. 12. Additionally, to compare the surface waviness and roughness, a Bruker Profile GT-K optical profiler with a maximum lateral resolution of 3.8×10^{-4} mm was used. To avoid the edge effect, the sampling was conducted in the middle of indentation. The measurement length was 1 mm, and the lateral resolution was 1.5×10^{-3} mm. The typical results are shown in Figures 18–20, with arrows indicating the measuring location, direction, and length.

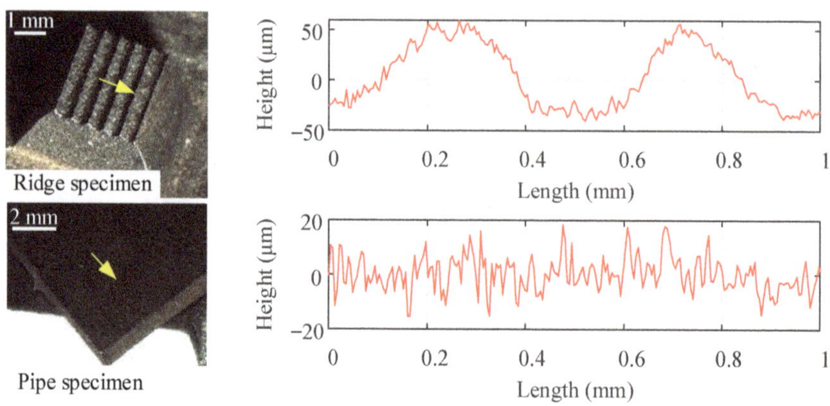

Figure 18. The measured surface topography under initial conditions.

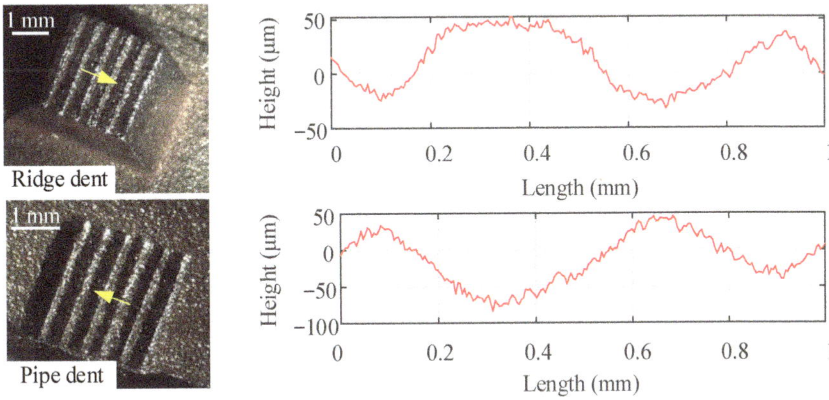

Figure 19. The measured surface topography after vertical compression.

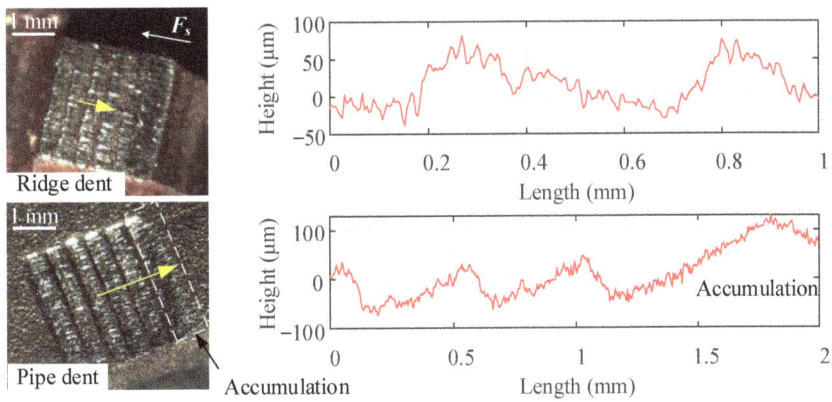

Figure 20. The measured surface topography after axial overload.

The initial condition of the specimens is shown in Figure 18, where we selected 1 mm as the evaluation length. For the pipe, the surface roughness, *Ra*, was grade 6.3 [54], and the total profile height, $Rt = 37.78$ mm; for the ridge, $Rt = 135.59$ mm, and *Ra* was grade 6.3 when sampled parallel to the streaks. For streaks, the mean height variation over all evaluation lengths was 2.64%, which was mainly due to the edge effect and the machining error.

For the vertical compacted surfaces shown in Figure 19, the streak peaks were slightly flattened. The surface roughness increased due to hardening, ranging from grades 6.3 to 12.5. For the waviness, the mean deviation between the test and the FEA results was 4.28% for the ridge and 4.62% for the pipe. The error was acceptable, and the deformation trend between the test and the FEA was consistent.

Figure 20 shows the topography after the overload by averaging multiple results. Both surfaces exhibited severe plastic deformation with a distinct bright contact area, suggesting the presence of the high-pressure zones. The amplitude of surface waviness further decreased, and there was a ploughing phenomenon [52] along the overload direction, resulting in an uneven increase in roughness. For the waviness, the mean deviation between the test and the FEA results was 5.93% for the ridge and 6.85% for the pipe. Accumulation was also observed on the colliding side of the specimens. The deformation trends of the test and the FEA were consistent. The area suffered severe plastic deformation and ploughing, and the width of the contact increased.

In order to examine the sealing pressure during the overload, the mean sealing pressure at each streak as a function of pipe displacement is shown in Figure 21 based on the FEA results.

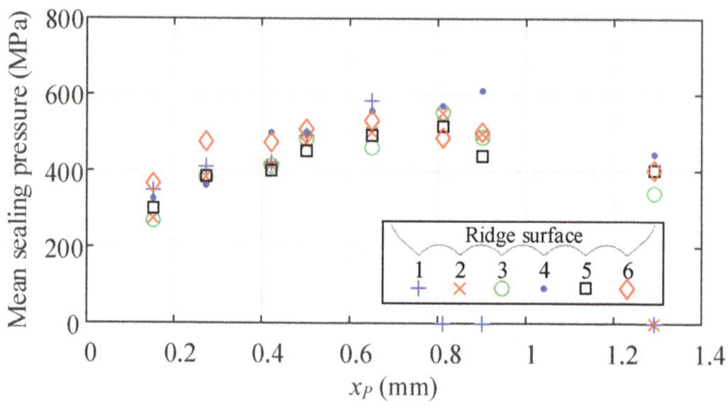

Figure 21. The mean sealing pressure at each streak as a function of pipe displacement, x_P.

The following conclusions were drawn from Figure 21:

- Mean sealing pressures rose and then fell within a narrow range, with the reason being that the contact area decreased due to the x_r, and then increased due to wear and deformation.
- During overload, the pressures at streaks 1 and 2 approached zero, indicating that separation occurred. Conversely, high pressure was consistently found on streaks 5 and 6, where the accumulation also occurred.

It can be inferred that the metal accumulation can cause a blockage at the end of the helical channel, allowing the connector to remain sealed, and this needs to be verified by a prototype test.

6.2.2. Tension and Compression Overload Test

Based on ISO 21329:2004 [55], full-scale prototypes of 3 to 8 inches were tested to assess the sealing performance under overload conditions. The sealing mediums used were water and air. Due to safety concerns, air tests were limited to compression tests, with restrictions on the air pressure and pipe diameter. Leakage detection was performed by the pressure drop and bubbling test (1×10^{-3} L/h), meeting the standard requirement (between 5.34×10^{-2} and 0.4 L/h). All tests were conducted with the room temperature of 20 °C. The schematic diagram and experimental setup are shown in Figure 22.

The typical test results for these prototypes are presented in Tables 2–4. Samples from the prototypes were cut and observed, and the typical surface roughness and waviness are shown in Figure 23.

Table 2. Typical results of the tension overload test of the water medium.

Type (Inch)	Nominal Diameter (mm)	Nominal Thickness (mm)	Medium Pressure (MPa)	Theoretical Tension Joint Strength (N)			Experimental Mean Strength (N)	Relative Error	Leakage (≥1 mL/h)	
			Water	Pipe	Seal	Total			Initial	Final
3	89	2	15	6.29×10^2	1.04×10^5	1.05×10^5	1.11×10^5	6.02%	×	×
4	114	3	7	9.30×10^3	2.01×10^5	2.10×10^5	2.18×10^5	3.82%	×	×
6	168	3	6	1.84×10^4	2.96×10^5	3.15×10^5	3.31×10^5	5.23%	×	×
8	219	4	4.5	6.59×10^4	5.16×10^5	5.82×10^5	6.25×10^5	7.38%	×	×

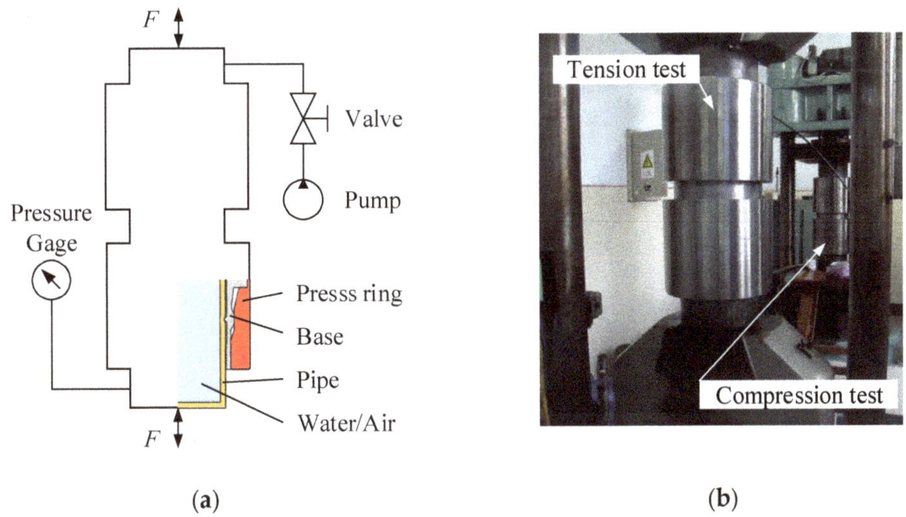

Figure 22. (**a**) Schematic diagram and (**b**) experimental setup of the tension and compression overload tests.

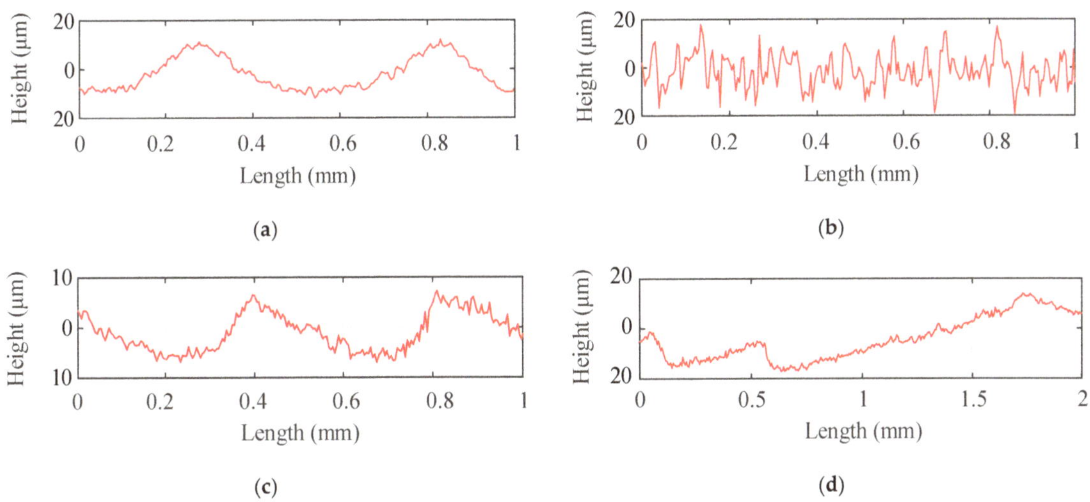

Figure 23. Turning topography of the ridge: (**a**) before the test and (**b**) after the test, (**c**) the pipe roughness before the test, and (**d**) the accumulation after the test.

Table 3. Typical results of the compression overload test of the water medium.

Type (Inch)	Nominal Diameter (mm)	Nominal Thickness (mm)	Medium Pressure (MPa)	Theoretical Compression Joint Strength (N)			Experimental Mean Strength (N)	Relative Error	Leakage (≥1 mL/h)	
			Water	Pipe	Seal	Total			Initial	Final
3	89	2	15	4.72×10^2	1.04×10^5	1.05×10^5	1.11×10^5	5.22%	×	×
4	114	3	7	9.40×10^3	2.01×10^5	2.10×10^5	2.19×10^5	4.24%	×	×
6	168	3	6	1.86×10^4	2.96×10^5	3.15×10^5	3.27×10^5	3.88%	×	×
8	219	4	4.5	6.73×10^4	5.16×10^5	5.84×10^5	6.22×10^5	6.59%	×	×

Table 4. Typical results of the compression overload test of the air medium.

Type (Inch)	Nominal Diameter (mm)	Nominal Thickness (mm)	Medium Pressure (MPa)	Theoretical Compression Joint Strength (N)			Experimental Mean Strength (N)	Relative Error	Leakage (≥1 mL/h)	
			Air	Pipe	Seal	Total			Initial	Final
3	89	2	0.5	8.34×10^3	1.04×10^5	1.12×10^5	1.17×10^5	4.09%	×	×
4	114	3	0.5	1.62×10^4	2.01×10^5	2.17×10^5	2.28×10^5	5.14%	×	×

Comparing the results, both the joint strength and deformation trends of the sealing surface well-agreed with the analysis. The error was caused by the simplification of the methods, allowable geometrical and installation errors, etc. No leakage (≥1 mL/h) was observed under overload. The test results showed that the compression-type connector is capable of maintaining a sealed state even under maximum axial overload, setting it apart from other connection types, such as pipe flanges.

7. Conclusions

This paper presented two semi-analytical methods for analyzing the joint strength and the overload sealing performance of the subsea compression-type connector. Specimen and prototype tests were carried out. The following conclusions were drawn:

1. A semi-analytical method for predicting the joint strength of the subsea pipeline compression connector was proposed. On the one hand, the axial strengths of the connected pipe under the axial tension and compression conditions were separately deduced based on membrane theory, taking into account the effects of hardening and bending stress. On the other hand, the joint strength of the radial seal was analyzed based on the FEA method. Using the model, the effects of the critical parameters on the joint strength were analyzed.
2. A semi-analytical method of the leakage ratio under axial overload was derived based on the 2D axisymmetric FEA model combined with the Reynolds equation of the laminar flow. Using the model, the overload sealing performance was obtained.
3. The evolution of the sealing interface during the axial overload was divided into three stages: Follow, Separate, and Slide. The laws governing the joint strength, relative displacement, and the cross-section of the leakage channel, and their impacts on the leakage rate, were analyzed. The maximum leakage ratio occurred after reaching the maximum overload. Besides, the accumulation of the metal can cause an obstruction at the terminus of the helical leakage channel.
4. The axial shear test and axial overload test were carried out. Using the shear test, the characteristics of the surface roughness and the deformation of the interface under overload conditions were obtained. With the tension and compression overload test, the overload sealing performance of the prototype was tested. Through the experiments, the feasibility of the proposed methods was verified.
5. The study showed that the internally turned sealing surface and the deflection of the connected pipe can improve the joint strength of the compression-type connector. Moreover, a distinctive feature of this connection type is the ability to remain sealed even under the maximum axial load.

The proposed methods allowed the prediction and identification of the joint strength and overload sealing condition of the compression-type connector and provided a better understanding of the radial metal seal. The results can help optimize the FEA design process and improve the test efficiency in engineering applications.

The current study is only focused on the unidirectional load, and it is necessary to further determine the joint strength and sealing performance against impacts, alternating loads, wear, etc. In addition, a more precise and comparable test method and device are needed to quantify the leakage ratio.

Author Contributions: Conceptualization, Z.L., G.W. and L.W.; methodology, Z.L.; software, Z.L.; validation, Z.L. and L.W.; formal analysis, Z.L.; investigation, G.W.; resources, L.W. and X.W.; data curation, G.W. and S.Y.; writing—original draft preparation, Z.L.; writing—review and editing, S.Y.; visualization, Z.L.; supervision, P.J. and F.Y.; project administration, G.W.; funding acquisition, L.W. All authors have read and agreed to the published version of the manuscript.

Funding: This research was funded by the National Natural Science Foundation of China, grant numbers 51779064 and 52001089, and the China Postdoctoral Science Foundation, grant number 2020M670889.

Institutional Review Board Statement: Not applicable.

Informed Consent Statement: Not applicable.

Data Availability Statement: The data used to support the findings of this study are included within the article.

Conflicts of Interest: The authors declare no conflict of interest.

Nomenclature

A_c	Cross-sectional area of the leakage channel (mm^2)
b	Axial length of the streak (mm)
C	Integration constant
D	Flexural rigidity of the pipe (N·mm)
E	Young's modulus (MPa)
F	Total joint strength (N)
F_A	Joint strength of the anchor (N)
F_L, F_P	Joint strength of the pipe under tension and compression overloads, respectively (N)
F_s	Joint strength of the seal (N)
h	Radial height of the leakage channel (mm)
$\overline{h}$	Height of the hypothetical rectangular channel (mm)
K_1	Hardening modulus (MPa)
K_2	Defection coefficient
K_3	Tresca criterion coefficient
L_c	Total length of the leakage channel (mm)
M	Bending moment (N·mm)
n	Number of the turns of the streaks
p_i, p_o	Inlet and outlet pressures of the leakage channel (MPa)
P	Sealing pressure (MPa)
Q_c	Leakage ratio (L/h)
R	Radius of mid-surface of the pipe in the deflected section (mm)
R_a, R_b	Radius of mid-surface of the pipe before and after deflection, respectively (mm)
R_{r1}, R_{r2}	Radius of bottom of the dent and pitch radius of the streaks, respectively (mm)
Ra	Grade of the surface roughness (mm)
Rt	Total profile height of the surface roughness (mm)
w	Deflection of the pipe (mm)
x_r	Relative displacement between the sealing surfaces (mm)
x_r'	Change rate of the relative displacement
x_P	Axial displacement of the midpoint of the pipe sealing area (mm)
x_R	Axial displacement of the midpoint of the ridge surface (mm)
α	Average half-cone angle of the deflection (°)
γ, θ, φ	Rotations of cylindrical coordinates (°)
δ	Thickness of the pipe (mm)
ε^T	Accumulated plastic strain
μ	Friction coefficient
ρ	Radius of curvature (mm)
σ^T	True stress (MPa)
σ_L, σ_P	Axial stresses considering hardening under tension and compression overload, respectively (MPa)

σ_{LM}, σ_{PM}	Axial stresses considering both hardening and bending, under tension and compression overload, respectively (MPa)
σ_M	Secondary stress caused by bending (MPa)
σ_n	Normal stress (MPa)
σ_s	Yield stress of the pipe (MPa)
σ_θ	Circumferential stress (MPa)
τ_p	Allowable shear stress (MPa)
υ	Poisson's ratio
ψ	Half-cone angle of the pipe (°)
ζ	Axial width of the leakage channel (mm)

References

1. Chirkov, Y.A.; Kushnarenko, V.M.; Repyakh, V.S.; Chirkov, E.Y. Analysis of the Causes of Failure of a Pipeline Welded Joint. *Met. Sci. Heat Treat.* **2018**, *59*, 669–672. [CrossRef]
2. Yun, F.; Liu, D.; Xu, X.; Jiao, K.; Hao, X.; Wang, L.; Yan, Z.; Jia, P.; Wang, X.; Liang, B. Thermal–Structural Coupling Analysis of Subsea Connector Sealing Contact. *Appl. Sci.* **2022**, *12*, 3194. [CrossRef]
3. Geoffroy, S.; Prat, M. On the Leak through a Spiral-Groove Metallic Static Ring Gasket. *J. Fluids Eng.* **2004**, *126*, 48–54. [CrossRef]
4. Robbe-Valloire, F.; Prat, M. A Model for Face-Turned Surface Microgeometry. Application to the Analysis of Metallic Static Seals. *Wear* **2008**, *264*, 980–989. [CrossRef]
5. Okada, H.; Itoh, T.; Suga, T. The Influence of Surface Profiles on Leakage in Room Temperature Seal-Bonding. *Sens. Actuators A Phys.* **2008**, *144*, 124–129. [CrossRef]
6. Lebeck, A.O. Hydrodynamic Lubrication in Wawy Contacting Face Seals-a Two Dimensional Iodei. *J. Tribol.* **1981**, *103*, 578–586. [CrossRef]
7. Matsuzaki, Y.; Kazamaki, T. Effect of Surface Roughness on Compressive Stress of Static Seals. *JSME Int. J. Ser. III* **1988**, *31*, 99–106. [CrossRef]
8. Yanagisawa, T.; Sanada, M.; Koga, T.; Hirabayashi, H. The Influence of Designing Factors on the Sealing Performance of C-Seal. In *Proceedings of the SAE Technical Papers*; SAE International: Warrendale, PA, USA, 1991.
9. Nakamura, T.; Funabashi, K. Effects of Directional Properties of Roughness and Tangential Force on Pressure Flow between Contacting Surfaces. *Lubr. Sci.* **1991**, *4*, 13–23. [CrossRef]
10. Polycarpou, A.A.; Etsion, I. A Model for the Static Sealing Performance of Compliant Metallic Gas Seals Including Surface Roughness and Rarefaction Effects. *Tribol. Trans.* **2000**, *43*, 237–244. [CrossRef]
11. Arghavani, J.; Derenne, M.; Marchand, L. Effect of Surface Characteristics on Compressive Stress and Leakage Rate in Gasketed Flanged Joints. *Int. J. Adv. Manuf. Technol.* **2003**, *21*, 713–732. [CrossRef]
12. Nitta, I.; Matsuzaki, Y.; Ito, Y. Observation of Real Contact Area at Gasket Surfaces Using the Thin PC Film. *Trans. Jpn. Soc. Mech. Eng. Ser. C* **2005**, *71*, 265–271. [CrossRef]
13. Nitta, I.; Matsuzaki, Y. Experimental Study of the Performance of Static Seals Based on Measurements of Real Contact Area Using Thin Polycarbonate Films. *J. Tribol.* **2010**, *132*, 022202. [CrossRef]
14. Ledoux, Y.; Lasseux, D.; Favreliere, H.; Samper, S.; Grandjean, J. On the Dependence of Static Flat Seal Efficiency to Surface Defects. *Int. J. Press. Vessel. Pip.* **2011**, *88*, 518–529. [CrossRef]
15. Bourniquel, J.; Lasseux, D.; Rit, J.F. Prediction and Measurement of Sealing Properties of Joints Between Wavy Metal Surfaces. *J. Press. Vessel. Technol.* **2018**, *140*, 051203. [CrossRef]
16. Beheshti, A.; Khonsari, M.M. Asperity Micro-Contact Models as Applied to the Deformation of Rough Line Contact. *Tribol. Int.* **2012**, *52*, 61–74. [CrossRef]
17. Jackson, R.L.; Green, I. On the Modeling of Elastic Contact between Rough Surfaces. *Tribol. Trans.* **2011**, *54*, 300–314. [CrossRef]
18. Lorenz, B.; Persson, B.N.J. On the Dependence of the Leak Rate of Seals on the Skewness of the Surface Height Probability Distribution. *Europhys. Lett.* **2010**, *90*, 38002. [CrossRef]
19. Zhang, F.; Liu, J.; Ding, X.; Yang, Z. An Approach to Calculate Leak Channels and Leak Rates between Metallic Sealing Surfaces. *J. Tribol.* **2017**, *139*, 011708. [CrossRef]
20. Marie, C.; Lasseux, D.; Zahouani, H.; Sainsot, P. An Integrated Approach to Characterize Liquid Leakage through Metal Contact Seal. *Eur. J. Mech. Environ. Eng.* **2003**, *48*, 81–86.
21. Liao, Y.; Stephenson, D.A.; Ni, J. Multiple-Scale Wavelet Decomposition, 3D Surface Feature Exaction and Applications. *J. Manuf. Sci. Eng.* **2012**, *134*, 011005. [CrossRef]
22. Deltombe, R.; Bigerelle, M.; Jourani, A. Analysis of the Effects of Different Machining Processes on Sealing Using Multiscale Topography. *Surf. Topogr. Metrol. Prop.* **2016**, *4*, 015003. [CrossRef]
23. Yan, Y.; Fan, Y. Influence of Fluid on Seal and Assembly of Pipeline Fittings Based on the Multiscale Finite Element Model. *Complexity* **2020**, *2020*, 4960734. [CrossRef]
24. So, H.; Liu, D.C. An Elastic-Plastic Model for the Contact of Anisotropic Rough Surfaces. *Wear* **1991**, *146*, 201–218. [CrossRef]

25. Shao, Y.; Du, S.; Xi, L. 3D Machined Surface Topography Forecasting with Space-Time Multioutput Support Vector Regression Using High Definition Metrology. In Proceedings of the ASME Design Engineering Technical Conference, Cleveland, OH, USA, 6–9 August 2017; Volume 1.
26. Tang, L.; He, W.; Zhu, X.; Zhou, Y. Sealing Performance Analysis of an End Fitting for Marine Unbonded Flexible Pipes Based on Hydraulic-Thermal Finite Element Modeling. *Energies* **2019**, *12*, 2198. [CrossRef]
27. Ernens, D.; Pérez-Ràfols, F.; van Hoecke, D.; Roijmans, R.F.H.; van Riet, E.J.; Vande Voorde, J.B.E.; Almqvist, A.; de Rooij, M.B.; Roggeband, S.M.; van Haaften, W.M.; et al. On the Sealability of Metal-to-Metal Seals with Application to Premium Casing and Tubing Connections. *SPE Drill. Complet.* **2019**, *34*, 382–396. [CrossRef]
28. Wei, Z.; Wang, L.; Guan, Y.; Yao, S.; Li, S. Static Metal Sealing Mechanism of a Subsea Pipeline Mechanical Connector. *Adv. Mech. Eng.* **2016**, *8*, 1–16. [CrossRef]
29. Wang, L.Q.; Wei, Z.L.; Yao, S.M.; Guan, Y.; Li, S.K. Sealing Performance and Optimization of a Subsea Pipeline Mechanical Connector. *Chin. J. Mech. Eng. (Eng. Ed.)* **2018**, *31*, 18. [CrossRef]
30. Weddeling, C.; Demir, O.K.; Haupt, P.; Tekkaya, A.E. Analytical Methodology for the Process Design of Electromagnetic Crimping. *J. Mater. Process. Technol.* **2015**, *222*, 163–180. [CrossRef]
31. Weddeling, C.; Walter, V.; Haupt, P.; Tekkaya, A.E.; Schulze, V.; Weidenmann, K.A. Joining Zone Design for Electromagnetically Crimped Connections. *J. Mater. Process. Technol.* **2015**, *225*, 240–261. [CrossRef]
32. Henriksen, J.; Hansen, M.R.; Thrane, F.C. Axial Load Capacity of Cold Formed Pipe Flange Connection. *J. Press. Vessel. Technol.* **2017**, *139*, 051201. [CrossRef]
33. Agrawal, A.K.; Narayanan, R.G. Experimental and Numerical Studies on Joining Steel Tubes by End Forming. *J. Constr. Steel Res.* **2020**, *167*, 105792. [CrossRef]
34. Onyegiri, I.; Kashtalyan, M. Threaded Connectors for Sandwich Pipes—Part 1: Parametric & Comparative Studies. *Int. J. Press. Vessel. Pip.* **2018**, *168*, 117–124. [CrossRef]
35. Quispe, J.L.P.; Pasqualino, I.P.; Estefen, S.F.; de Souza, M.I.L. Structural Behavior of Threaded Connections for Sandwich Pipes under Make-up Torque, External Pressure, and Axial Load. *Int. J. Press. Vessel. Pip.* **2020**, *186*, 104156. [CrossRef]
36. Liu, Z.; Zhang, L.; Wang, F.; Li, S.; Wang, P.; Cai, M.; Han, L.; Ma, Y.; Ma, Z.; Yan, B. Study on Optimization Design of Permanent Packer Slip Structure. *J. Fail. Anal. Prev.* **2020**, *21*, 50–60. [CrossRef]
37. Li, Z.; Wang, G.; Yao, S.; Yun, F.; Jia, P.; Li, C.; Wang, L. A Semi-Analytical Method for the Sealing Performance Prediction of Subsea Pipeline Compression Connector. *J. Mar. Sci. Eng.* **2023**, *11*, 854. [CrossRef]
38. Wu, L.; Guo, X.; Chen, H.; Liu, J.; Su, Y. Calculation Method for the Axial Load-Bearing Capacity of Steel Pipe-to-Sleeve Grouted Connections. *Constr. Build. Mater.* **2022**, *314*, 125621. [CrossRef]
39. *ASTM F1387*; Standard Specification for Performance of Piping and Tubing Mechanically Attached Fittings. ASTM International: West Conshohocken, PA, USA, 2012.
40. *ISO 8434-1*; Metallic Tube Connections For Fluid Power and General Use Part 1:24° Compression Fittings. International Organization for Standardization: Geneva, Switzerland, 2018.
41. *ASTM E8/E8M*; Standard Test Methods for Tension Testing of Metallic Materials. American Society of Mechanical Engineers: New York, NY, USA, 2021.
42. Rojíček, J.; Čermák, M.; Halama, R.; Paška, Z.; Vaško, M. Material Model Identification from Set of Experiments and Validation by DIC. *Math. Comput. Simul.* **2021**, *189*, 339–367. [CrossRef]
43. Roostaei, A.A.; Jahed, H. Fundamentals of Cyclic Plasticity Models. In *Cyclic Plasticity of Metals: Modeling Fundamentals and Applications*; Elsevier: Amsterdam, The Netherlands, 2022; pp. 23–51.
44. Yu, T.; Xue, P. Yield Criteria. In *Introduction to Engineering Plasticity*; Elsevier: Amsterdam, The Netherlands, 2022; pp. 67–87.
45. Weddeling, C.; Gies, S.; Khalifa, N.B.; Tekkaya, A.E. Analytical Methodology for the Process and Joint Design of Form-Fit Joining by Die-Less Hydroforming. In Proceedings of the 9th ASME International Manufacturing Science and Engineering Conference (MSEC2014), Detroit, MI, USA, 9–13 June 2014; Web Portal ASME (American Society of Mechanical Engineers): New York, NY, USA, 2014.
46. *ASME B16.5*; Pipe Flanges and Flanged Fittings. American Society of Mechanical Engineers: New York, NY, USA, 2020.
47. Duan, W.; Joshi, S. Failure Analysis of Threaded Connections in Large-Scale Steel Tie Rods. *Eng. Fail. Anal.* **2011**, *18*, 2008–2018. [CrossRef]
48. *Abaqus*, version 6.14. Software for Technical Simulation. Dassault Systemes: Vélizy-Villacoublay, France, 2014.
49. Georgescu, C.; Cristea, G.C.; Dima, C.; Deleanu, L. Evaluating Lubricating Capacity of Vegetal Oils Using Abbott-Firestone Curve. *IOP Conf. Ser. Mater. Sci. Eng.* **2017**, *174*, 12057. [CrossRef]
50. Prat, M.; Plouraboué, F.; Letalleur, N. Averaged Reynolds Equation for Flows between Rough Surfaces in Sliding Motion. *Transp. Porous Media* **2002**, *48*, 291–313. [CrossRef]
51. Kumar, A.; Lal Das, S.; Wahi, P. Instabilities of Thin Circular Cylindrical Shells under Radial Loading. *Int. J. Mech. Sci.* **2015**, *104*, 174–189. [CrossRef]
52. Wojciechowski, S.; Krajewska-Śpiewak, J.; Maruda, R.W.; Krolczyk, G.M.; Nieslony, P.; Wieczorowski, M.; Gawlik, J. Study on Ploughing Phenomena in Tool Flank Face—Workpiece Interface Including Tool Wear Effect during Ball-End Milling. *Tribol. Int.* **2023**, *181*, 108313. [CrossRef]

53. Hoelz, K.; Kleinhans, L.; Matthiesen, S. Wood Screw Design: Influence of Thread Parameters on the Withdrawal Capacity. *Eur. J. Wood Wood Prod.* **2021**, *79*, 773–784. [CrossRef]
54. *ISO 468*; Surface Roughness—Parameters, Their Values and General Rules for Specifying Requirements. International Organization for Standardization: Geneva, Switzerland, 1982.
55. *ISO 21329:2004*; Petroleum And Natural Gas Industries—Pipeline Transportation Systems—Test Procedures for Mechanical Connectors. International Organization for Standardization: Geneva, Switzerland, 2004.

Journal of
Marine Science and Engineering

MDPI

Article

Analysis of the Reliability of Subsea Clamp Connector Based on Multiple Response Surface Methodology

Weizheng An [1,2], Yi Wang [1,*], Baijiang Chen [1], Menglan Duan [1] and Xiyang Zhang [1]

1 College of Safety and Ocean Engineering, China University of Petroleum (Beijing), Beijing 102249, China
2 CNOOC Research Institute, Beijing 100028, China
* Correspondence: wangyizyn@sina.com

Abstract: The subsea clamp connector is susceptible to sealing and locking failures over its lifetime in harsh marine environments and complex loading conditions, posing a serious challenge to the safe development of subsea oil and gas fields. Accurately predicting the reliability of the subsea clamp connector under realistic and complex operating conditions is therefore an important guarantee of its safe operation. Considering the main structural characteristic parameters of the subsea clamp connector, this paper conducts a reliability analysis using finite element numerical simulation combined with multiple response surface methodology (MRSM), based on the seal failure and yield failure criteria. The applicability has been verified through the application of subsea clamp connector in the Bohai Sea. The results show that the failure probability of the system is mainly affected by the radius of the seal, the contact angle of the upper and lower flanges and internal pressure. Considering the influence of various factors, the reliability of the connector was calculated to be 98.73%, and the reliability was verified by the sealing performance test. This paper provides a practical method for the reliability analysis of the subsea clamp connector structure under the comprehensive consideration of multiple factors, and provides a new technology to ensure the safe operation of subsea oil and gas fields.

Keywords: subsea clamp connector; reliability; failure criteria; multiple response surface methodology

Citation: An, W.; Wang, Y.; Chen, B.; Duan, M.; Zhang, X. Analysis of the Reliability of Subsea Clamp Connector Based on Multiple Response Surface Methodology. *J. Mar. Sci. Eng.* **2023**, *11*, 1378. https://doi.org/10.3390/jmse11071378

Academic Editor: Erkan Oterkus

Received: 30 May 2023
Revised: 27 June 2023
Accepted: 3 July 2023
Published: 6 July 2023

1. Introduction

With the rapid development of the world's industry, the demand for oil and gas resources is constantly increasing, and the exploration and exploitation of offshore oil and gas has become the main development direction of the petroleum industry [1–3]. In the development of offshore oil and gas fields, subsea connectors are important connecting components for subsea production system, such as the Christmas tree and manifold [4]. Subsea clamp connectors (as shown in Figure 1) are widely used due to the advantages of simple structure, rapid connection, and wide applicability, and their reliability directly affects the safety of subsea oil and gas field development.

At present, research on subsea connectors is mainly focused on the theoretical relationship between the load transfer of the contact surface, the locking force and the pre-tightening force, the optimization of the sealing ring structure, and related experiments [5–8]. In terms of sealing and leakage research, Yun et al. [9] established the contact mechanics model of the metal lens-type sealing gasket at the macroscopic scale based on the Hertz theory. They also analyzed the contact characteristics between the metal lens seal and the flange structure at the microscopic scale by equivalently replacing the peak cutting coefficient of a one-dimensional sinusoidal wave. Li et al. [10] analyzed the influence of pre-tightening force, contact width, preloading compression, and operating pressure on contact stress through theoretical calculations and finite element methods. Liu et al. [11] derived a calculation expression for the size ratio of a spherical sealing groove based on the spherical structure of the connector's sealing ring. They determined the width and depth of the groove, and used finite element methods to study the influence of different pressures and compression ratios on the sealing performance of O-ring seals from various aspects, such as von mises stress, contact pressure, and contact width of different contact surfaces.

Figure 1. Subsea clamp connectors (https://www.fogt.com/subsea-connector (accessed on 3 May 2023)).

In the field of subsea connector reliability research, the main approaches are fault tree analysis and dynamic Bayesian network analysis. Wan et al. [12] analyzed the failure modes and causes of the subsea connector drive ring, and established a fault tree model. Zhang et al. [13] obtained the failure probability magnitude of subsea connector installation through fuzzy quantitative analysis of the fault tree. Chen et al. [14] proposed a fault diagnosis method based on a three-layer dynamic Bayesian network to diagnose faults and predict failures for the mechanical structure of vertical collet subsea connectors. Torfinn et al. [15] applied structural reliability analysis (SRA) to wellhead connectors, fatigue induced by overload and plastic collapse are two failure modes to estimate the fatigue failure probability. Bhardwaj et al. [16] used the first-order second-moment method and the Monte Carlo method to evaluate the structural reliability of the pipeline system in deep water under high internal pressure and high temperature. Pang et al. [17] proposed a fuzzy Markov method that integrates risk, reliability, availability, and uncertainty analysis based on fault tree, fuzzy comprehensive evaluation, and Markov methods. The method pre-processes input fault data using fault tree and fuzzy theory to improve the reliability of the input fault data. Wang et al. [18] proposed a reliability analysis method based on a dynamic Bayesian network using the Monte Carlo simulation to evaluate the failure probability of subsea wellhead connectors during their service life. Tsai et al. [19] considered wind speed probability, fatigue strength of mast arm base, miner summation, and other uncertain factors to perform the Monte Carlo simulation and generate failure probability curve. Simon et al. [20] described a verification method for subsea pipeline maintenance to confirm that the product design meets DNV standards.

The MRSM is a reliability analysis method that can simultaneously consider two or more failure factors, and it is commonly used in reliability analysis under multi-factor coupling by establishing multiple response surface functional functions. Das et al. [21] proposed an improved response surface method and applied it to the reliability analysis of stiffened plate structures. The response surface functions are formed cumulatively so that second-order effects in the response surface are properly taken into account using acceptable computational work in the evaluation of state functions. Somdatta et al. [22] investigates moving least squares (MLSM) to construct response surface functions. The advantage of MLSM over the least square method (LSM) is that it reduces the number of iterations required to obtain the updated central point of the experimental design (DOE). The final response surface was constructed for effective structural reliability analysis. Zhang et al. [23]

conducted a reliability study of aircraft engine blades using the multiple response surface method by establishing four response surface models for overall deformation, radial deformation, stress, and temperature of the blade. Lu [24] proposed a structural reliability analysis method to study the dynamic reliability of turbine disk structures. Liu et al. [25] used finite element numerical simulation combined with the multiple response surface method and the Monte Carlo method to analyze the reliability of tunnel structural systems. Zhai [26] calculated the reliability of the fan connectors according to the MRSM method.

This paper proposed an MRSM by combining the failure criterion of yield and minimum sealing ratio pressure to analyze the reliability of the subsea clamp connector under multi-factors. The proposed MRSM provides a feasible approach for the reliability prediction of subsea clamp connectors.

2. Structure Characteristics of Subsea Connection System

The subsea clamp connectors are mainly used to connect and fix a variety of subsea equipment, pipelines, and structural parts, including oil and gas pipelines, subsea oil wells, isolators, oil-water separators, Christmas tree, etc. The subsea clamp connector studied in the paper is a kind of structure to ensure the smooth, simple, and fast connection between the mobile end flange and the fixed end flange. It has the characteristics of simple operation, low requirements for installation environment, and reduces the influence of various uncontrollable factors. The subsea clamp connector is a rotary body axisymmetric structure, mainly including the upper flange, lower flange, clamp jaws, metal seals, and other components (as shown in Figure 2). The three-lobe clamp form makes the force of the flange more uniform. When the axial external force is applied, the overall structure is stressed evenly, and the fixed constraint and load are symmetrical along the central axis. It should maintain structural locking and sealing performance under axial force, bending moment and pressure etc.

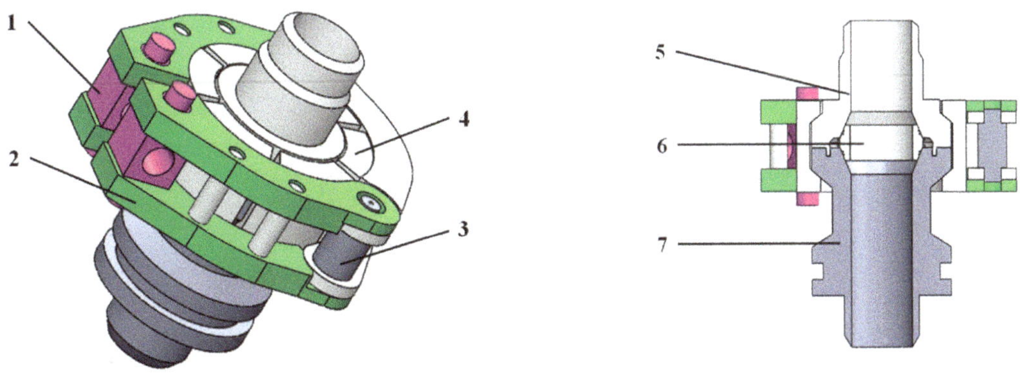

Figure 2. Schematic of the overall structure of the subsea clamp connector. 1—Locking bolt hole; 2—Three-flap clamp piece; 3—Axial support plate; 4—Clamp jaws; 5—Upper flange; 6—Seal ring; 7—Bottom flange.

The working condition of subsea clamp connection is very complicated. On the one hand, the sealing structure of the subsea clamp connector should play a sealing role in the internal oil and gas medium. On the other hand, the external seawater pressure caused by the water depth of the pipeline should also play a sealing role. Therefore, the lens seal and O-type composite seal (as shown in Figure 3) are commonly used in subsea clamp connectors. The metal seal is the primary seal for sealing oil and gas media, and the O-type is the secondary seal for sealing external seawater. The combination of these two sealing methods can effectively improve the reliability and service life of the seal.

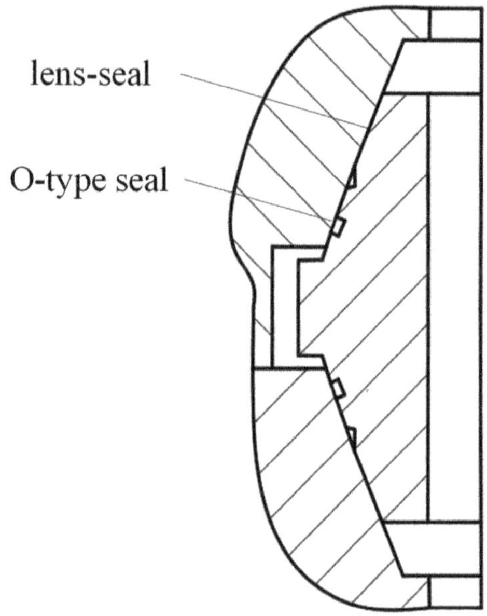

Figure 3. Dual sealing structure with lens seal and O-type seal.

3. The MRSM for Subsea Clamp Connector

The MRSM is a reliability analysis method based on the response surface method, which can consider multiple failure factors of the system by establishing multiple response surface function functions. The response surface function is transformed into a specific mathematical polynomial expression by an implicit performance function, which can be fitted to obtain a global approximation of the output variable, and then goes to replace the true response surface while proving its correctness. The response surface function of a quadratic polynomial is often used in practical applications, and its basic expressions are as follows:

$$y = a_0 + \sum_{i=1}^{n} a_i x_i + \sum_{i=1}^{n} a_{ii} x_i^2 + \sum_{i>1}^{n} a_{ij} x_i x_i \quad (1)$$

The process of the newly proposed MRSM for the subsea clamp connector is shown in Figure 4 and its main steps include:

(1) Selection of suitable failure criteria;
(2) sensitivity analysis of structural parameters and internal pressure of subsea clamp connectors after parametric modeling;
(3) acquisition of test points based on initial data using the Latin hypercube sampling method;
(4) numerical calculations were performed on the test point data, and the response surface function was obtained by fitting the test point data and the resultant data using the least squares method;
(5) the obtained response surface function is sampled using the Monte Carlo method linkage;
(6) and the probability curve of the normal distribution of sample data is obtained by using the kernel density method, and finally the reliability is calculated combining the stress-strength interference theory.

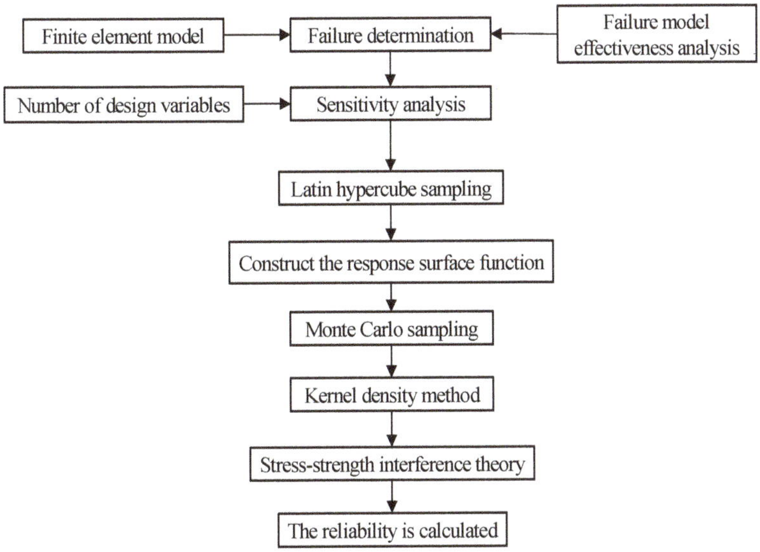

Figure 4. Multiple response surface method calculation process.

3.1. Reliability Failure Criteria

3.1.1. Seal Failure Criteria

In the operating condition, the subsea clamp connector needs the flange to provide sufficient preload, so that the residual contact force can ensure sufficient sealing, and thus avoid the occurrence of leakage. A schematic of the preload force transfer is shown in Figure 5.

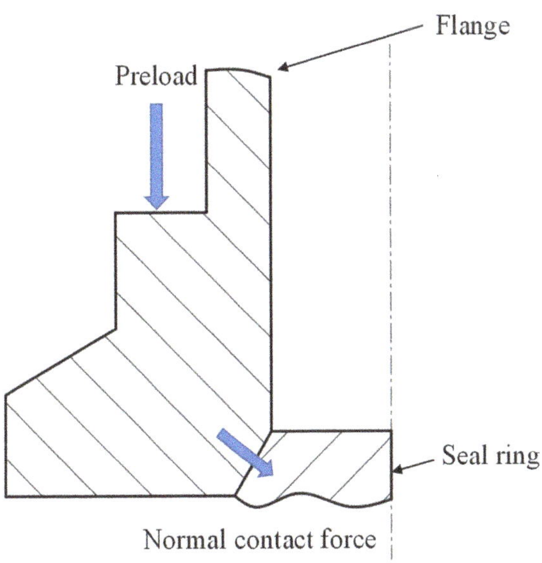

Figure 5. Schematic of preload force transmission.

Engineering has found a linear relationship between the sealing pressure and the contact stress, and defined the ratio of the mean contact stress applied to the unit effective area to the internal pressure, called the gasket coefficient, denoted by m [27], as follows:

$$\overline{p} = mp_n \tag{2}$$

where: $\overline{p}$ is the sealing specific pressure and p_n is the internal medium pressure.

Therefore, to ensure reliable sealing of subsea connectors, it is necessary to satisfy:

$$p_{Max} \geq \overline{p} \tag{3}$$

where: p_{max} is the maximum contact pressure on the seal ring.

3.1.2. Yield Failure Criteria

Yield failure refers to the phenomenon that occurs when a material reaches its yield limit. When the subsea clamp connector is subjected to preload, the seal ring and other related parts will generate a certain amount of stress to resist small plastic deformation, when the external force is too large, the parts will undergo permanent plastic deformation, resulting in structural failure. The yield failure criterion adopts the fourth strength theory. Therefore, to ensure that no structural failure occurs in each part of the subsea connector, it is necessary to satisfy:

$$\sigma_\varepsilon = \sqrt{\frac{1}{2}(\sigma_1 - \sigma_2)^2 + (\sigma_2 - \sigma_1)^2 + (\sigma_3 - \sigma_1)^2} < [\sigma] \tag{4}$$

where: σ_ε is the calculated equivalent stress.

3.2. Kernel Density Estimation Method

The probability density distribution function of equivalent stress and maximum contact pressure needs to be obtained before reliability analysis of the subsea clamp connector. The sample set of equivalent stress and maximum contact pressure obtained from the Monte Carlo method of sampling can only be expressed as a histogram. Therefore, the histogram of the equivalent stress and maximum contact pressure is converted into the probability density distribution function of the sample set using the kernel density estimation method.

Using the idea of differentiation, the group distance of the frequency histogram is further reduced. As the group distance decreases, the width of the rectangle becomes smaller, so that in the limit the frequency histogram becomes a curve, and this curve is the probability density curve [28]. The formula is as follows:

$$f(x) = \frac{1}{2nh}\sum_{i=1}^{n} 1_{x-h \leq x_i \leq x+h} \tag{5}$$

where: $f(x)$ is probability density distribution function; h is bandwidth; and n is the number of samples.

According to this formula, in the actual calculation, the value of h must be given. The value of h cannot be too large or too small, too large does not satisfy the condition of $h \rightarrow \rightarrow 0$, too small uses too few sample data points, and the error will be large. Therefore, there is more research on the choice of the value of h. Bandwidth h is usually selected by the AMISE rule.

3.3. Stress-Strength Interference Theory

For the subsea clamp connector related parts, when the material strength is greater than the stress, it is in a safe state. When the material strength is less than the stress, it fails. As shown in Figure 6, a schematic diagram of stress-strength interference theory, $f_s(s_0)$, is the probability density function of stress, and $f_S(S)$ is the probability density function of strength. The shadow where two normal distribution curves intersect is the subsea connector failure area.

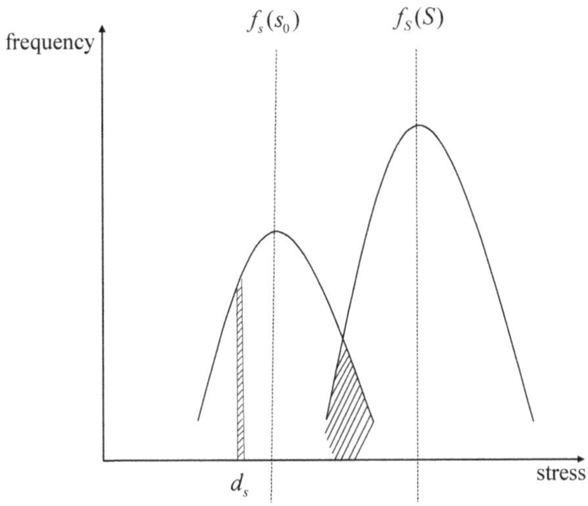

Figure 6. Schematic of stress-strength interference theory.

Assuming that both strength and stress conform to the law of normal distribution, when the mean value of stress is less than the mean value of strength, the distribution area of the two will produce an overlapping interference region. When the stress is greater than the strength that occurs in this region, the structure is a failure. Therefore, the model for calculating the probability of failure in the interference region is the stress-strength interference model [29]. The probability of failure is as follows:

$$R = \int dR = \int_{-\infty}^{\infty} f(s) \cdot \left[\int_{s}^{\infty} f(S) dS \right] ds \quad (6)$$

where: R is probability of failure.

When the failure probability of the subsea clamp connector is calculated, its reliability can be obtained.

4. Case Study of Subsea Clamp Connector Based on MRSM

4.1. Finite Element Model of Subsea Clamp Connector

In the paper, the subsea clamp connector used in the Bohai Sea in China is selected as the object of study, and its reliability analysis is based on two response surfaces: equivalent stress and maximum contact pressure.

The structural parameters of metal seals and flanges affect the locking and sealing performance of subsea clamp connectors, which mainly contain seal ring thickness (x_1), seal ring contact angle (x_2), seal ring radius (x_3), upper flange thickness (x_4), upper and lower flange contact angle (x_5), and lower flange thickness (x_6) (as shown in Figure 7).

Its structure is shown in Figure 2, which mainly includes the locking block, flanges, seal ring, clamp flap, and other parts. Due to the large number of parts and complex contact relationships, the three core components of the upper and lower flanges and seal are selected for analysis in this numerical simulation. The one-eighth model is chosen for its axisymmetric structure for modeling, which can effectively improve the computational efficiency while ensuring the computational accuracy. In the numerical calculation model as shown Figure 8, the mesh all use hexahedral mesh, and the seal ring and the upper and lower flanges contact location of the grid size is set to 1 mm. The upper and lower flanges and other parts of the grid size is set to 3 mm, the number of grid is about 170,000. The constraints of the numerical model are set as follows: fixed constraints are applied to the lower flange position to limit all degrees of freedom; the upper and lower flange sides and seal ring sides are set as axisymmetric constraints and normal constraints to simulate the effect of the 1/8 model as an overall force analysis. The model load is set to apply a concentrated force on the upper flange end face to simulate an axial preload of 200 kN and a bending moment of 30 kN·m. A 35 MPa pressure load is applied to the internal surface to simulate the fluid pressure of the medium inside the connector. The initial sample data are shown in Table 1.

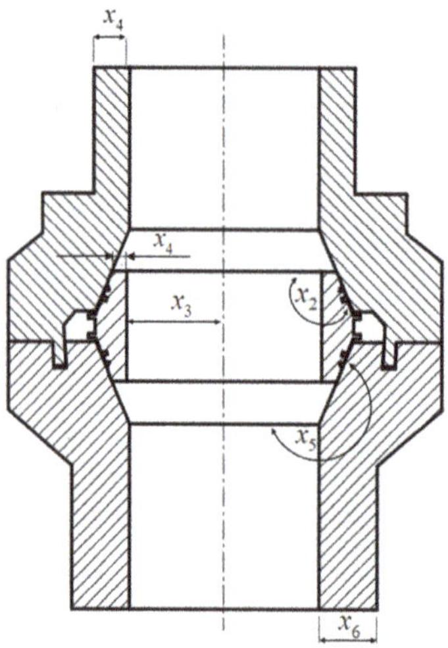

Figure 7. Subsea clamp connector structure parameterization.

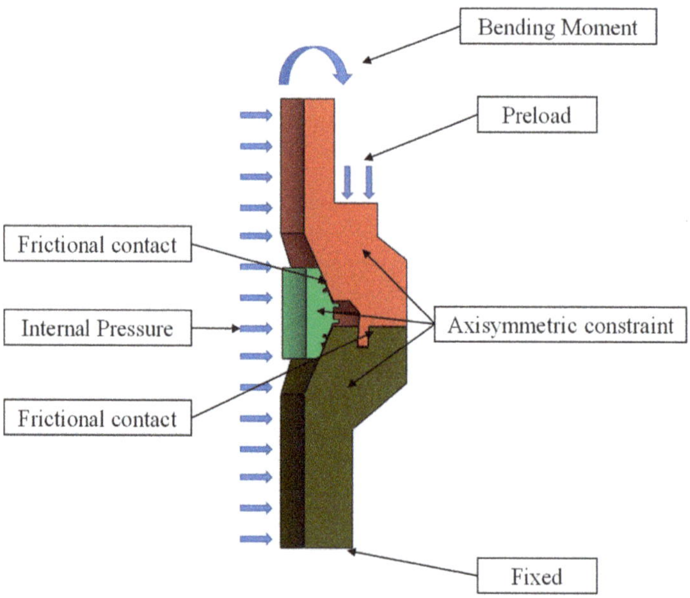

Figure 8. Constraints and loads of the numerical model.

Table 1. Initial sample data.

Random Variable Parameters	Symbols	Distribution Type	Average Value (mm)	Coefficient of Variation	Standard Deviation
Seal thickness	x_1	Normal distribution	11	0.02	0.22
Sealing ring contact angle	x_2	Normal distribution	107	0.02	2.14
Seal radius	x_3	Normal distribution	74	0.02	1.48
Upper flange thickness	x_4	Normal distribution	26	0.02	0.52
Upper flange contact angle	x_5	Normal distribution	145	0.02	2.9
Lower flange thickness	x_6	Normal distribution	38	0.02	0.76
Internal pressure	x_7	Normal distribution	35	0.02	0.7

The material parameters for each component of the model are set as shown in Table 2.

Table 2. Material parameters.

Part Name	Materials	Strength Limit (MPa)	Yield Limit (MPa)	Elastic Modulus (GPa)	Poisson Ratio	Density (g/cm^3)
Upper and lower flange	12Cr2Mo1	540	340	211	0.29	7.85
Seals	316	515	310	211	0.3	8.0

4.2. Reliability Analysis of Subsea Clamp Connector

The finite element analysis allows the equivalent stress cloud of the subsea clamp connector (shown in Figure 9) and the maximum contact pressure cloud of the seal (shown in Figure 10) to be obtained. As can be seen from the graph, the maximum equivalent stress on the upper and lower flanges is 238.07 MPa and the yield strength of the seal is 330 MPa, both of which are less than the yield strength of the material. The maximum contact pressure on the seal is 246.96 MPa, which is greater than the minimum sealing specific pressure of 227.5 MPa (according to the ASME standard, the gasket factor *m* is selected as 6.5). Therefore, it is clear from the two failure criteria that the sealing structure of the subsea clamp connector meets the design requirements. The performance of the subsea clamp connector is good under the rated conditions. The analysis of the reliability under the relevant parameters can be continued.

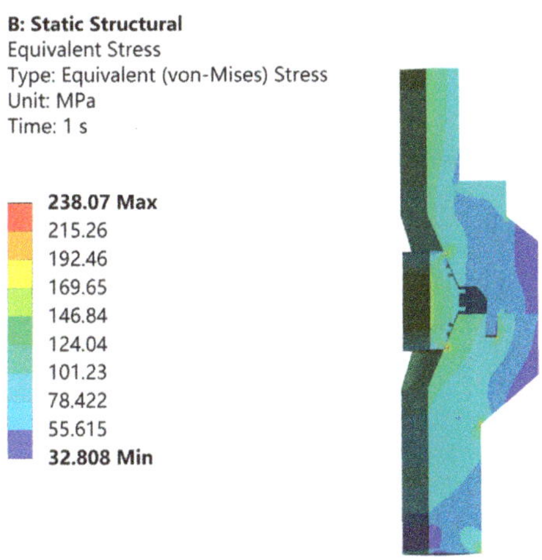

Figure 9. Equivalent stress cloud diagram.

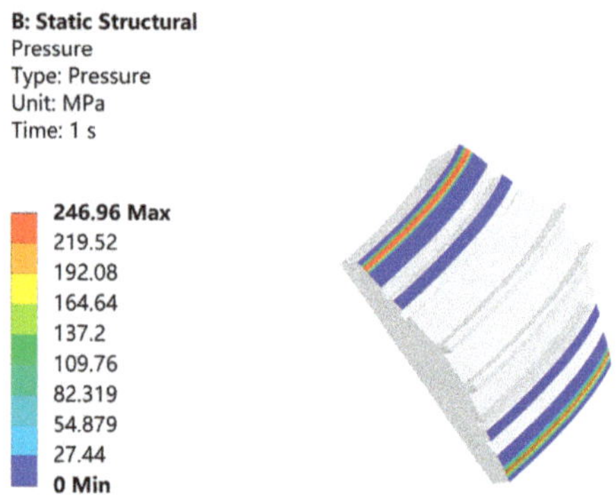

Figure 10. Maximum contact pressure cloud diagram.

By varying the main structural parameters of the subsea clamp connector, a sensitivity diagram expressing the effect of different parameters can be obtained, as shown in Figure 11, identifying the seal radius (x_3), the upper flange contact angle (x_5), and the internal pressure of the medium (x_7) as the main influencing factors in the subsequent analysis.

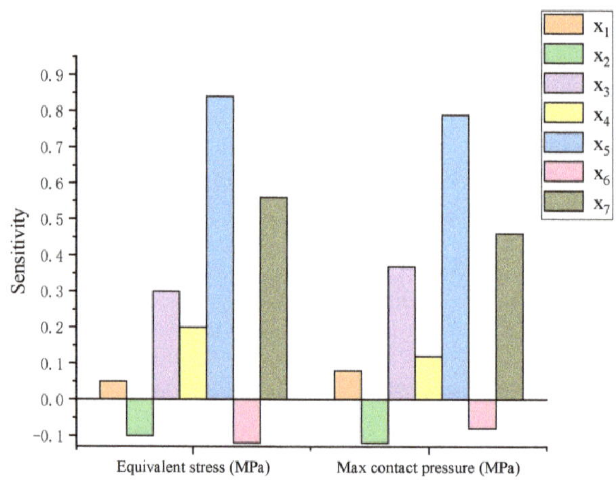

Figure 11. Sensitivity of input parameters to the response surface.

In the paper, the Latin hypercube sampling method was chosen to obtain the test data. The Latin hypercube sampling technique was a stratified sampling method, which means that the characteristics of the overall sample can be reflected by a smaller sample size, which can greatly reduce the number of test points and save calculation time. The test data obtained from the sampling were then analyzed using finite element software, and the response surface functions for the equivalent stress S and the maximum contact pressure L were fitted separately based on the least squares method.

$$\begin{aligned} S = & -66190.32642 - 375.23899x_3 + 929.50191x_5 + \\ & 761.76991x_7 - 0.914321x_3 \cdot x_5 - 5.04020x_3 \cdot x_7 - \\ & 2.65118x_5 \cdot x_7 + 4.69174\ x_3{}^2 - 2.68596\ x_5{}^2 - 0.076828x_7{}^2 \end{aligned} \tag{7}$$

$$\begin{aligned} L = & 105347.587 - 108.7357x_3 - 1066.17x_5 - 1417.9x_7 \\ & -11.165x_3 \cdot x_5 - 0.341x_3 \cdot x_7 + 8.4x_5 \cdot x_7 + 12.21x_3{}^2 \\ & +5.438x_5{}^2 + 3.29541x_7{}^2 \end{aligned} \tag{8}$$

The response surface function obtained from the above fit was subjected to the Monte Carlo sampling method to obtain the sampling iterations shown in Figures 12 and 13. The figure shows that after 2000 cycles, subsea clamp connector equivalent stress and maximum contact pressure sampling curves gradually plateau, which means that the data sampling requirements can be met using 2000 sampling cycles.

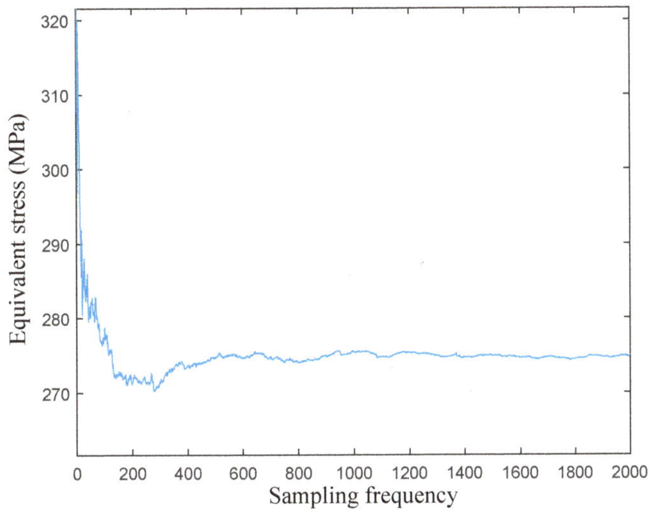

Figure 12. Iteration diagram of the response surface *S*.

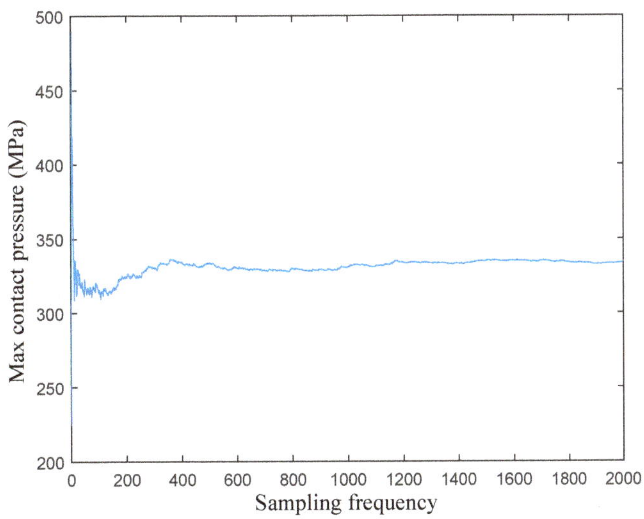

Figure 13. Iteration diagram of the response surface *L*.

The mean value of the response surface *S* is 276.56 MPa with a variance of 226. The mean value of the response surface *L* is 330.03 MPa with a variance of 205. The red line in the figure is the normal distribution curve obtained by fitting, as shown in Figures 14 and 15.

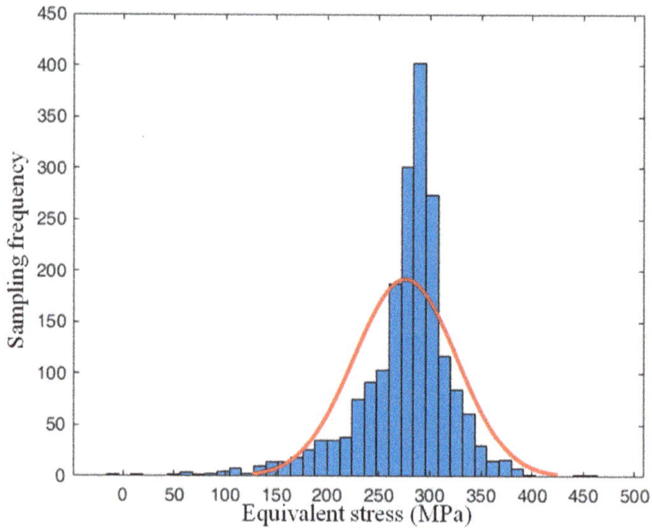

Figure 14. Normal distribution plot fitted to response plane *S*.

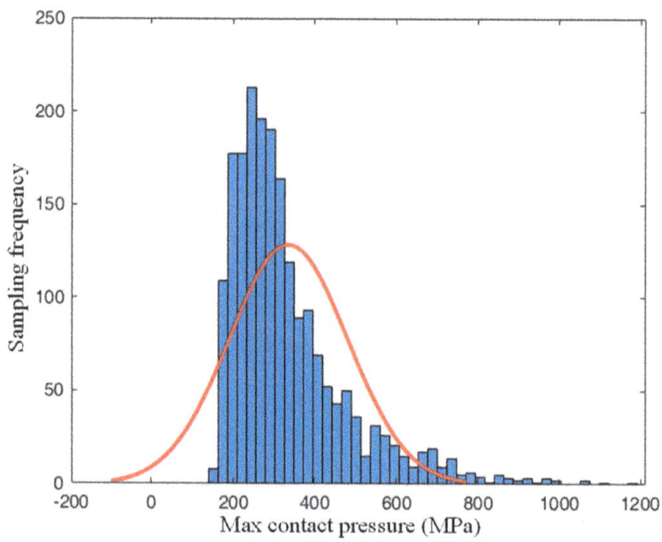

Figure 15. Normal distribution plot fitted to response plane *L*.

First, the probability density expression of the actual response surface is obtained. Then, the intensity probability density expression is obtained. Finally, the reliability of subsea clamp connector under each single response can be obtained by using the stress-strength interference theory:

$$P_t(S) = \int_{-\infty}^{\infty} f_s(s) \int_{s}^{\infty} f_{S}(S) ds dS = 0.0048 \tag{9}$$

$$P_t(L) = \int_{-\infty}^{\infty} f_l(l) \int_{L}^{\infty} f_L(L) dl dL = 0.008 \tag{10}$$

$P_t(S)$ and $P_t(L)$ are the failure probabilities obtained from the yield failure criterion and the seal failure criteria. The combined probability of failure is:

$$P(S \cup L) = P(S) + P(L) - P(S) \cdot P(L) = 0.0127 \quad (11)$$

Therefore, the reliability of the subsea clamp connector is:

$$1 - P(S \cup L) = 98.73\% \quad (12)$$

The reliability is the reliability performance of the subsea clamp connector in the initial state of production. According to a literature study [30], this reliability belongs to the low-level fault and meets its working requirements.

5. Model Test Verification

The test system shown in Figure 16 is used to check the sealing reliability of the subsea clamp connector studied in this paper. The test system is mainly composed of connector model, intelligent pressure test machine, internal sensor, and fixed brackets, etc. The test was carried out using a stepwise pressurization method, where the internal pressure of the connector model was gradually increased to 52.5 MPa. The pressure is loaded in 12 steps with a 5-min holding operation at each stage and 15 min at the last step. During the holding period measure, if the pressure drop is less than 5% of the test pressure, observe for leaks and abnormal sounds.

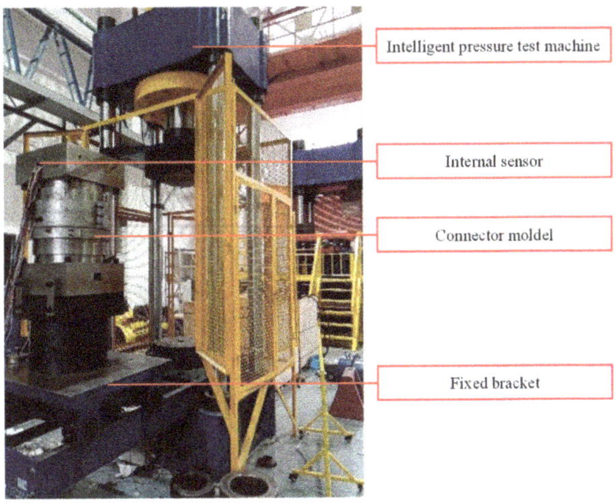

Figure 16. Reliability performance verification test model.

The pressure drop rate formed from the measured pressure drop data is shown in Figure 17. The maximum pressure drop rate during the whole pressurization process is 2%, which is lower than the standard requirement of 5%. As pressure increases, the pressure drop rate decreases and the internal pressure retention performance becomes stable. After the first pressurization step, the pressure is released and then re-pressurized in two steps to verify the accuracy of the above test procedure. The results of the model tests show that the subsea clamp connector has good sealing performance to meet the design requirements during both the pressurization and pressure-hold periods. It validates the applicability of the MRSM for the safe design of subsea clamp connectors.

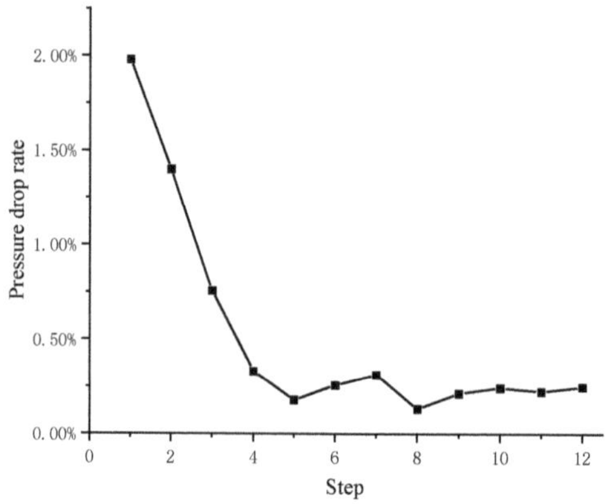

Figure 17. Diagram of pressure drop rate for each stage during the test.

6. Conclusions

The paper proposes a reliability prediction method based on the multiple response surface method for subsea clamping connectors and provides a specific analysis process for the structural characteristics and load bearing of subsea clamping connectors. The reliability analysis and model test validation are carried out with a model of subsea clamp connector, and the main conclusions are formed as follows:

(1) The multiple response surface method can accurately predict the reliability of underwater clamp connectors under different structural characteristic parameters and load conditions due to its advantages of simultaneously considering multiple failure factors of the system.
(2) The reliability of the subsea clamp connector analyzed in this paper is mainly affected by the radius of the sealing ring, the contact angle of the upper and lower flanges, and the internal pressure of the medium, and the reliability is 98.73% by calculation.
(3) The maximum pressure drop rate of the internal pressure test of the underwater clamp connector analyzed in this paper is 2%, the pressure drop rate decreases with increasing pressure, and the internal pressure retaining performance tends to be stable. The applicability of the MRSM in the safety design of subsea clamp connectors is verified.

Author Contributions: Conceptualization, M.D.; methodology, W.A.; validation, B.C.; formal analysis, B.C.; investigation, X.Z.; resources, W.A.; data curation, Y.W.; writing—original draft preparation, W.A.; writing—review and editing, Y.W.; All authors have read and agreed to the published version of the manuscript.

Funding: This research was funded by National Natural Science Foundation, grant number 52222111.

Institutional Review Board Statement: Not applicable.

Informed Consent Statement: Not applicable.

Data Availability Statement: Not applicable.

Conflicts of Interest: The authors declare no conflict of interest.

References

1. Chen, S. Current Situation and Improvement Method of Offshore Oil Drilling Platform Equipment Management. *Chem. Manag.* **2021**, *8*, 191–192.
2. Li, Q.; Zhu, H.; Li, X. The Current State and Future of Deep Water Subsea Production Technology. *Chin. Eng. Sci.* **2016**, *18*, 76–84.
3. Wang, L. The development status and trend of Chinese offshore petroleum engineering industry. *China Pet. Chem. Stand. Rain Qual.* **2022**, *42*, 142–144.
4. Wang, Y. *Subsea Production Systems and Engineering*; Publishing House of China University of Petroleum: Dongying, China, 2017.

5. Peng, F.; Duan, M.; Wang, J. Optimisation method for mathematical model of deepwater collet connector locking mechanism. *Ships Offshore Struct.* **2015**, *11*, 575–590. [CrossRef]
6. Zhao, H.; Chen, R.; Luo, X.; Duan, M.; Lu, Y.; Fu, G.; Tian, H.; Ye, D. Metal sealing performance of subsea X-tree wellhead connector sealer. *Chin. J. Mech. Eng.* **2015**, *28*, 649–656. [CrossRef]
7. Hu, X.; Yun, F.; Shi, L. Structural Optimization of Metal Sealing Ring of Underwater Clamp Connector. *Lubr. Seal.* **2022**, *47*, 117–124.
8. Wang, R. *Sealing Structure Optimization Analysis and Performance Research of Deep Sea Horizontal Clamp Connector*; Harbin Engineering University: Harbin, China, 2017.
9. Yun, F.; Wang, G.; Yan, Z.; Jia, P.; Xu, X.; Wang, L.; Sun, H.; Liu, W. Analysis of Sealing and Leakage Performance of the Subsea Collet Connector with Lens-Type Sealing Structure. *J. Mar. Sci. Eng.* **2020**, *8*, 444. [CrossRef]
10. Li, Y.; Zhao, H.; Wang, D.; Xu, Y. Metal sealing mechanism and experimental study of the subsea wellhead connector. *J. Braz. Soc. Mech. Sci. Eng.* **2020**, *42*, 26. [CrossRef]
11. Liu, D.; Yun, F.; Jiao, K.; Wang, L.; Yan, Z.; Jia, P.; Wang, X.; Liu, W.; Hao, X.; Xu, X. Structural Analysis and Experimental Study on the Spherical Seal of a Subsea Connector Based on a Non-Standard O-Ring Seal. *J. Mar. Sci. Eng.* **2022**, *10*, 404. [CrossRef]
12. Wan, B.; Sun, Z.; Jiang, Y.; Cai, Y.; Chen, J. Reliability analysis of underwater connectors based on fault tree. *China Shipp. Surv.* **2015**, *12*, 85–87.
13. Zhang, K.; Duan, M.; Luo, X.; Hou, G. A fuzzy risk matrix method and its application to the installation operation of subsea collet connector. *J. Loss Prev. Process Ind.* **2017**, *45*, 147–159. [CrossRef]
14. Chen, Z.; Liu, G.; Wang, Y. Fault diagnosis of subsea collet connector based on dynamic Bayesian network. *Chin. J. Saf.* **2020**, *30*, 81–87.
15. Hørte, T.; Reinås, L.; Wormsen, A.; Aardal, A. Structural Reliability Analysis Method for Assessing the Fatigue Capacity of Subsea Wellhead Connectors. In Proceedings of the ASME 2020 39th International Conference on Ocean, Offshore and Arctic Engineering, Online, 3–7 August 2020. [CrossRef]
16. Bhardwaj, U.; Teixeira, A.P.; Soares, C.G. Reliability assessment of a subsea pipein-pipe system for major failure modes. *Int. J. Pres. Ves. Pip.* **2020**, *188*, 104177. [CrossRef]
17. Pang, N.; Jia, P.; Liu, P.; Yin, F.; Zhou, L.; Wang, L.; Yun, F.; Wang, X. A Fuzzy Markov Model for Risk and Reliability Prediction of Engineering Systems: A Case Study of a Subsea Wellhead Connector. *Appl. Sci.* **2020**, *10*, 6902. [CrossRef]
18. Wang, Y.; Liu, S.; Chen, Z. Dynamic Bayesian networks for reliability evaluation of subsea wellhead connector during service life based on Monte Carlo method. *J. Loss Prev. Process Ind.* **2021**, *71*, 104487. [CrossRef]
19. Tsai, L.W.; Alipour, A. Structural health monitoring and fatigue life reliability assessment of a flexible structure in extreme wind. *J. Civ. Struct. Health Monit.* **2023**, *13*, 677–691. [CrossRef]
20. Simon, O.; Fei, T.; Mohsen, S.; Chen, C. Risk Based Verification for Subsea Pipeline Repair Connectors and a Case Study. In Proceedings of the Offshore Technology Conference, Houston, TX, USA, 1–4 May 2023; p. 32412. [CrossRef]
21. Das, P.K.; Zheng, Y. Cumulative formation of response surface and its use in reliability analysis. *Probabilistic Eng. Mech.* **2000**, *15*, 309–315. [CrossRef]
22. Goswami, S.; Ghosh, S.; Chakraborty, S. Reliability analysis of structures by iterative improved response surface method. *Struct. Saf.* **2016**, *60*, 56–66. [CrossRef]
23. Zhang, C.; Liu, L.; Sun, X. Reliability Analysis for Aero-engine Blades with Multiple Response Surface Method. *J. Harbin Eng. Univ.* **2016**, *21*, 22–27.
24. Cheng, L. *Research on Structural Reliability Analysis Method Based on Multiple Response Surfaces of Coupling Failure*; Harbin Engineering University: Harbin, China, 2016.
25. Liu, Z.; Lin, H.; Li, L. Analysis on the Reliability of Tunnel System Based on Multiple Response Surface Methodology and Monte Carlo Method. *Mod. Tunn. Technol.* **2022**, *59*, 78–87.
26. Zhai, X. *Structural Reliability Analysis and Optimization of Offshore Fan Connectors Based on Multiple Response Surface Method*; China University of Petroleum: Beijing, China, 2022.
27. Li, H. Factor Analysis of the Gasket Seal Failure. *Liaoning Chem. Ind.* **2013**, *42*, 1007–1009.
28. Węglarczyk, S. Kernel density estimation and its application. *XLVIII Semin. Appl. Math.* **2018**, *23*, 8. [CrossRef]
29. Niu, L.; Dong, H.; Zhao, X.; Li, X.; Yan, N. Reliability of Separation Nuts Based on Stress-Intensity Interference Model. *Equip. Environ. Eng.* **2022**, *19*, 8–13.
30. Pang, N.; Jia, P.; Wang, L. Reliability analysis of subsea connector structure. *J. Harbin Eng. Univ.* **2021**, *42*, 68–73.

Journal of
Marine Science and Engineering

MDPI

Article

A Semi-Analytical Method for the Sealing Performance Prediction of Subsea Pipeline Compression Connector

Zhenyu Li [1], Gang Wang [2,*], Shaoming Yao [1], Feihong Yun [1], Peng Jia [1], Chao Li [1] and Liquan Wang [1]

1 College of Mechanical and Electrical Engineering, Harbin Engineering University, Harbin 150001, China; jade10623@sina.com (Z.L.); yao.sm@hotmail.com (S.Y.); yunfeihong@hrbeu.edu.cn (F.Y.); 13633605161@139.com (P.J.); 54529880@hrbeu.edu.cn (C.L.); wangliquan@hrbeu.edu.cn (L.W.)
2 College of Shipbuilding Engineering, Harbin Engineering University, Harbin 150001, China
* Correspondence: wanggang@hrbeu.edu.cn

Abstract: To predict the sealing performance of the subsea pipeline compression connector, a semi-analytical method is proposed and verified. The leakage condition is obtained as a function of the minimum radial deflection. The semi-analytical method consists of three parts: a macroscopic analytical model for the interference process deduced based on the membrane theory, a mesoscopic two-dimensional finite element analysis of the internally turned sealing surface according to scanning electron microscope observation, and a formulation of the leakage ratio according to the Abbott–Firestone curve. The influences of the geometry parameters are analyzed. The prediction and identification of the leakage condition near the minimum preloaded deflection are investigated. Experimental tests are also carried out to verify the proposed method. The results show that, within five times the pipeline threshold of the thin wall thickness, compression connectors designed by the proposed method can create a reliable seal with a rough internally turned surface, saving 57% of the design time while reducing the machining time and costs.

Keywords: semi-analytical method; leakage ratio; compression connector; metal static seal; membrane theory; Abbott–Firestone curve

Citation: Li, Z.; Wang, G.; Yao, S.; Yun, F.; Jia, P.; Li, C.; Wang, L. A Semi-Analytical Method for the Sealing Performance Prediction of Subsea Pipeline Compression Connector. *J. Mar. Sci. Eng.* **2023**, *11*, 854. https://doi.org/10.3390/jmse11040854

Academic Editor: Bruno Brunone

Received: 20 March 2023
Revised: 13 April 2023
Accepted: 17 April 2023
Published: 18 April 2023

1. Introduction

Non-welded connections are increasingly used for the maintenance of subsea pipelines in marine industries [1]. The compression connector is one of the most efficient connectors that belongs to the mechanically attached connection (ASTM F1387 [2], ISO 8434-1 [3]). This technology has been developed by many well-known companies, including the Phastite® connector from Parker Hannifin (Columbus, OH, USA), the Tube-Mac® connector from PYPLOK® Corp. (Stoney Creek, ON, Canada), the compression connector from HAELOK® (Schlieren, Switzerland), etc. The connector is directly attached to the pipe by the mechanical deformation and creates a seal and a permanent joint by elastic–plastic deformation. The sealing principle is a metal static seal that uses the plastic deformation of the metal to block the leakage channel.

In studies on the leak ratio on randomly rough surfaces, the topography formed by the turning texture shows unique sealing properties [4]. This can be traced back to groundbreaking studies on the surface waviness theory [5] and further developments on surface roughness [6] and sealing design factors [7]. In the aspect of the gasket, Nakamura and Funabashi [8] observed the leakage on a spiral-like topography. Then, by establishing a simplified semi-empirical leakage model, Polycarpou and Etsion [9] demonstrated the importance of micro-profile parameters for the sealing performance of the metal gasket seal. To further understand the sealing performance of the turning topography, Geoffroy and Prat [10] established the seal leakage model of the face-turned metal gasket and found the transition from the radial leakage to the spiral leakage by studying the radial and circumferential leakage paths. Liao et al. [11] first presented the simplification of the radial leakage path and the radial leakage model to clarify the effects of the micro-profile parameters on the sealing behaviors. More recently, the influence of wavy deformations and the formation of synthetic "turned-like" surfaces have been extensively studied. Scholars have also conducted much research on sealing topography related to the studies of asperity [12] and the probabilistic models [13] of rough surfaces. By establishing a 2D stochastic model, Marie et al. [14] developed a practical model and applied the multiscale calculation method based on the leakage ratio of the sealing structure. Liao et al. [15] proposed a method to decompose 3D surfaces into multi-scale surfaces by using wavelets to predict surface functions and detect machining errors. Deltombe et al. [16] used multi-scale analysis to characterize sealing performance. By comparing the differences between machining and superfinishing, it was shown that micro-roughness plays a major role in the leakage. In addition, Bricaud et al. [17] studied the leakage between firmly pressed rigid surfaces by measuring mass flow under different contact geometries using leakage tests.

Scholars have also conducted much research on compression connections. Weddeling et al. [18] derived an analytical model to determine the charging energy to a clamped connection with a mandrel inside the pipe. In another study, Weddeling et al. [19] developed a formula for the transferable load of connection with different groove shapes using membrane theory. Wei et al. [20] deduced the formula of the sealing pressure of a compression connector based on the superposition theorem of elasticity, and obtained the sealing condition using the finite element method (FE). Henriksen et al. [21] used experimental data and FE simulations to describe and understand the forces and deformations during the connection process and identified the increase radial stiffness as the best indicator of a clamped pipe flange connection while also predicting the load capacity of the connection. Wang et al. [22] proposed an optimization method for a compression connector based on a static metal sealing mechanism and optimized the structure using an FE model with the zero-order method. Onyegiri et al. [23] investigated a compression connector used on sandwich pipes combining 2D axisymmetric and 3D FE models. Quispe et al. [24] proposed a 2D FE model for threaded compression connection with square teeth and a metal static seal to join subsea sandwich pipes. Yan and Fan [25] employed a multiscale model to study the effect of the fluid pressure on the sealing performance of a pipeline connection. Agrawal et al. [26] developed a new method for a compression connector and carried out the extrusion test and compression test, concluding that the connection strength is considerable with welding. Liu et al. [27] optimized the anchoring structure by using a nonlinear FE analysis method. Wu et al. [28] introduced an axial load-bearing capacity calculation method for steel pipe-to-sleeve grouted connections, which also showed the significance of the compression connection.

For the compression connection, there is no simple analytical method to determine the achievable joint strength based on the process and joint characteristics [19,29]. In practice, the FE method is still the only option in many cases, resulting in a long design cycle and high test costs. Better knowledge and understanding of the elasticity and plastic deformation of the compression connection is required. In addition, most of the research on metal seals are about face-turned seals. Few studies can be found on the internally turned seal. The purpose of this study is to propose an effective semi-analytical method to predict the leakage ratio on the internally turned sealing surface of the compression connector, and, by analyzing the sealing behavior on the mesoscopic scale, to predict the leakage ratio for all standard pipe series.

The paper is organized as follows: In Section 2, the structure and fitting process of the compression connector are explained. In Section 3, an analytical model of the radial interference process is derived based on the membrane theory and verified by a macroscopic FE model. In Section 4, the SEM observation of the internally turned sealing surface is performed, and a mesoscopic FE model for the seal is developed. Then, the change law of the sealing surface during the compression is analyzed. In Section 5, by transferring the deflection into the mesoscopic FE model, and using the Abbott–Firestone curve, the leakage ratio formulation is derived. In Section 6, a parameter analysis and experimental tests are presented.

2. Connector Fitting Process

The pipeline compression connector, the fitting tools, and the fitting process are shown in Figure 1. The connector is designed to maintain and/or repair typical single-wall subsea pipes with a diameter between 2 and 12 inches.

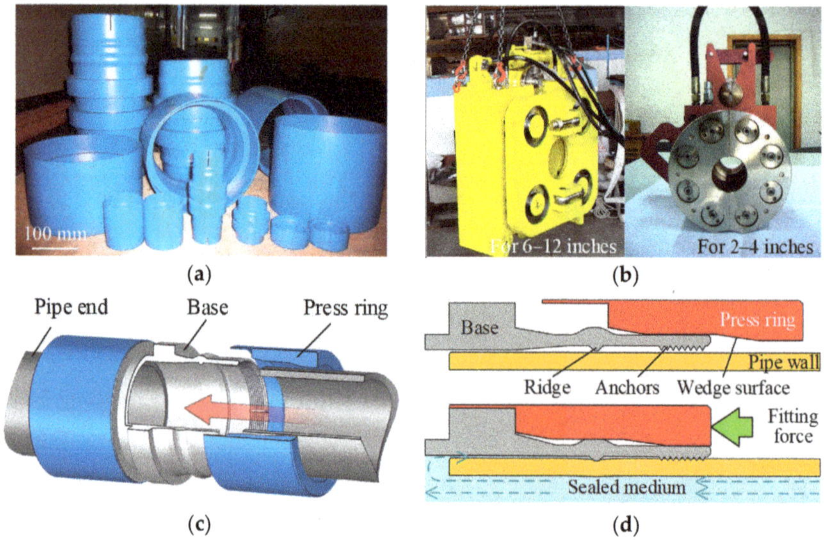

Figure 1. (**a**) Compression connector parts; (**b**) fitting tools; (**c**) connector assembly; (**d**) fitting process.

As shown in Figure 1, the connector consists of a base and a pair of press rings, which are made of metal and are symmetrically arranged along the pipe. Inside the base, there are one or more ring-shaped ridges for sealing and anchors (optional configuration) for gripping. A hydraulic fitting tool is used to drive the press rings to create a static compression metal seal [30].

3. Analytical Model of the Fitting Process

During the fitting process, external loads are transferred through the press ring, the shaped ridges are compressed on the sealing surface of the pipe and partially press into the pipe surface, creating a form-fit [31] sealing condition. The geometries can be idealized as 2D axisymmetric thin-walled cylinders [32] and the membrane theory of symmetrically loaded cylinders based on Kirchhoff's hypothesis [33] is applicable. The internal forces can be solved by the physical relations between stress and strain, the differential equations of the deflections, and the relations between strain and displacement.

3.1. Formulation

A shell element of the thin-walled long cylinder with mid-surface radius R and thickness δ is shown in Figure 2. The mid-surface of the shell is located at the cylindrical coordinates system with angle φ, axial position x, and radius $\rho(= R + z)$. The corresponding displacements on the mid-surface are $u, v,$ and w. P_x is the shearing force, N_φ and N_x are the membrane forces, and M_φ and M_x are the bending moments. N_φ and M_φ are constant in the circumference.

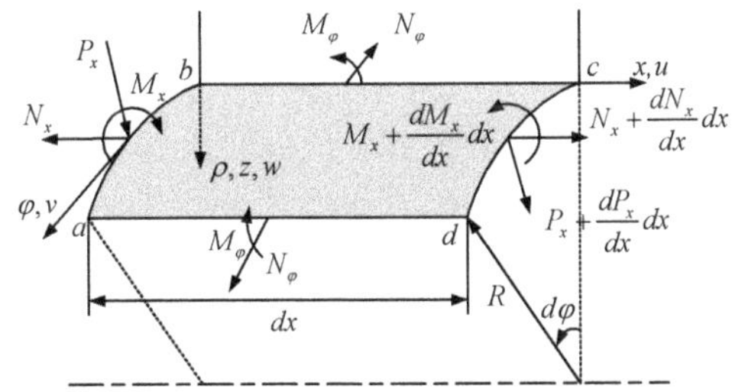

Figure 2. A thin-walled cylindrical shell in the cylindrical coordinates.

Assuming that external forces only contribute to the normal pressure on the surface, the following geometrical and equilibrium equations apply:

$$\frac{dN_x}{dx} = 0, \ \frac{dP_x}{dx} + \frac{N_\varphi}{R} + P_z = 0, \ \frac{dM_x}{dx} - P_x = 0 \tag{1}$$

The expressions for the strains can be written as

$$\varepsilon_x = \frac{du}{dx}, \varepsilon_\varphi = -\frac{w}{R}, \tag{2}$$

and, when applying Hook's law and yield,

$$\begin{cases} N_x = \frac{E\delta}{1-v^2}(\varepsilon_x + \nu\varepsilon_\varphi) = \frac{E\delta}{1-v^2}\left(\frac{du}{dx} - \nu\frac{w}{R}\right) \\ N_\varphi = \frac{E\delta}{1-v^2}(\varepsilon_\varphi + \nu\varepsilon_x) = \frac{E\delta}{1-v^2}\left(-\frac{w}{R} + \nu\frac{du}{dx}\right) \end{cases}. \tag{3}$$

With the load of P_x applied, the basic differential equation of the symmetrically loaded cylinder can be written as

$$\frac{d^4w}{dx^4} + 4\beta^4 w = \frac{P_x}{D} + \frac{\nu}{RD}N_x, \tag{4}$$

where

$$\beta = \left(\frac{E\delta}{4R^2D}\right)^{\frac{1}{4}}, D = \frac{E\delta^3}{12(1-v^2)}. \tag{5}$$

The general solution of the constant coefficients equation is

$$w = e^{\beta x}(C_1\cos\beta x + C_2\sin\beta x) + e^{-\beta x}(C_3\cos\beta x + C_4\sin\beta x) + f(x), \tag{6}$$

in which $f(x)$ is a particular solution of Equation (4) and C_1 to C_4 are the constants of integration that need to be determined in each special case by the conditions at the cylinder ends.

The forces and moments for w are

$$\left.\begin{aligned} N_\varphi &= -E\delta\frac{w}{R} + \nu N_x \\ M_x &= -D\frac{d^2w}{dx^2} \\ M_\varphi &= v M_x \\ P_x &= -D\frac{d^3w}{dx^3} \end{aligned}\right\}. \tag{7}$$

Both the base and the pipe can be considered long cylinders and subject to the radial external load, P_x, which is symmetrically distributed about the cylinder central line. Fixing the origin in the cross-section where P_x is applied, the constants of Equation (6) are

$$C_1 = C_2 = 0, C_3 = -\frac{1}{2\beta^3 D}\left(\beta M_0 + \frac{P_x}{2}\right), C_4 = \frac{M_0}{2\beta^3 D}, \tag{8}$$

and the radial deflection w can be derived as

$$w = \frac{e^{-\beta x}}{2\beta^3 D}\left(\beta M_0(\sin\beta x - \cos\beta x) + \frac{P_x}{2}\cos\beta x\right). \tag{9}$$

Then, the deflection, slope, shearing load, and bending moment can be rewritten as

$$\left.\begin{aligned} w &= \tfrac{P_x}{8\beta^3 D}e^{-\beta x}(\cos\beta x + \sin\beta x) \\ \tfrac{dw}{dx} &= -\tfrac{P_x}{4\beta^2 D}e^{-\beta x}\sin\beta x \\ M_x &= \tfrac{P_x}{4\beta}e^{-\beta x}(\cos\beta x - \sin\beta x) \\ Q_x &= -\tfrac{P_x}{2}e^{-\beta x}\cos\beta x \end{aligned}\right\}. \tag{10}$$

According to Equation (9), the influenced axial range of the cylinder can be predicted as

$$\left.\begin{aligned} \chi_{p+} &= \tfrac{3}{4}\pi\left(\tfrac{E_p\delta_p}{4R_p^2 D_p}\right)^{-\frac{1}{4}} \\ \chi_{p-} &= -\chi_{p+} \end{aligned}\right\}. \tag{11}$$

The half of the 2D axisymmetric connector–pipe assembly with the radial equivalent load, P_z, of the press ring is shown in Figure 3 with regard to the symmetry about the central plane of the connector. The axial distance of the ridge and the anchor is much greater than the half influence range, $L_1 \gg \chi_{p+}$; hence, the coupling effect between the ridge and the anchor can be neglected in the following analysis. In the fitting process, the applied equivalent load, P_z, can be divided into two components: one is for eliminating the maximum radial clearance, h, and the other is for compressing the ridge into the pipe. Meanwhile, the ridge is subjected to the reaction load, P_r, of the pipe. Due to the wedge surfaces inside the press ring, there is a short axial distance, L_0, between P_z and P_r.

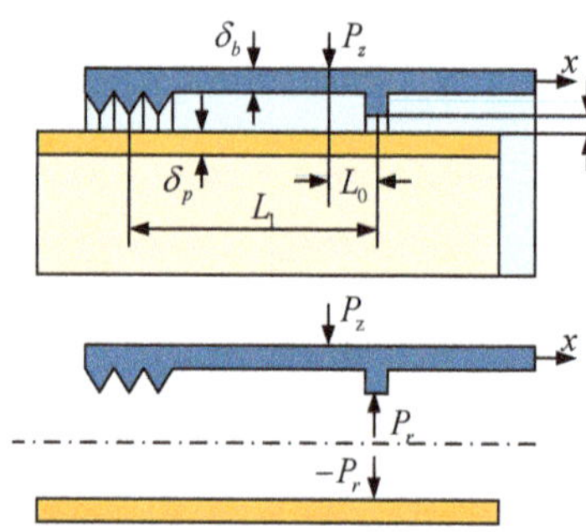

Figure 3. Free-body diagram of the assembly subjected to the radial equivalent load, P_z.

When $\psi(\beta_b x) = e^{-\beta_b x}(\cos\beta_b x + \sin\beta_b x)$, the deflection of the mid-surface of the cylindrical base, w_{b1}, in the radial direction can be expressed as

$$\left.\begin{aligned} w_{b1}(x) &= \tfrac{P_{b1}}{8\beta_b^3 D_b}\psi(\beta_b x) \\ w_{b1}(L_0) &= h \end{aligned}\right\}. \tag{12}$$

When $\chi_0 = \beta_b L_0$, P_{b1} can be rewritten as:

$$P_{b1} = \frac{8\beta_b^3 D_b}{\psi(\chi_0)}h. \tag{13}$$

Once the ridge contacts the pipe surface, a sealing pressure occurs on the pipe surface. When the pressure reaches the pipe's elastic limit, σ_{sp}, the reaction load, P_r, can be given as

$$P_r = \zeta\sigma_{sp}. \tag{14}$$

In this case, according to the Equation (10), the pipe deflection is

$$w_p(0) = \frac{P_r}{8\beta_p^3 D_p}. \tag{15}$$

The deflection of the base is

$$w_r(0) = \frac{1}{8\beta_b^3 D_b}(P_{b2}\psi(\chi_0) - P_r). \tag{16}$$

Substituting P_r from Equation (15) into Equation (16) yields

$$P_{b2} = \left(\frac{1}{\beta_b^3 D_b} + \frac{1}{\beta_p^3 D_p}\right)\frac{\beta_b^3 D_b}{\psi(\chi_0)}P_r \tag{17}$$

The total equivalent load on the base is

$$P_z = \frac{\beta_b^3 D_b}{\psi(\chi_0)}\left[8h + \left(\frac{1}{\beta_b^3 D_b} + \frac{1}{\beta_p^3 D_p}\right)P_r\right]. \tag{18}$$

Finally, the deflection can be rewritten as

$$w_x(x) = \left[h + \frac{1}{8}\left(\frac{1}{\beta_b^3 D_b} + \frac{1}{\beta_p^3 D_p}\right)P_r\right]\frac{\psi(\beta_b|x|)}{\psi(\chi_0)} - \frac{P_r}{8\beta_b^3 D_b}\psi\Big(\beta_b(|L_0 - x|)\Big). \tag{19}$$

In particular, the radial deflection at the point of P_z is

$$w_x(0) = \left[h + \frac{1}{8}\left(\frac{1}{\beta_b^3 D_b} + \frac{1}{\beta_p^3 D_p}\right)P_r\right]\frac{1}{\psi(\chi_0)} - \frac{P_r}{8\beta_b^3 D_b}\psi(\chi_0). \tag{20}$$

The bending moment and slope can also be derived with the same method. Multiple ridges can be designed according to the superposition principle.

3.2. Numerical Modeling and Validation

The finite element method (FE) can be used in the design of the compression connector. In order to validate the proposed macroscopic analytical solutions, a macroscopic 2D axisymmetric FE model is created for the simulation using ABAQUS®/Standard [34].

The base and pipe are idealized as cylinders with a given thickness. An isotropic bilinear hardening model (BISO) with von Mises yield criterion [35] is adopted in which the yield plateau is ignored. The constitutive relationship of the plastic stage takes the form of

$$\sigma^T = \sigma_s + \xi \cdot \varepsilon^T, \tag{21}$$

where σ^T is the true stress, σ_s is the yield strength, and ε^T is the accumulated plastic strain. ξ is the hardening modulus, which can be determined by the tangent modulus, E_t, and Young's modulus, E, with the relation of

$$\xi = \frac{E_t E}{E - E_t}. \tag{22}$$

For general applicability, E_t is averaged and rounded to 5.0×10^2 Mpa by referring to the properties of wildly used pipe steels: mild steels, Q235, X60, X70, and X80. The geometrical parameters and material properties used in the paper are listed in Table 1.

Table 1. Geometrical parameters and material properties.

Name	Unit	Symbol	Value	
			Base	Pipe
Type	-	-	6-inch	DN 150
Radius	mm	R_b, R_p	215	168
Thickness	mm	δ_b, δ_p	20	20
Yield strength	Mpa	σ_{sb}, σ_{sp}	600	235
Axial distance	mm	L_0	0.2	
Radial clearance	mm	h	0.4	
Young's modulus	Mpa	E_b, E_p	2.10×10^5	
Tangent modulus	Mpa	E_t	5×10^2	
Poisson coefficient	-	υ	0.3	
Friction coefficient	-	μ	0.2	

The FE model is shown in Figure 4. To consider the large mesh distortion without twist, the four-node bilinear axisymmetric quadrilateral element with reduced integration and hourglass control (CAX4R) is used. The mesh sensitivity is examined. The right corners of the base and the pipe are fixed.

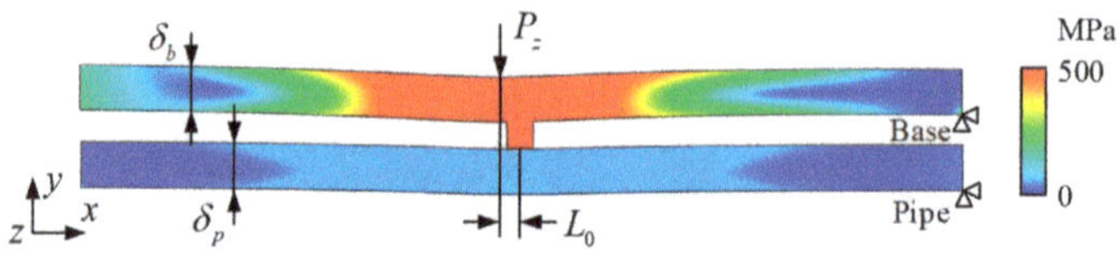

Figure 4. Macroscopic 2D axisymmetric FE model of the assembly.

Between the analytical solution and the FE model, a realistic assumption should be made. Because the analytical solution satisfies the Kirchhoff's hypotheses, one of the hypotheses is that the thickness of the cylinder, δ, is much smaller than the radius of the mid-surface, R (the threshold is $\delta / R \leq 0.05$). In practice, the δ_p / R_p of some pipes is higher than 0.05, e.g., for six-inch pipes, the range is $\delta_p / R_p \in [0.04, 0.54]$ (ASME B36 10M [36] and 19M [37]). To satisfy the maximum mean error of less than 10% between the analytical model and the FE model, in this study, it is assumed that the pipe thickness in the FE simulation is within five times the thin wall threshold; hence, $\delta_p = 20$ mm and $\delta_p / R_p \approx 0.25$. During the macroscopic FE analysis, an equivalent radial external load, P_z, is applied on the appropriate width of the external surface of the base part and is gradually loaded from one to two times the theoretical threshold. The mid-surface deflections of both the base, w_x, and the pipe, w_p, are observed, as shown in Figure 5.

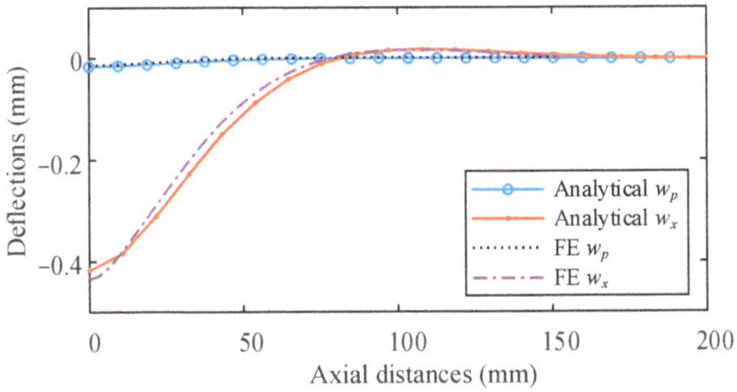

Figure 5. Comparison between the analytical and FE results.

The deflections of the analytical method are consistent with the FE results. In the axial direction, the mean error of w_x with the analytical method is 8.5% compared to that of the FE method, and the mean error of w_p is 5.7%. The error should usually be within 10% [38]. In particular, within an axial distance of 20 mm from the seal (at $x = 0$), the mean error of w_x is only 3.7%. However, the corresponding error of w_p is much higher due to the local plastic deformation on the sealing surface under mesoscopic interference. With the help of a mesoscopic FE model, the sealing process can be analyzed, as discussed in Section 4.

4. FE Model of the Sealing Surface

In this section, a mesoscale FE model is created to examine the plastic deformation of the internally turned surface during the fitting process.

4.1. Characterization of the Sealing Surface

The turning process is widely used in the manufacturing of flange surfaces, and these surfaces usually reveal spiral morphology [11,39]. Similarly, due to the difficulty of polishing a thin-walled cylindrical shell [40], the topography produced by the rough turning process is usually left on the ridge surface. The surface was observed using scanning electron microscopy (SEM), as shown in Figure 6.

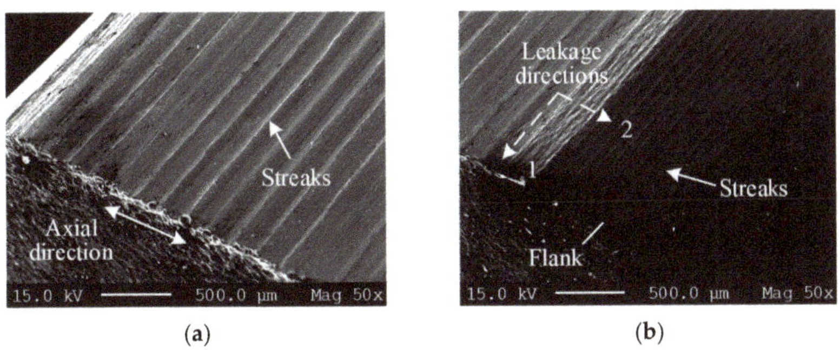

Figure 6. Observation of ridge surfaces with SEM: (**a**) sealing surface and (**b**) flank surface.

The internally turned surface has a distinctive anisotropic topography, which is characterized by regular helical and continuous streaks with slight defects and irregularities. Therefore, there are two potential leakage paths: the first one (Direction 1) is the longest helical path along the streak, and the second (Direction 2) is the shortest leak path in the axial direction.

4.2. FE Model of the Sealing Surface

According to the turning topography [41,42] and ISO recommendations (ISO 16610-41 [43] and 16610-85 [44]), the surface profile can be considered a periodic connection from the nose profile of the turning tool. Consistently with the SEM results, the mesoscale sealing surface profile is created, as shown in Figure 7.

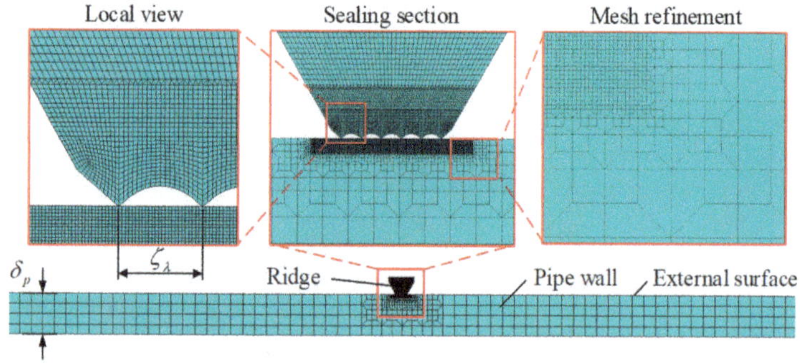

Figure 7. The mesoscale 2D axisymmetric FE model of the seal and local mesh views.

The turning process produces a helical profile on the inner surface in 3D. Chen and Shih [45] have demonstrated that the 2D axisymmetric model can be used to reduce the computational complexity of the helical 3D model while meeting the accuracy requirement. Therefore, a 2D axisymmetric FE model is used. According to Saint-Venant's principle [46], the computation domain is limited to the localized contact region with a portion of the base and is set as $L_0 = 0$ mm. On the sealing surface, since leakage is more of a concern than stress concentration, a groove is positioned at the center and streaks are arranged at both ends. The topography layout is set to an acceptable minimum arrangement of six streaks with five equally spaced grooves. The cross-section of the groove is assumed to be part of a constant circle according to the ISO recommendation. For consistency, additional FE parameters are inherited from the previous macroscopic FE model (in Section 3.2). Since the pipe surface is preprocessed before connection, the surface roughness is 1–2 orders of magnitude smaller than the ridge, and it is simplified to an even surface (no roughing) in the analysis. To show subtle deformation and reduce computation, the mesh is optimized. The sensitivity of the mesh is examined. The following parameters for the study case are provided: the groove radius, $r_c = 0.3$ mm; the pitch of the streaks, $\zeta_\lambda = 0.5$ mm; the sealing width, $\zeta = 3$ mm.

4.3. Change Laws of the Leakage Channel

The interference can be obtained by applying an equivalent deflection/displacement to the ridge in the simulation. According to the FE simulation, the stress contour plots for the corresponding deformation phases are shown in Figure 8a. The deflections of the pipe surface, w_s, the pipe mid-surface, w_p, the streaks, w_k, and the embedded depth of the streaks, h_b, are plotted in Figure 8b. The contact pressure on the sealing surface is shown in Figure 8c.

As shown in Figure 8a, both of the streaks of the ridge and the pipe surface are deformed during interference. The local upheaval of the pipe surface further fills the grooves and leads to the change in the cross-section of the leakage channels.

As shown in Figure 8b, the deflection of the pipe surface, w_s, which is distinct from other deflections, has a plateau phase during the embedding, and correspondingly, the embedded depth, h_b, levels off just at the end of the plateau. Hence, the plateau indicates the interference process. The process can be divided into three phases: synchronous (in this case, $w_{b2} \in (0, 0.06)$), interference ($w_{b2} \in (0.6, 0.16)$), and form-fit ($w_{b2} > 0.16$).

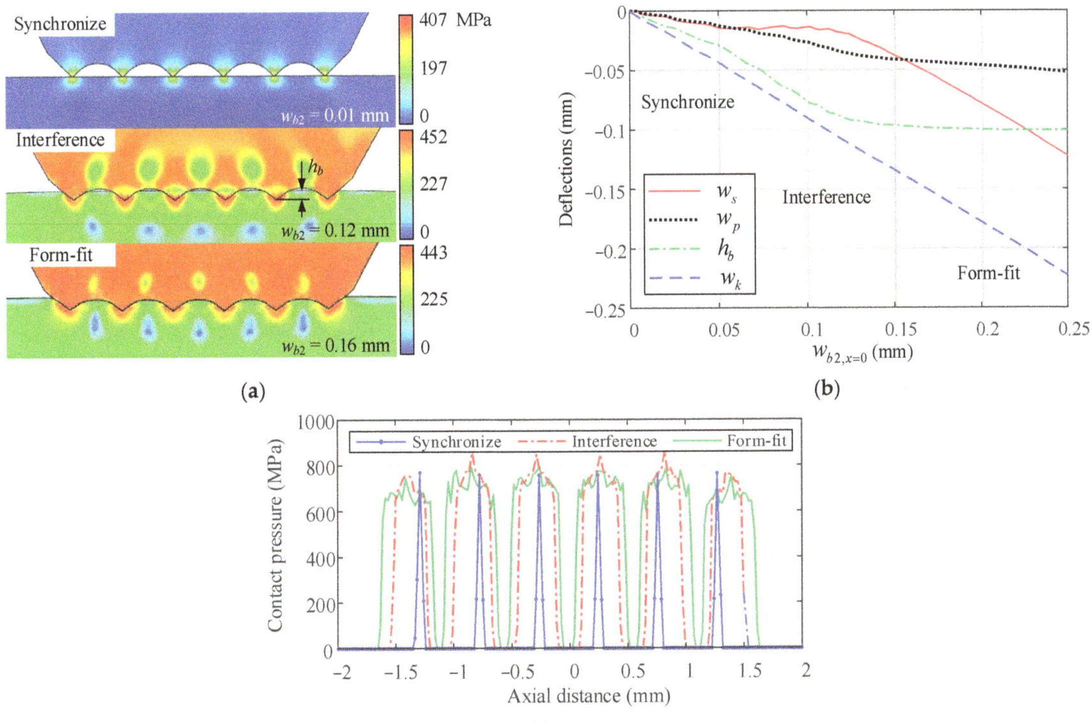

Figure 8. (**a**) Stress contour plots of corresponding deformation phases; (**b**) deflections and embedded depth of the streaks as a function of w_{b2}; (**c**) contact pressure on the sealing surface.

As shown in Figure 8c, the contact pressure is distributed axially with stress concentration areas (corresponding to streaks) and untouched areas (corresponding to the helical leakage channel).

The following change laws can be obtained:

- In the synchronize phase, the ridge and pipe deflect synchronously, the streaks are barely embedded in the pipe surface, and the stress concentration occurs at the contact points. Materials are not yielded;
- In the interference phase, streaks continue to be embedded in the pipe surface, the helical leakage channel is narrowed by the plastic deformations of the pipe, and the equispaced contact areas with high-contact pressure extend in the axial direction;
- In the form-fit phase, embedding is mainly performed with constant h_b; however, the plastic deformation of the seal area (the difference between w_s and w_p) continually increases as w_{b2} increases, which makes the seal tighter. The untouched areas narrow considerably and will eventually disappear with the increase in w_{b2}.

To verify the availability of the mesoscale FE model, the analytical values and the mesoscale FE results need to be compared. For w_x, the analytical preload value is 1.35×10^{-2} mm, while the FE range in Figure 8a is from 1.20×10^{-2} to 1.40×10^{-2} mm; for w_p, the analytical value is 1.54×10^{-2} mm, while the FE range is from 0.26×10^{-2} to 3.66×10^{-2} mm. Therefore, the analytical variables of both the base and the pipe are included in the mesoscale FE model. The availability of the FE model is verified.

5. Semi-Analytical Model of the Leakage

In this section, the leakage ratio is derived from the macroscopic analytical model and the mesoscopic FE model using the Abbott–Firestone curve.

According to the FE results in Section 4.3, the leakage channel and the relevant dimensions are shown in the Figure 9. The height of the leakage paths in the radial direction is h_a, and the embedded depth of the streaks is h_b. Since the axial leakage vanishes during the interference, the direction of the leakage (in Section 4.1) is different before and after the interference [11,47]. Therefore, for the compression connector, the leakage ratio of the helical leak dominates and the boundary condition is limited to radial clearance $h(w_x) < h_a$.

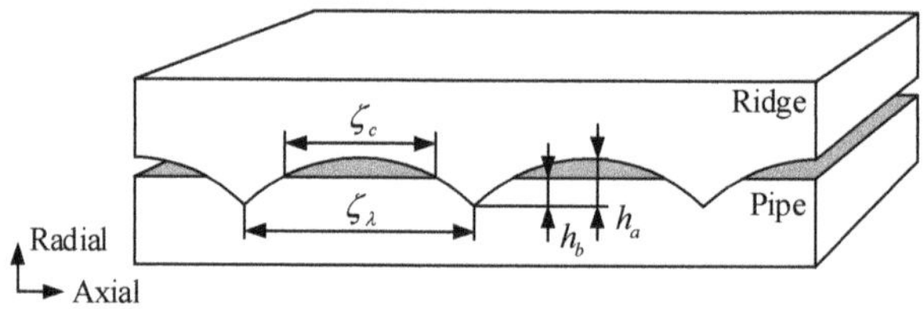

Figure 9. Illustration of the helical streaks embedding into the pipe surface.

To attribute the mesoscale profile parameters for calculating the seal leakage ratio, the Abbott–Firestone curve [48] is used. The effectiveness of the curve has been proved in describing the turning surface. The curve is shown in Figure 10.

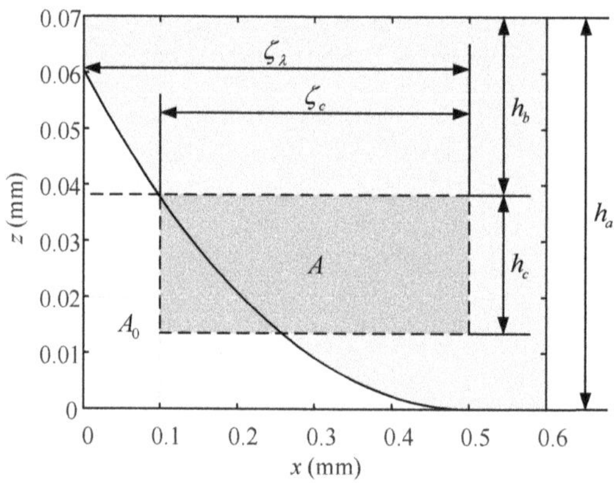

Figure 10. The Abbott–Firestone curve of the sealing surface.

The mathematical formula of the Abbott–Firestone curve can be represented by $z(x)$, which can be mathematically acquired by the fitting method of a polynomial. However, according to the geometric characteristics of the FE model and the parameters in Table 1, $z(x)$ can be given as

$$\frac{1}{4}(\zeta_\lambda - x)^2 + z^2(x) = r_c, x \in (0, \zeta_\lambda). \tag{23}$$

A hypothetical 3D rectangular channel can be used as the leakage path in the mathematical model to calculate the leakage ratio, as shown in Figure 11.

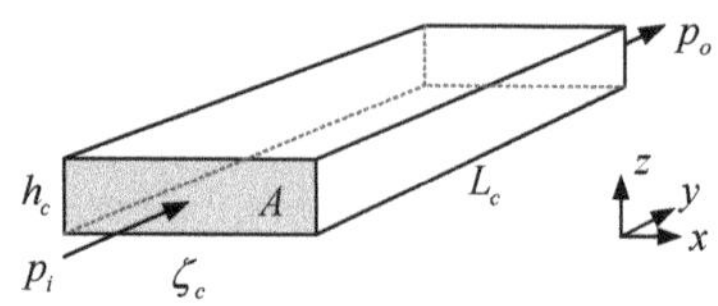

Figure 11. Hypothetical rectangular leakage channel.

The equivalent dimensions are width ζ_c, height h_c and length L_c, of which ζ_c and h_c are regarded as definite values along L_c. A represents the cross-sectional area of the hypothetical rectangular channel. Given the uncertainty of the pressure, the magnitude of the inlet pressure, p_i, and the outlet pressure, p_o, is assumed as $p_i \geq p_o$.

The integral value of the cross-sectional area of the streaks A_o is

$$A_o = \int_0^{\zeta_\lambda} z(x)dx. \tag{24}$$

Subjected to $h(w_x)$, the true cross-section of the leakage channel can be calculated by the following relations:

$$A = \zeta_\lambda \left(h_a - h_b(w_x)\right) - A_0. \tag{25}$$

The equivalent height of the leakage channel can be expressed as

$$h_c(w_x) = \frac{A}{\zeta_c(w_x)}. \tag{26}$$

The radius of the leakage path is given by

$$R_c = R_p + \frac{\delta_p + h_a - h_b(w_x)}{2} - w_x. \tag{27}$$

The length of the leakage channel can be defined as

$$L_c = 2\pi R_c \left(\frac{\zeta}{\zeta_\lambda} - 1\right). \tag{28}$$

The governing equation of the flow field based on the laminar flow can be defined using the Reynolds equations [49]:

$$\begin{aligned} &\nabla \cdot \left(\vec{q}_v\right) = 0.; \vec{q}_v = -\tfrac{h^3}{12\eta}\Delta p; \\ &p = p_i, \quad at\ x = x_i; \\ &p = p_o, \quad at\ x = x_o, \end{aligned} \tag{29}$$

where $\vec{q}_v$ is the volume flow ratio per unit width, η is the viscosity of the fluid, and p is the fluid pressure. The circumferential leakage ratio, Q_c, can be represented in the following form:

$$Q_c = \int_0^{\zeta_c} \int_0^{h_c} \frac{1}{2\eta}\frac{dp}{dy}(z^2 - h_c z)dz\, dx. \tag{30}$$

The leakage ratio, Q_c, as a function of the radial deflection, w_x, is given as:

$$Q_c(w_x) = \frac{(p_i^2 - p_o^2)\,\zeta_c(w_x)h_c(w_x)^3}{24\eta L_c p_o}. \tag{31}$$

The following equation introduces the notation:

$$K = \frac{(p_i^2 - p_o^2)}{24\eta L_c p_o}, \quad J(w_x) = \zeta_c(w_x)h_c(w_x)^3. \tag{32}$$

The leakage ratio can be represented in the simplified form:

$$Q_c(w_x) = K \cdot J(w_x). \tag{33}$$

6. Results and Discussion

6.1. Parameter Analysis

The proposed semi-analytical method can reveal the relationship between the deflection and the leakage condition. Therefore, the key parameters are those that have an impact on the relationship as they influence deflection and are easy to implement in design. Via multi-scale parameter analysis, key design and performance rules can be identified, and the minimum preloaded deflection can be obtained.

On the macroscopic level, the effects of the radius and clearance on deflections are first investigated. Because R_b is determined by R_p, δ, and h, by taking R_p as an independent variable while others are constant, the influences of R_p on w_x and w_p at $x = 0$ are plotted with logarithmic coordinates using Equations (15) and (20), as shown in Figure 12a. However, according to Equations (12) and (13), h is positively correlated with the deflection component, w_{b1}, and its influence on the total deflection, w_x, varies. As a reference for the design, with the increase in h, the corresponding change in h/w_x and its gradient is calculated, as shown in Figure 13b. Note that w_p is not shown in Figure 12b since it is irrelevant to h from Equation (15).

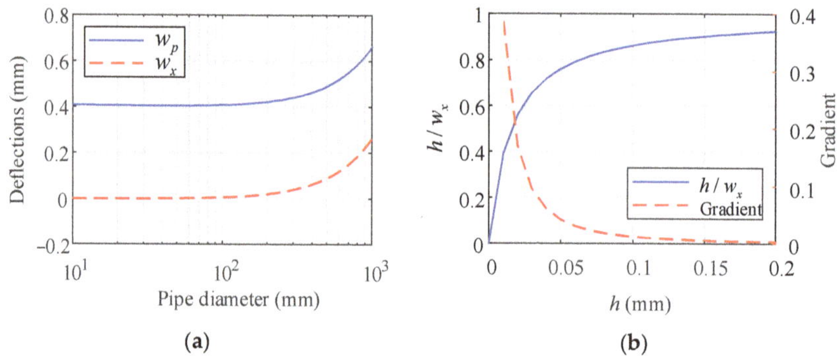

Figure 12. (**a**) Deflections as a function of pipe diameter and (**b**) ratio h/w_x as a function of radial clearance, h, at x = 0.

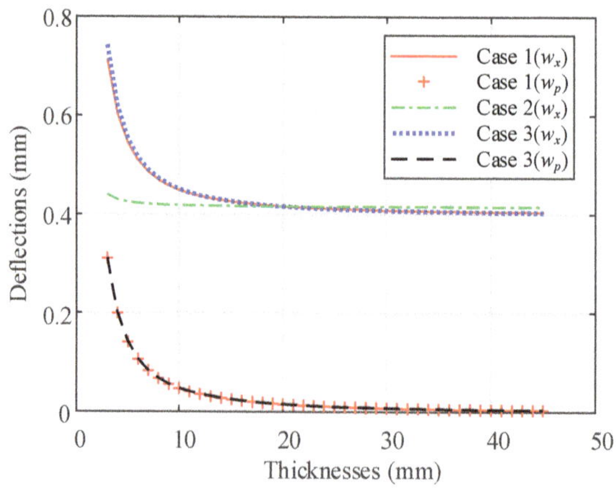

Figure 13. Deflections as a function of thickness at $x = 0$.

The following conclusions are drawn from Figure 12a:

- Compared to w_p, pipe diameter shows almost exact influence trends on w_x; hence R_p is positively correlated with both deflections;

- For all standard diameters in the ratio range of $\delta_p/R_p < 0.25$, within four-inch diameters ($R_p \leq 50$ mm), changes of w_x and w_p can be considered independent of R_p.

The following conclusions are drawn from Figure 12b:

- The increment of w_x is gradually related to h, and, in this instance, when $h > 0.2$ mm, the gradient of h/w_x approaches zero;
- There is a threshold of h (when $h = 0.2$) at which the growth of w_x is directly proportional to the growth of h.

Similarly to the radius, R, the thickness, δ, also shows a prominent influence on the deflections. The thickness of the base, δ_b, and the pipe, δ_p, are analyzed under three conditions: in Case 1, δ_b is constant and δ_p is variable; Case 2 has the opposite conditions to Case 1; in Case 3, $\delta_b = \delta_p$ is considered. Note that, according to Equation (15), δ_b is irrelevant to w_p. The influences of δ on the deflections are shown in Figure 13. The curve in Case 1 (w_x) is the influence of δ_p on w_x while $\delta_b = 20$ mm, and the w_p in Case 1 is the influence on w_p. The w_x in Case 2 is the influence of δ_b on w_x while $\delta_p = 20$ mm. The w_x in Case 3 is the influence of δ_p on w_x while $\delta_b = \delta_p$, and the w_p in Case 3 is the influence on w_p.

The following conclusions are drawn from Figure 13:

- As the δ_p increases, all deflections decrease with a decreased change rate;
- δ_b has a minimum influence on w_x (the curve w_x in Case @), and is irrelevant to w_p, so the change in δ_b has a negligible impact on the interference. Therefore, consideration may be given to appropriately increasing or decreasing the thickness, δ_b, as needed.

In addition to the above parameters, deflections are also affected by the sealing width. For generality, a ratio of the pipe thickness to the sealing width, δ_p/ζ, is used. The deflection as a function of δ_p/ζ is shown in Figure 14 with logarithmic coordinates. Moreover, since the ratio also affects flexural rigidity, the flexural rigidity of the pipe, D_p, is also shown in the figure.

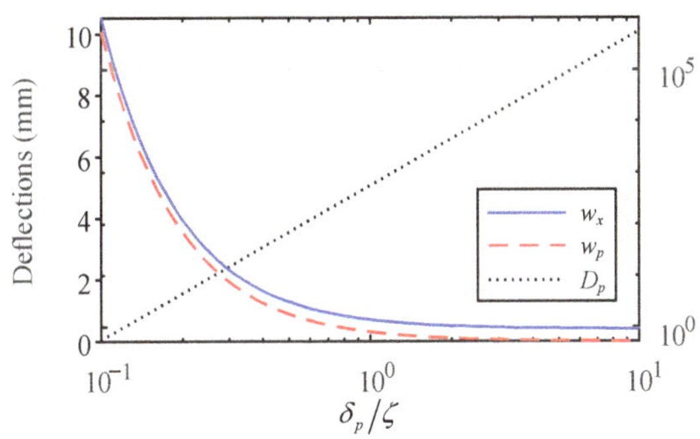

Figure 14. Deflections as a function of ratio δ_p/ζ.

As shown in Figure 14, deflections and D_p have different responses:

- For the flexural rigidity, a linear relationship exists between the ratio and D_p in logarithmic coordinates;
- For deflections, they are influenced by a threshold of the ratio: when $\delta_p/\zeta \leq 1$, deflections decrease as the ratio increases; when $\delta_p/\zeta > 1$, the deflection change is practically unrelated to the ratio. This shows that when the thickness exceeds the sealing width, δ_p/ζ hardly affects deflection.

Next, the effects of D_b and D_p on deflections are also investigated as depicted in the logarithmic coordinates in Figure 15. Since w_p is irrelevant to D_b, its curve is not shown.

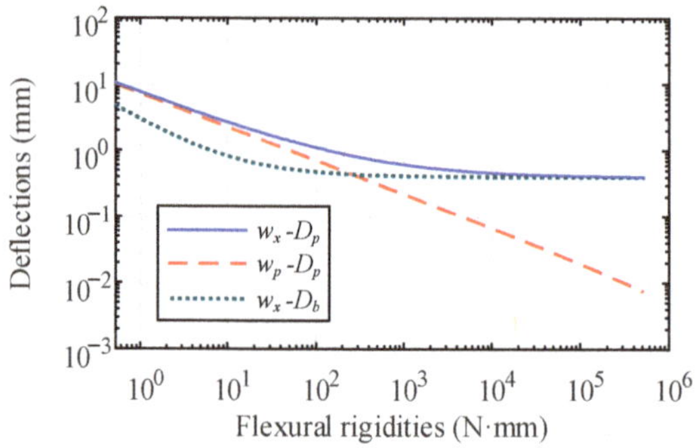

Figure 15. Deflections as a function of flexural rigidities.

As shown in Figure 15, D_p exhibits distinct effects on w_p and w_x:

- D_p exhibits an almost linear relationship with w_p (the curve w_p-D_p) in logarithmic coordinates but a decreasing effect on w_x (w_x-D_p);
- When $D_p < 10^3$, all effects on deflections are approximately linear, but when $D_p \geq 10^3$, the effect on w_x and w_p is significantly different;
- When $D_p \geq 10^4$, the only effect of the flexural rigidity is that of D_p on w_p;
- The effect on w_x of the flexural rigidity of D_b (w_x-D_b) is comparable to that of the pipe (w_x-D_p).

On the mesoscopic level, the leakage condition can be analyzed. The influence of w_x on the nonlinear change in the cross-sectional area of the leakage channel A is investigated first. Based on the mesoscopic FE model (in Section 4), the true width, ζ_c, and the equivalent height, h_c, of A are shown in Figure 16 with w_x increments, where the corresponding radius of the leakage path, R_c, is also depicted to indicate the relationship between macroscopic and mesoscopic changes.

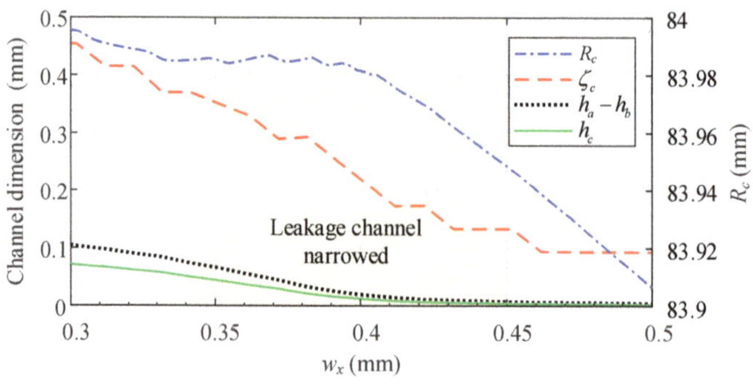

Figure 16. Leakage channel dimension and sealing radius as a function of base deflection.

The following conclusions can be drawn from Figure 16:

- There is a strong correlation between w_x and ζ_c: as w_x increases, both ζ_c and h_c decrease;
- Coincidentally with the macroscopic results in Section 4.3, the true height of the channel ($h_a - h_b$) and h_c also have a notable reduction during the interference within the range $w_x \in (0.34, 0.45)$, while h_c shows a clear descent stage;
- When $w_x > 0.45$ mm, ζ_c and h_c are virtually unchanged, indicating that the leakage ratio is stable.

Based on the above findings, the minimum preloaded deflection can be predicted. Since Q_c varies with different sealed mediums, the typical leakage ratios of water

($\eta = 1.77 \times 10^5$ Pa·s), air ($\eta = 1.01 \times 10^3$ Pa·s), and oil ($\eta = 1.50$ Pa·s), with inlet pressure, $p_i = 1$ MPa, and outlet pressure, $p_o = 101$ KPa, can be calculated as a function of w_x, as shown in Figure 17 in the logarithmic coordinates. The corresponding R_c is also shown. Other unlisted parameters are consistent with the above analyses.

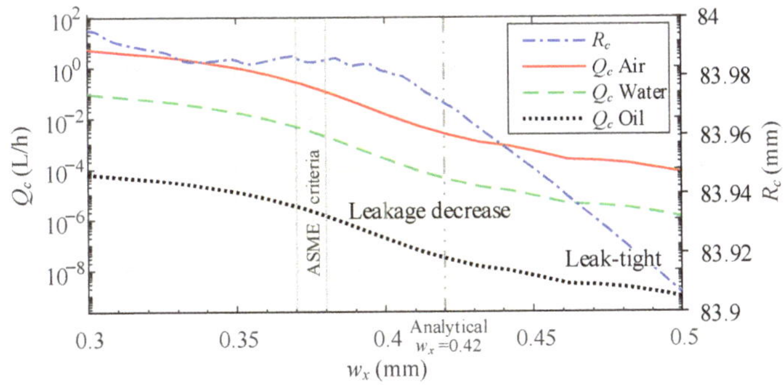

Figure 17. Leakage ratios as a function of base deflection.

As shown in Figure 17, the air leakage ratio is distinct from other mediums, which makes a great difference to the sealing performance of connectors. Accordingly, the following conclusions are drawn:

- For air medium, according to the highest Class VI of ASME B16.104 [50], the connector can be considered leak-tight below, $Q_c = 0.24$ L/h in the study case, while the corresponding minimum w_x is in the range from 0.37 to 0.38 mm (0.25 to 0.10 L/h, shadowed area in the figure);
- Similar to the equivalent height curve, h_c, in Figure 16, there is also an accelerated descent range of the leakage ratio and all leakage ratios are reduced to a minimum with $w_x > 0.45$ mm;
- The macroscopic minimum, w_x, calculated from the proposed analytical model is 0.42 mm (the dash-double-dotted line in the figure), which is covered by the ASME criteria;
- The ASME criteria and the theoretical minimum, w_x, are included within the leakage decrease range (from 0.37 to 0.45). Therefore, the results of the semi-analytical method are self-consistent, and the case connector is theoretically leak-tight to all typical mediums. The theoretical feasibility of the proposed semi-analytical method to predict the minimum preloaded deflection is verified.

At the end of the parameter analysis, considering the negligible effect of the thickness, the leakage ratio as a function of the rate, δ_p/ζ, is explored in Figure 18 with logarithmic coordinates. The corresponding R_c and ASME criteria of the air leakage ratio (from 0.53 to 0.20 L/h, shadowed area in the figure) are also shown.

The following conclusions can be drawn from Figure 18:

- When $\delta_p/\zeta \in (1, 3]$, w_x and Q_c begin descending with the increase in δ_p/ζ;
- When $\delta_p/\zeta \in (3, 5]$, the leakage is noticeably decreasing;
- When $\delta_p/\zeta > 5$, the effects on w_x, Q_c, and r_c are moderated, the leakage is stable, and the sealing state needs to be further verified by prototype tests;
- When $\delta_p/\zeta \leq 3$, the leakage may occur on the internally turned topography, the recommended design value is $\delta_p/\zeta \geq 5$.

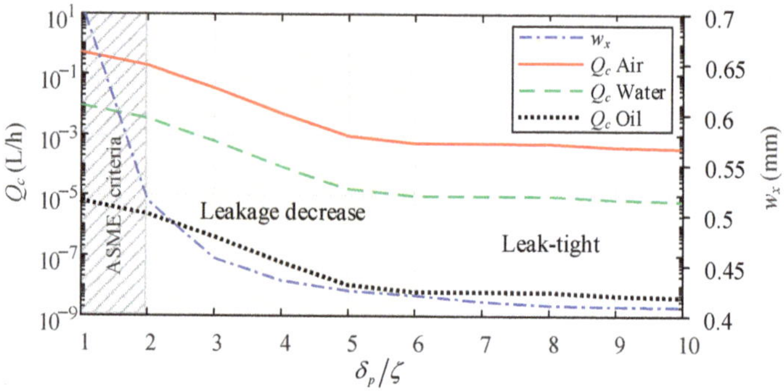

Figure 18. Leakage ratios as a function of rate δ_p/ζ.

6.2. Experimental Test

In order to verify the feasibility of the proposed semi-analytical method on the sealing performance prediction, according to ISO 21329:2004 [51], a working pressure test and a burst pressure test are carried out. All tests are conducted under the temperature of 25 °C and the maximum gauge pressure is 60 MPa with an accuracy rating of 0.4 (0.24 MPa). Furthermore, a customized 80 MPa manual hydraulic pump and an air compressor are used. The schematic diagram of the experimental setup is shown in Figure 19.

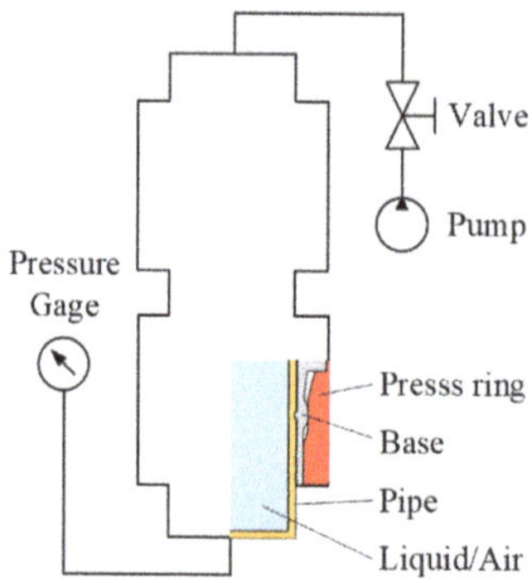

Figure 19. Schematic diagram of the experimental setup.

Prototypes are developed based on the proposed semi-analytical approach using seamless wrought steel pipes (ASME B36 10M [36]) of 3 to 8 inches. Geometrical parameters are listed in Table 2. A few redesigned prototypes are shown in Figure 20.

Table 2. Geometric and material parameters for prototype types.

Type		Nominal Diameter (mm)	Nominal Thickness (mm)	Medium		Yield Strength (MPa)	
DN	Inch			Water	Air	Pipe	Connector
80	3	89	5	15	0.2	235	345
100	4	114	7	7	0.1		
150	6	168	8	6	0.1		
200	8	219	10	4.5	0.08		

(a) (b)

Figure 20. Prototypes: (**a**) DN 200 type and (**b**) DN 100 type.

6.2.1. Working Pressure Test

The purpose of the test is to verify whether the designed prototype can reach the designed working pressure. The mediums are clean air and water. For the air pressure test, the prototype is pressurized with air at the required depth of an experimental tank. During the working pressure test, the internal pressure gradually increases to the required working pressure, rising below a maximum rate of 100 MPa/min. When the working pressure is reached, to stabilize the system and periodically check the assembly for the leakage it is necessary to maintain constant pressure for a sufficient time (10 min for air pressure and 24 h for hydraulic pressure in the study). Neither structural failure nor visual indications of the leakage are permitted. The setups of the air pressure test and the hydraulic pressure test are shown in Figure 21. A sectional view of the tested prototype is shown in Figure 22.

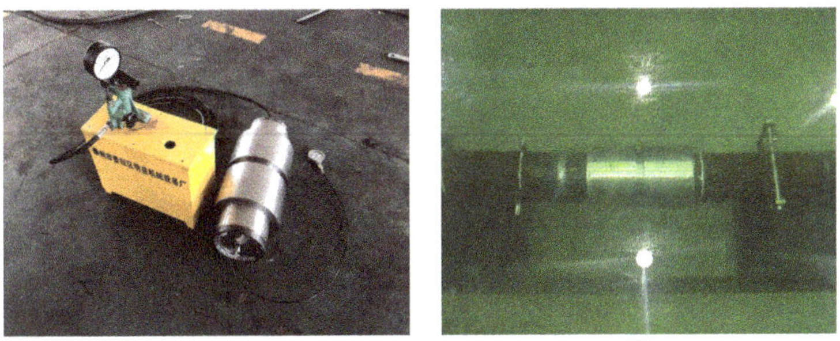

(a) (b)

Figure 21. Experiment setups with assemblies of (**a**) hydraulic pressure test and (**b**) air pressure test.

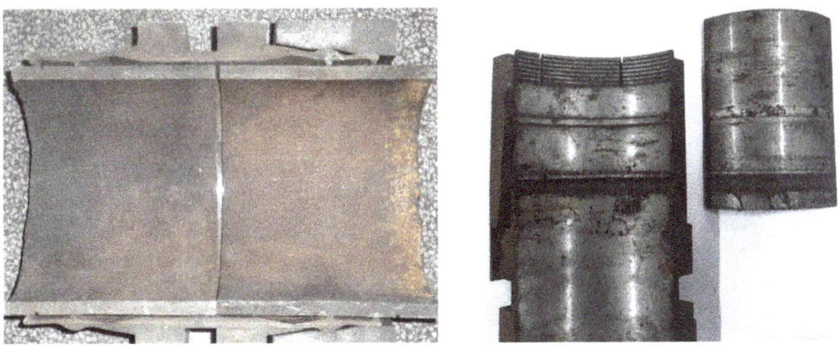

(a) (b)

Figure 22. Section view of assemblies of (**a**) DN 200 and (**b**) DN 80.

From Figure 22, the internal deflection of the prototypes conforms to the design expectation, there are no visual indications of leakage during sampling inspection with different sealing widths under variable structure configurations, and specimens are in good contact with seals. The sealing surface condition is not detailed here since it has already been discussed in our previous studies [20,22].

Typical test results for the prototypes are shown in Table 3, indicating that all prototypes are leak tight. Therefore, the prototypes developed by the proposed semi-analytical method meet the working pressure requirement.

Table 3. Typical results of the working pressure test.

Type		Pressure (MPa)		Pressure Drop		Leakage		Structural Change	
DN	Inch	Water	Air	Water	Air	Water	Air	Water	Air
80	3	15	0.2	×	×	×	×	×	×
100	4	7	0.1	×	×	×	×	×	×
150	6	6	0.1	×	×	×	×	×	×
200	8	4.5	0.08	×	×	×	×	×	×

6.2.2. Burst Pressure Test

The purpose of the test is to find the failure pressure of the designed prototypes. To avoid the risk of the compression air, only the hydraulic pressure was applied. Following the leak test, the pressure is continuously raised below the maximum pressurization rate, and the pressure is maintained for 5 min per 1 MPa until there is evidence of the structural failure or leakage. The mean failure pressures for the prototypes are listed in Table 4. For comparison, the results of the same prototype types with polished sealing surfaces from our previous design by FE methods [20,22] are also listed.

Table 4. Failure pressures of burst test with failure types and comparison with previous designs.

Type		Mean Burst Pressure (MPa)		Pressure Drop	Leakage	Structural Change
DN	Inch	Current	Previous			
80	3	45	36	√	×	√
100	4	33	30	√	×	√
150	6	20	20	×	×	√
200	8	12	10	×	×	√

According to Table 4, there is no significant difference between the analytical design and the previous FE design, indicating that the semi-analytical solution does not have a considerable effect on the previous solution. In addition, the experiments did not show any leakage. All failures are structural, and most are manifested as the axial pipe movement. In the failure, the internal pressure no longer increased and decreased slightly in some tests. This type of failures involves axial anchoring failure and pull-out deformation, which will be discussed in a follow-up study.

To summarize, the results show that even with a rough radial sealing surface, the compression connector designed by the proposed method can quickly create a reliable radial seal. On the other hand, the computational time has been reduced, saving more than 57% design time in practice. Meanwhile, the internal finish turning machining is no longer necessary, which further reduces the time and cost of the manufacturing process.

7. Conclusions

A semi-analytical method for the sealing performance prediction of the subsea pipeline compression connector is proposed in this paper. Prototypes are fabricated on the basis of the design results with this method and the failure pressures are tested in lab. The following conclusions can be drawn:

An analytical model of the fitting process of the compression connector is derived, based on the membrane theory, in which the assembly is idealized as pairs of thin-walled cylinders under uniformly distributed radial load. The expressions of the deflections with the contact pressure are obtained;

1. A macroscopic 2D FE model is created to verify the analytical model. An agreement is found between the analytical and FE results with a mean error of less than 8.5%. The analytical model is proven to be good enough with the thickness–radius ratio being no more than 0.25 ($\delta/R \leq 0.25$);
2. The geometric parameter effects on the interference process are investigated. The influence and thresholds of the geometric parameters on sealing performance are presented, which provide the design guidance for the subsea pipeline compression connector;
3. The radial sealing surface of the connector is observed by SEM. The surface is an anisotropic topography characterized by regular helical and continuous streaks with slight defects and irregularities. Potential leak paths in two orthogonal directions are identified as axial and circumferential;
4. A mesoscopic 2D FE model is created based on the SEM result and the macroscopic FE model. The effects of the leakage channel in the sealing process are analyzed. Then, the analytical minimum deflections and those of FE are compared and shown to be consistent with each other. Since the macroscopic analytical minimum deflections of both the base and the pipe are included in the mesoscale results, it is theoretically verified that the semi-analytical method can identify and predict the leakage condition of the connector;
5. The Abbott–Firestone curve of the sealing surface is introduced based on SEM observation and the FE model. A semi-analytical leakage model was obtained. Since the true height of the leakage channel shows a clear descent stage, the minimum preloaded deflection of the leak-tight range is identified. On the rough internally turned sealing surface, leakage may occur when the ratio of the pipe thickness to the sealing width is $\delta_p/\zeta < 3$, and the recommended leak-tight design value is $\delta_p/\zeta > 5$;
6. The results with the proposed semi-analytical method correlate well with the working pressure test and the burst pressure test. The designed connectors can quickly create a reliable radial metal static seal with the rough turning topography and save more than 57% of design time compared to the common FE method, while reducing machining time and costs.
7. The proposed semi-analytical method can be used to optimize the overall structural performance of compression connectors with a low computational cost. It allows the prediction and identification of the leakage condition of all standard pipe series with $\delta_p/R_p \leq 0.25$. The model also provides a good understanding of the radial sealing process. These results are helpful to highlight the different sealing performances with other type of seals and guide the further improvements.

In order to perform further quantitative analyses of the leakage ratio, it is necessary to explore the tangential contact mechanics and the deflection behaviors of the connector in subsequent studies.

Author Contributions: Conceptualization, Z.L., G.W. and L.W.; methodology, Z.L.; software, Z.L.; validation, Z.L. and L.W.; formal analysis, Z.L.; investigation, G.W.; resources, L.W. and C.L.; data curation, G.W. and S.Y.; writing—original draft preparation, Z.L.; writing—review and editing, S.Y.; visualization, Z.L.; supervision, P.J. and F.Y.; project administration, G.W.; funding acquisition, L.W. All authors have read and agreed to the published version of the manuscript.

Funding: This research was funded by the National Natural Science Foundation of China, grant number 51779064, 52001089 and the China Postdoctoral Science Foundation, grant number 2020M670889.

Institutional Review Board Statement: Not applicable.

Informed Consent Statement: Not applicable.

Data Availability Statement: The data used to support the findings of this study are included within the article.

Conflicts of Interest: The authors declare no conflict of interest.

Nomenclature

A	Cross-sectional area of the hypothetical rectangle channel (mm^2)
A_o	Integral value of the total Abbott–Firestone curve (mm^2)
D, D_b, D_p	Flexural rigidity of cylindrical shell, base, and pipe, respectively (N·mm)
E	Young's modulus (MPa)
E_t	Tangent modulus (MPa)
h	Radial clearance between sealing surfaces (mm)
h_a	Height of streaks (mm)
h_b	Depth of embedded streaks (mm)
h_c	Equivalent height of the hypothetical rectangular channel (mm)
L_0	Axial distance between radial load and ridge (mm)
L_1	Axial distance between anchors and ridge (mm)
L_c	Length of the leakage channel (mm)
M_φ, M_x	Bending moments of axial section and a section perpendicular to the axis of a cylindrical shell, respectively (N·mm)
N_φ, N_x	Membrane forces of axial section and a section perpendicular to the axis of a cylindrical shell, respectively (N)
p_i, p_o	Inlet the outlet pressures of the leakage channel (MPa)
P_{b1}, P_{b2}	Components of radial load P_z (N/mm)
P_r	Reaction load (N/mm)
P_x	Shearing force of intensity parallel to z axis of a section perpendicular to the axis of a cylindrical shell (N)
P_z	Total equivalent radial load on the base (N/mm)
Q_c	Helical leakage ratio (L/h)
r_c	Corner radius of the tool nose (mm)
R, R_b, R_p	Radius of mid-surface of cylinder, base, and pipe, respectively (mm)
R_c	Radius of leakage path (mm)
u, v, w	Displacements of the mid-surface in the cylindrical coordinate (mm)
w_{b1}, w_{b2}	Deflection components of P_z (mm)
w_x, w_p	Radial deflection of the base and pipe, respectively (mm)
w_k	Deflection of the streak peaks (mm)
w_r	Deflection component of P_r (mm)
w_s	Deflection of pipe surface at the seal (mm)
x, y, z	Global rectangular coordinates of 2D axisymmetric FE model
$\delta, \delta_b, \delta_p$	Thickness of cylinder, base, and pipe, respectively (mm)
$\varepsilon_\varphi, \varepsilon_x$	Strain components of cylindrical coordinates
ε^T	True strain
ζ	Sealing width (mm)
ζ_c	Width of the leakage channel (mm)
ζ_λ	Pitch of the streaks (feed rate) (mm)
η	Kinematic viscosity (Pa·s)
μ	Friction coefficient
υ	Poisson's ratio
ξ	Hardening modulus (MPa)
σ_{sb}, σ_{sp}	Yield strength of the base and pipe, respectively (MPa)
σ^T	True stress (MPa)
φ, x, ρ(=R+z)	Cylindrical coordinates

References

1. Yun, F.; Liu, D.; Xu, X.; Jiao, K.; Hao, X.; Wang, L.; Yan, Z.; Jia, P.; Wang, X.; Liang, B. Thermal–Structural Coupling Analysis of Subsea Connector Sealing Contact. *Appl. Sci.* **2022**, *12*, 3194. [CrossRef]
2. *ASTM F1387*; Standard Specification for Performance of Piping and Tubing Mechanically Attached Fittings. ASTM International: West Conshohocken, PA, USA, 2012.
3. *ISO 8434-1*; Metallic Tube Connections For Fluid Power and General Use Part 1:24° Compression Fittings. International Organization for Standardization: Geneva, Switzerland, 2018.
4. Okada, H.; Itoh, T.; Suga, T. The Influence of Surface Profiles on Leakage in Room Temperature Seal-Bonding. *Sens. Actuators A Phys.* **2008**, *144*, 124–129. [CrossRef]
5. Lebeck, A.O. Hydrodynamic Lubrication in Wawy Contacting Face Seals-a Two Dimensional Iodei. *J. Tribol.* **1981**, *103*, 578–586. [CrossRef]
6. Matsuzaki, Y.; Kazamaki, T. Effect of Surface Roughness on Compressive Stress of Static Seals. *JSME Int. J. Ser. 3* **1988**, *31*, 99–106. [CrossRef]
7. Yanagisawa, T.; Sanada, M.; Koga, T.; Hirabayashi, H. The Influence of Designing Factors on the Sealing Performance of C-Seal. *SAE Trans.* **1991**, *100*, 651–657.
8. Nakamura, T.; Funabashi, K. Effects of Directional Properties of Roughness and Tangential Force on Pressure Flow between Contacting Surfaces. *Lubr. Sci.* **1991**, *4*, 13–23. [CrossRef]
9. Polycarpou, A.A.; Etsion, I. A Model for the Static Sealing Performance of Compliant Metallic Gas Seals Including Surface Roughness and Rarefaction Effects. *Tribol. Trans.* **2000**, *43*, 237–244. [CrossRef]
10. Geoffroy, S.; Prat, M. On the Leak through a Spiral-Groove Metallic Static Ring Gasket. *J. Fluids Eng. Trans. ASME* **2004**, *126*, 48–54. [CrossRef]
11. Liao, C.; Xu, X.; Fang, H.; Wang, H.; Man, M. A Leakage Model of Metallic Static Seals Based on Micromorphology Characteristics of Turning Flange Surface. *Ind. Lubr. Tribol.* **2015**, *67*, 572–581. [CrossRef]
12. Beheshti, A.; Khonsari, M.M. Asperity Micro-Contact Models as Applied to the Deformation of Rough Line Contact. *Tribol. Int.* **2012**, *52*, 61–74. [CrossRef]
13. Jackson, R.L.; Green, I. On the Modeling of Elastic Contact between Rough Surfaces. *Tribol. Trans.* **2011**, *54*, 300–314. [CrossRef]
14. Marie, C.; Lasseux, D.; Zahouani, H.; Sainsot, P. An Integrated Approach to Characterize Liquid Leakage through Metal Contact Seal. *Eur. J. Mech. Environ. Eng.* **2003**, *48*, 81–86.
15. Liao, Y.; Stephenson, D.A.; Ni, J. Multiple-Scale Wavelet Decomposition, 3D Surface Feature Exaction and Applications. *J. Manuf. Sci. Eng.* **2012**, *134*, 011005. [CrossRef]
16. Deltombe, R.; Bigerelle, M.; Jourani, A. Analysis of the Effects of Different Machining Processes on Sealing Using Multiscale Topography. *Surf. Topogr. Metrol. Prop.* **2016**, *4*, 015003. [CrossRef]
17. Bricaud, C.; Schulz, O.; Zierer, T.; Peltier, V.; Schwitzke, C.; Bauer, H.J. Experimental Investigations Into the Effect of Surface Roughness and Contact Force on Leakage Between Two Rigid Metallic Surfaces. *J. Eng. Gas Turbines Power* **2022**, *144*, 061006. [CrossRef]
18. Weddeling, C.; Demir, O.K.; Haupt, P.; Tekkaya, A.E. Analytical Methodology for the Process Design of Electromagnetic Crimping. *J. Mater. Process. Technol.* **2015**, *222*, 163–180. [CrossRef]
19. Weddeling, C.; Walter, V.; Haupt, P.; Tekkaya, A.E.; Schulze, V.; Weidenmann, K.A. Joining Zone Design for Electromagnetically Crimped Connections. *J. Mater. Process. Technol.* **2015**, *225*, 240–261. [CrossRef]
20. Wei, Z.; Wang, L.; Guan, Y.; Yao, S.; Li, S. Static Metal Sealing Mechanism of a Subsea Pipeline Mechanical Connector. *Adv. Mech. Eng.* **2016**, *8*, 1–16. [CrossRef]
21. Henriksen, J.; Hansen, M.R.; Thrane, F.C. Axial Load Capacity of Cold Formed Pipe Flange Connection. *J. Press. Vessel Technol.* **2017**, *139*, 051201. [CrossRef]
22. Wang, L.Q.; Wei, Z.L.; Yao, S.M.; Guan, Y.; Li, S.K. Sealing Performance and Optimization of a Subsea Pipeline Mechanical Connector. *Chin. J. Mech. Eng. Engl.* **2018**, *31*. [CrossRef]
23. Onyegiri, I.; Kashtalyan, M. Threaded Connectors for Sandwich Pipes—Part 1: Parametric & Comparative Studies. *Int. J. Press. Vessel. Pip.* **2018**, *168*, 117–124. [CrossRef]
24. Quispe, J.L.P.; Pasqualino, I.P.; Estefen, S.F.; de Souza, M.I.L. Structural Behavior of Threaded Connections for Sandwich Pipes under Make-up Torque, External Pressure, and Axial Load. *Int. J. Press. Vessel. Pip.* **2020**, *186*, 104156. [CrossRef]
25. Yan, Y.; Fan, Y. Influence of Fluid on Seal and Assembly of Pipeline Fittings Based on the Multiscale Finite Element Model. *Complexity* **2020**, *2020*, 4960734. [CrossRef]
26. Agrawal, A.K.; Narayanan, R.G. Experimental and Numerical Studies on Joining Steel Tubes by End Forming. *J. Constr. Steel Res.* **2020**, *167*, 105792. [CrossRef]
27. Liu, Z.; Zhang, L.; Wang, F.; Li, S.; Wang, P.; Cai, M.; Han, L.; Ma, Y.; Ma, Z.; Yan, B. Study on Optimization Design of Permanent Packer Slip Structure. *J. Fail. Anal. Prev.* **2020**, *21*, 50–60. [CrossRef]
28. Wu, L.; Guo, X.; Chen, H.; Liu, J.; Su, Y. Calculation Method for the Axial Load-Bearing Capacity of Steel Pipe-to-Sleeve Grouted Connections. *Constr. Build. Mater.* **2022**, *314*, 125621. [CrossRef]
29. Weddeling, C.; Gies, S.; Khalifa, N.B.; Tekkaya, A.E. Analytical Methodology for the Process and Joint Design of Form-Fit Joining by Die-Less Hydroforming. In Proceedings of the ASME 2014 International Manufacturing Science and Engineering Conference,

MSEC 2014 Collocated with the JSME 2014 International Conference on Materials and Processing and the 42nd North American Manufacturing Research Conference, Web Portal ASME (American Society of Mechanical Engineers), Detroit, MI, USA, 9–13 June 2014; Volume 2.
30. Ernens, D.; Pérez-Ràfols, F.; van Hoecke, D.; Roijmans, R.F.H.; van Riet, E.J.; Vande Voorde, J.B.E.; Almqvist, A.; de Rooij, M.B.; Roggeband, S.M.; van Haaften, W.M.; et al. On the Sealability of Metal-to-Metal Seals with Application to Premium Casing and Tubing Connections. *SPE Drill. Complet.* **2019**, *34*, 382–396. [CrossRef]
31. Weber, F.; Gebhard, J.; Gitschel, R.; Goyal, S.; Kamaliev, M.; Wernicke, S.; Tekkaya, A.E. Joining by Forming–A Selective Review. *J. Adv. Join. Process.* **2021**, *3*, 100054. [CrossRef]
32. Orynyak, I.; Bai, Y. Coupled Approximate Long and Short Solutions versus Exact Navier and Galerkin Ones for Cylindrical Shell under Radial Load. *Thin-Walled Struct.* **2022**, *170*, 108536. [CrossRef]
33. Kumar, A.; Das, S.L.; Wahi, P.; Żur, K.K. On the Stability of Thin-Walled Circular Cylindrical Shells under Static and Periodic Radial Loading. *J. Sound Vib.* **2022**, *527*, 116872. [CrossRef]
34. *Abaqus*, Version 6.14. Software for Technical Simulation. Dassault Systemes: Vélizy-Villacoublay, France, 2014.
35. Rojíček, J.; Čermák, M.; Halama, R.; Paška, Z.; Vaško, M. Material Model Identification from Set of Experiments and Validation by DIC. *Math. Comput. Simul.* **2021**, *189*, 339–367. [CrossRef]
36. *ASME B36.10M*; Welded and Seamless Wrought Steel Pipe. American Society of Mechanical Engineers: New York, NY, USA, 2018.
37. *ASME B36.19M*; Stainless Steel Pipe. American Society of Mechanical Engineers: New York, NY, USA, 2018.
38. Yang, Y.; Zhu, H.; He, D.; Du, C.; Xu, L.; He, Y.; Zheng, Y.; Ye, Z. Contact Mechanical Behaviors of Radial Metal Seal for the Interval Control Valve in Intelligent Well: Modeling and Theoretical Study. *Energy Sci. Eng.* **2020**, *8*, 1337–1352. [CrossRef]
39. Liu, Y.; Du, H.; Ren, X.; Li, B.; Qian, J.; Yan, F. A Leakage Rate Model for Metal-to-Metal Seals Based on the Fractal Theory of Porous Medium. *Aerospace* **2022**, *9*, 779. [CrossRef]
40. Manikandan, H.; Chandra Bera, T. Modelling of Dimensional and Geometric Error Prediction in Turning of Thin-Walled Components. *Precis. Eng.* **2021**, *72*, 382–396. [CrossRef]
41. Abdalrahman, R.M.; Sure, S.A.H. The Influence of Cutting Edge Angles Included Angle and Nose Radius on Surface Finish of Aluminum Alloy 1050 in Turning. *Zanco J. Pure Appl. Sci.* **2020**, *32*, 31–38. [CrossRef]
42. Shao, Y.; Yin, Y.; Du, S.; Xia, T.; Xi, L. Leakage Monitoring in Static Sealing Interface Based on Three Dimensional Surface Topography Indicator. *J. Manuf. Sci. Eng.* **2018**, *140*, 101003. [CrossRef]
43. *ISO 16610-41-2015*; Geometrical Product Specifications (GPS)-Filtration-Part 41: Morphological Profile Filters: Disk and Horizontal Line-Segment Filters. International Organization for Standardization: Geneva, Switzerland, 2015.
44. *ISO 16610-85-2013*; Geometrical Product Specifications (GPS)-Filtration-Part 85: Morphological Areal Filters: Segmentation. International Organization for Standardization: Geneva, Switzerland, 2015.
45. Chen, J.J.; Shih, Y.S. Study of the Helical Effect on the Thread Connection by Three Dimensional Finite Element Analysis. *Nucl. Eng. Des.* **1999**, *191*, 109–116. [CrossRef]
46. Toupin, R.A. Saint-Venant's Principle. *Arch. Ration. Mech. Anal.* **1965**, *18*, 83–96. [CrossRef]
47. Nitta, I.; Matsuzaki, Y. Experimental Study of the Performance of Static Seals Based on Measurements of Real Contact Area Using Thin Polycarbonate Films. *J. Tribol.* **2010**, *132*, 022202. [CrossRef]
48. Georgescu, C.; Cristea, G.C.; Dima, C.; Deleanu, L. Evaluating Lubricating Capacity of Vegetal Oils Using Abbott-Firestone Curve. *IOP Conf. Ser. Mater. Sci. Eng.* **2017**, *174*, 012057. [CrossRef]
49. Prat, M.; Plouraboué, F.; Letalleur, N. Averaged Reynolds Equation for Flows between Rough Surfaces in Sliding Motion. *Transp. Porous Media* **2002**, *48*, 291–313. [CrossRef]
50. *ASME B16.104;* Control Valve Seat Leakage. American Society of Mechanical Engineers: New York, NY, USA, 1998.
51. *ISO 21329:2004;* Petroleum And Natural Gas Industries-Pipeline Transportation Systems-Test Procedures for Mechanical Connectors. International Organization for Standardization: Geneva, Switzerland, 2004.

Journal of
Marine Science and Engineering

MDPI

Article

Enhancing Power Transmission Stability of AUV's Wireless Power Transfer System with Compact Planar Magnetic Coupler

Haibing Wen [1], Jiayuan Li [1], Kehan Zhang [2,*], Jinying Ye [3], Zhengchao Yan [4], Baowei Song [2] and Xiangqian Tong [1]

1 School of Electrical Engineering, Xi'an University of Technology, Xi'an 710048, China
2 School of Marine Science and Technology, Northwestern Polytechnical University, Xi'an 710072, China
3 School of Astronautics, Northwestern Polytechnical University, Xi'an 710072, China
4 School of Electrical Engineering, Xi'an Jiaotong University, Xi'an 710049, China
* Correspondence: zhangkehan210@163.com

Abstract: In this paper, a wireless power transfer (WPT) system with a compact planar magnetic coupler for an autonomous underwater vehicle (AUV) is proposed. A passive induction (PI) coil is integrated into the circular transmitter (Tx) coil to build a uniform magnetic field (UMF), which can guarantee the stable output of the WPT system under uncertain radial and axial misalignments for AUV. Based on normalized magnetic induction intensity distribution analysis, a UMF constructing method with a PI coil is given, aiming to eliminate the fluctuation of magnetic field intensity, and the PI coil design principles and flow chart are obtained. The theoretical analysis shows the proposed integrated coil can effectively enhance the radial misalignment tolerance compared with a conventional circular spiral coil. The zero-phase angle (ZPA) input condition can be achieved by adjusting the series capacitor connected with the Tx coil in S-S compensation topology. Experimental results show that the proposed magnetic coupler containing an integrated coil significantly improves the stability of output power and power transfer efficiency within the possible radial and axial misalignments compared with a conventional coupler. It was demonstrated that the output power changes less than 5.5% and the power transfer efficiency maintains at approximately 84.5% in arbitrary radial positions within the possible working region with an axial transfer distance of 50 mm in saltwater.

Keywords: uniform magnetic field (UMF); autonomous underwater vehicle (AUV); wireless power transmission (WPT); zero phase angle (ZPA)

Citation: Wen, H.; Li, J.; Zhang, K.; Ye, J.; Yan, Z.; Song, B.; Tong, X. Enhancing Power Transmission Stability of AUV's Wireless Power Transfer System with Compact Planar Magnetic Coupler. *J. Mar. Sci. Eng.* **2023**, *11*, 566. https://doi.org/10.3390/jmse11030566

Academic Editor: Yassine Amirat

Received: 13 February 2023
Revised: 28 February 2023
Accepted: 2 March 2023
Published: 6 March 2023

1. Introduction

AUVs play an increasingly important role in the development and utilization of marine resources, scientific exploration, and military applications. Power supply is the key problem that restricts the long-term continuous operation of AUVs. The last decade has witnessed rapidly growing extensive research on WPT technology; recently, WPT has been widely used in electric vehicles, household appliances, and industrial devices [1]. Compared to conventional energy supplement approaches for AUVs, such as battery swapping [2] and wet-mate contact charging [3], which suffer from limited maneuverability, and insulation and worn conductor issues, WPT technology has attracted increasing attention to becoming a reliable and safe underwater charging method for AUV [4,5].

The magnetic coupler is the key component that determines the performance of a WPT system, including output power, power transfer efficiency, misalignment tolerance, and electromagnetic compatibility. Various magnetic couplers for AUVs with different structures have been studied [6–13]. A three-phase WPT system for a lightweight AUV was proposed in [6], which enhanced the system's misalignment tolerance to rotational offsets, but the axial and radial misalignments were not considered. Wu et al. [7] developed a magnetic coupler structure with a quadruple Tx coil and crossed dipole Rx coil to prevent rotational and axial misalignment. Zeng et al. [8] proposed a novel hybrid transmitter composed of conical and planar spiral coils. The experimental results showed that the proposed WPT system's output power changes were within 5.7%, and the power transfer efficiency was maintained at approximately 86% in an excessive misalignment area with a transfer distance of 2 cm. A magnetic coupler structure with a spiral tube type is adopted in [10] to be compatible with the structure of AUV. In addition, Wang et al. [14] aimed at proposing an omnidirectional and positioning-tolerant planar-type AUV docking and charging platform, which has no constraints on AUV structures. However, two main shortcomings restricted their application—on the one hand, some couplers are unconsolidated and space-consuming; furthermore, some magnetic couplers are incompatible with an AUV's hull, and modifications are inevitable, which have negative effects on the hydrodynamic or pressure-resistant performance of AUVs. On the other hand, in most couplers' design procedures, radial, axial, and rotational misalignments, which may be caused by docking errors or ocean current impact, were not taken into consideration simultaneously.

In order to guarantee AUVs obtain a stable power supply from the WPT system under uncertain misalignments, a UMF in a specified charging area should be formed. Some scholars have conducted relevant research on the construction of a UMF with a planar electromagnetic coupling coil. Wang et al. proposed a method of using a multiple-antiparallel square spiral structure transmitting coil to form a UMF [15]. In the literature [16], the geometric parameters of the anti-parallel square spiral transmitting coil were optimized by genetic algorithm, and the magnetic field uniformity factor was defined; the measurement results showed that the magnetic field uniformity factor was reduced from 0.154 to 0.089 after optimization. Zhang et al. added a compensation structure to the circular array coil, and extended the range of UMF [17]. A novel hybrid structure composed of a coil and a spiral winding was proposed for enhancing the uniform magnetic field distribution over the charging surface by Hui et al. [18]. A planar distributed multi-coil transmitter structure composed of a chief coil and booster coils to stretch the uniformity of the magnetic field was proposed in [19]. Transmitter arrays have also been proposed to build a UMF [20,21]. The above-mentioned UMF construction methods change the original coupling coil structure, resulting in variations in parameters such as self-inductance, which increases the complexity of WPT system modeling and analysis.

This paper proposes a WPT system with a compact planar magnetic coupler for an AUV. PI coil is incorporated into the circular Tx coil to generate a UMF, which enhances the radial and axial misalignment tolerance. The magnetic coupler is suitable for an AUV's WPT system because of its compact planar structure, and modifications to the AUV's hull are unnecessary. Additionally, the circular magnetic coupler has inherent insensitivity to rotational misalignments. Circuit analysis demonstrates that the addition of a PI coil changes the input impedance of the WPT system with S-S compensation topology. The ZPA input condition can be maintained by adjusting the value of compensation capacitor C_1, which allows the WPT system to remain with high efficiency when the load changes. Theoretical analysis reveals that a PI coil improves the stability of output power; meanwhile, it brings little effect to power transfer efficiencies of less than 3%. Experimental results verify the proposed magnetic coupler's performance in stability enhancement of output power and power transfer efficiency under different radial and axial misalignments in comparison with a conventional coupler. A comparison of some UMF construction methods and the proposed magnetic coupler of this work is obtained, as shown in Table 1.

Table 1. Comparison of some UMF construction methods.

References	[15]	[17]	[8]	[19]	This Work
Magnetic coupler structure	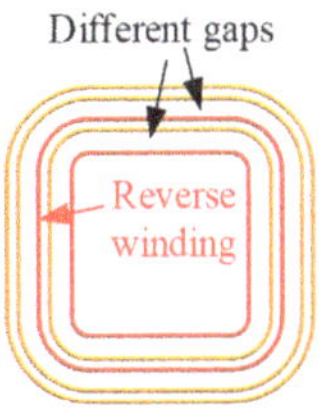	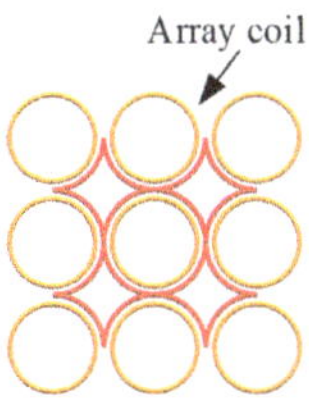	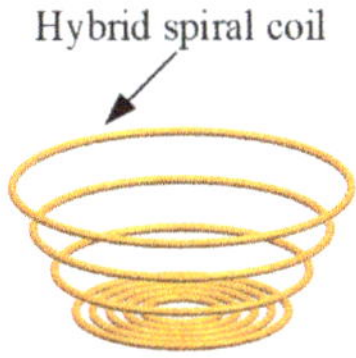	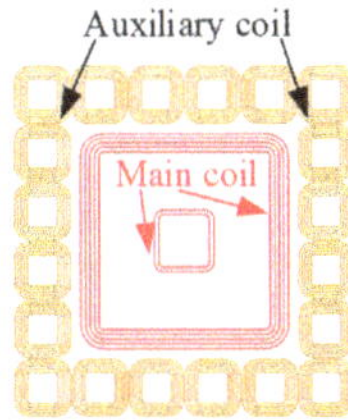	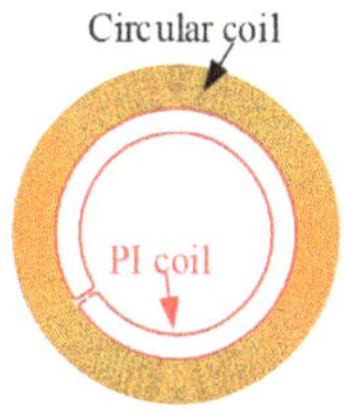
Compact of magnetic coupler (volume)	★★★★★	★★★☆☆	★★☆☆☆	★★☆☆☆	★★★★★
Misalignment tolerance	★★★☆☆	★★★★☆	★★★★★	★★★★☆	★★★★☆
UMF construction effect	★★★★☆	★★★★☆	★★★☆☆	★★★★★	★★★★☆
Easy to construct	★★☆☆☆	★★☆☆☆	★★☆☆☆	★☆☆☆☆	★★★★★

A higer "★" represents better performance on this item, ★★★★★ means best.

2. Magnetic Coupler Design

2.1. Discussion of Magnetic Field Distribution of Circular Coil

Suppose that the radius of a single-turn circular Tx coil is a; simultaneously, the receiver (Rx) coil is located in the circular charging plane S at the gap of h, and the radius of the S-plane is greater than the radius of the Rx coil, as shown in Figure 1.

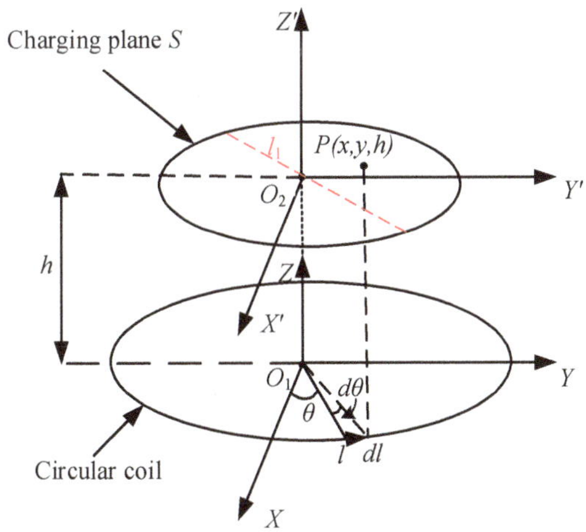

Figure 1. Schematic diagram of charging plane S formed by the circular coil.

Where θ is the included angle between $\overrightarrow{O_1 l}$ and the X-axis. When the Rx coil moves arbitrarily in the S-plane, the power "picked" by the Rx coil is determined by the magnetic induction intensity B_Z in the Z direction generated by the Tx coil [15]. According to

Biot–Savart's Law, assuming that the current flowing through the circular coil is I, the B_Z generated at any point P on the S-plane can be expressed as

$$B_z = \frac{\mu_0 I}{4\pi} \int_0^{2\pi} \frac{a^2 - ax\cos\theta}{\left(h^2 + a^2 + x^2 - 2ax\cos\theta\right)^{3/2}} d\theta \quad (1)$$

where μ_0 is the vacuum permeability. It can be seen from Equation (1) that the radius of the Tx coil a and the gap h determine the strength of B_Z. The outer diameter of the Rx coil should fit the dimension of the AUV as closely as possible. Set the outer diameter of the Rx coil as 200 mm, when the AUV is recharged, the maximum radial misalignment is 30 mm, which means that the S-plane is a circular plane with a radius of 130 mm. When the AUV drives into the underwater docking station for power replenishment, a misalignment between the Tx coil and the Rx coil will inevitably occur due to the influence of ocean current and docking error, which mainly includes the following three types: axial, radial, and rotational misalignments. Since the circular coil has good resistance to rotational offset, it will not be studied in this section.

To simplify the analysis, assume the current I flows through each circular coil as 1A, the distribution of B_Z in the S-plane at h = 50 mm and 70 mm, as shown in Figure 2.

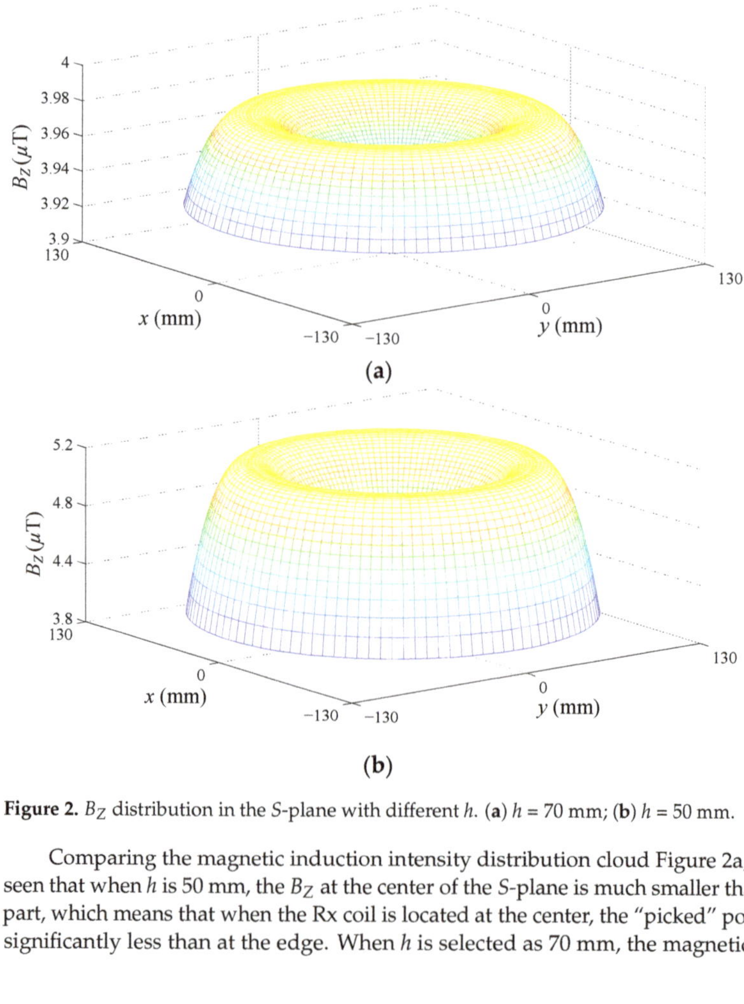

Figure 2. B_Z distribution in the S-plane with different h. (**a**) h = 70 mm; (**b**) h = 50 mm.

Comparing the magnetic induction intensity distribution cloud Figure 2a,b, it can be seen that when h is 50 mm, the B_Z at the center of the S-plane is much smaller than the edge part, which means that when the Rx coil is located at the center, the "picked" power will be significantly less than at the edge. When h is selected as 70 mm, the magnetic induction intensity distribution across the S-plane is more uniform, and the Rx coil is allowed to "dock" or move to any position in the area with stable output power.

The B_Z distribution produced by the circular coil is symmetrical, so the uniformity of B_Z distribution can be described by the degree of discretization of the magnetic induction intensity at each point on line l_1 with the center point in Figure 1. The deviation of B_Z distribution on line l_1 is defined as

$$\lambda(x) = \frac{B_Z(x) - B_Z(0)}{B_Z(0)}(x = 0, 1 \ldots 130) \tag{2}$$

Figure 3 describes the discretization of B_Z distribution on line l_1 when h = 50 mm and 70 mm, respectively, according to Equation (2).

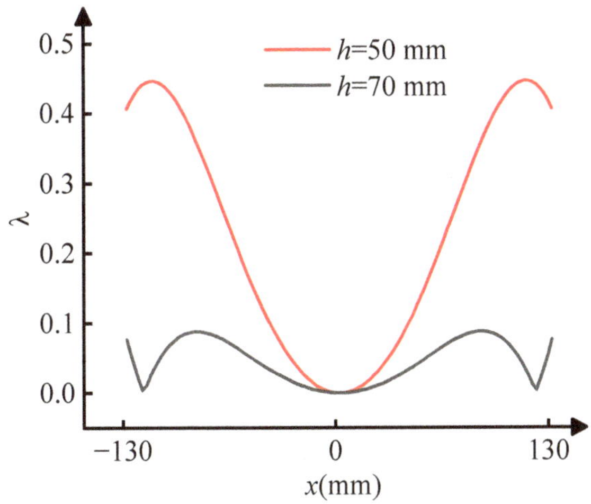

Figure 3. The discretization of B_Z distribution on line l_1.

According to previous design experience [16], the area that the deviation of B_Z distribution within 0.1 is specified as the UMF region and it is labeled as S_1. It can be seen from Figure 3 that when h = 70 mm, the distribution of B_Z in the entire S-plane meets the requirements of UMF, but when the Rx coil is located in the plane of h = 50 mm, the received power will fluctuate sharply with the change of position.

2.2. Analysis of the Circular Coil Turns

When the outer diameter of the Tx coil is determined, the average radius of the coil decreases as the turn numbers of the coil increase. In addition, the distribution of B_Z calculated by Equation (1) ignores the single-turn coil thickness, and the results may differ from the actual value. By using COMSOL simulation software, the influence of different circular coil turn numbers on the B_Z distribution in the S-plane can be more accurately analyzed. The maximum discretization of B_Z and the ratio of UMF area to the total area of the S-plane with h = 50 mm and h = 70 mm under different coil turn numbers are shown in Figure 4.

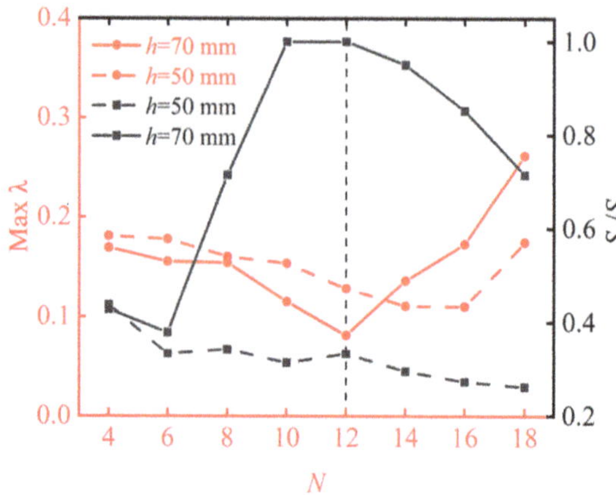

Figure 4. Influence of coil turn numbers N on B_Z distribution.

As can be seen from Figure 4, the distribution of B_Z and the area of the UMF region at h = 70 mm is related to the turn numbers N of the circular coil. When N = 12, the distribution of B_Z in the entire S-plane meets the conditions of a UMF; at the same time, the fluctuation of B_Z is minimal. However, for h = 50 mm, as shown in Figure 4, by adjusting the turn numbers of the circular coil, the area of the UMF region S_1 always stays at a low level. Some other methods need to be taken to increase the area of the UMF region when h = 50 mm.

2.3. Constructing UMF with a PI Coil

In the WPT system shielding measures, there is a method of reactive resonant shielding, which suppress the leakage magnetic field by passive coil [22]. Similarly, we can utilize the method of adding a PI coil to enhance or weaken the magnetic field in a specific area, and thus achieve the purpose of building a UMF.

The proposed magnetic coupler with an integrated PI coil is shown in Figure 5. When the alternating magnetic field generated by the Tx coil passes through the PI coil, an induced electromotive force is generated on the coil, and the induction coil will produce a magnetic field with opposite direction. Using this property, combined with an induced magnetic field distribution of circular coil, the parameters of the PI coil are designed to construct a UMF.

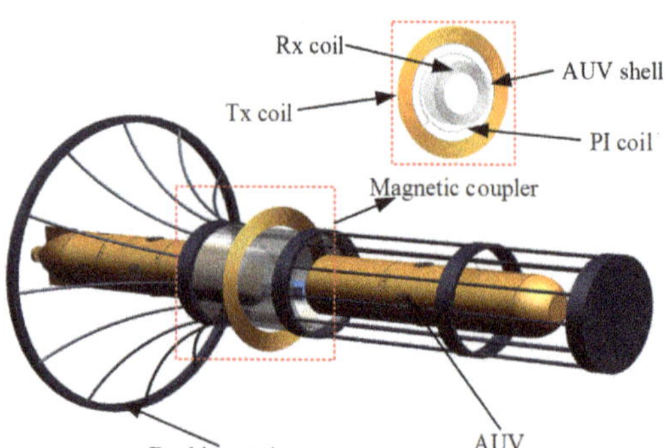

Figure 5. Proposed WPT system with a PI coil for AUV.

The induced voltage V_{ind} generated on the PI coil can be expressed as

$$V_{ind} = -\frac{d\phi}{dt} = -j\omega B_z e^{j\omega t} S_2 \tag{3}$$

where φ is the magnetic flux through the plane PI coil located, and S_2 is the area of the PI coil. The induced current I_{ind} in the PI coil can be expressed as

$$I_{ind} = \frac{V_{ind}}{Z_2} = \frac{V_{ind}}{j\omega L_2 + R_2} \tag{4}$$

where L_2 is the self-inductance of the PI coil and R_2 is the parasitic resistance of the PI coil. According to Equation (1), the magnetic induction intensity generated by the PI coil at any point P on the S-plane can be calculated as

$$B'_z = \frac{N\mu_0 I_{ind}}{4\pi} \int_L \frac{dl \times e_r}{r^2} \tag{5}$$

Since the magnetic induction intensity satisfies the law of superposition, the magnetic induction intensity at any point P on the S-plane is the sum of the magnetic induction intensity generated by Tx coil and the magnetic induction intensity generated by the PI coil, expressed as

$$B''_z = B'_z + B_z \tag{6}$$

According to (3)–(6), when the parameters of the Tx coil are fixed, the area S_2 of the PI coil determines the magnetic induction intensity at point P. The direction of the current in the PI coil can be determined by Lenz's law, as shown in Figure 6.

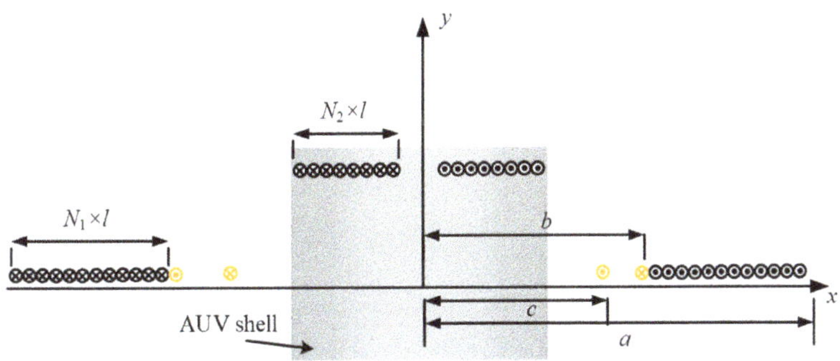

Figure 6. Cross-sectional view of the proposed magnetic coupler.

Since AUV needs to maintain a radial offset of at least 30 mm; in other words, the inner diameter of the PI coil should be greater than 130 mm, assuming c = 130 mm. Figure 7 shows the magnetic induction intensity distribution on line l_1 varying with the size of the PI coil.

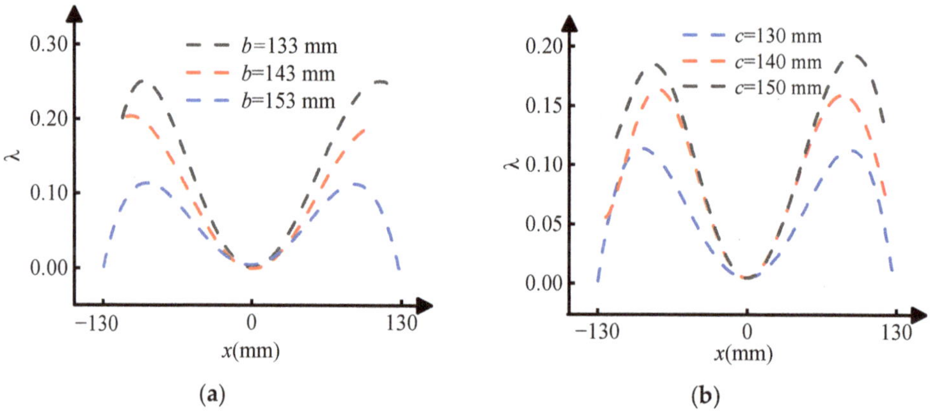

Figure 7. Magnetic induction intensity distribution varying with PI coil size: (**a**) When the outside diameter of PI coil changes; (**b**) When the inner diameter of PI coil changes.

The effect of PI coil on the B_Z distribution can be concluded as per the following:

1. By adjusting the parameters of the PI coil, the fluctuation of magnetic induction intensity can be reduced to extend the UMF area.
2. The deviation of B_Z distribution on line l_1 reduces with the increase in the outer diameter of PI coil b. The outer diameter of the PI coil can be fixed to the inner diameter of the Tx coil.
3. When the outer diameter of PI coil b is determined, the deviation of B_Z distribution on line l_1 reduces as the inner diameter of PI coil c decreases.

The above discussion shows that the introduction of a PI coil can improve the uniformity of the induced magnetic field distribution. The values of parameters b and c that meet the requirements can be obtained through the flowchart shown in Figure 8.

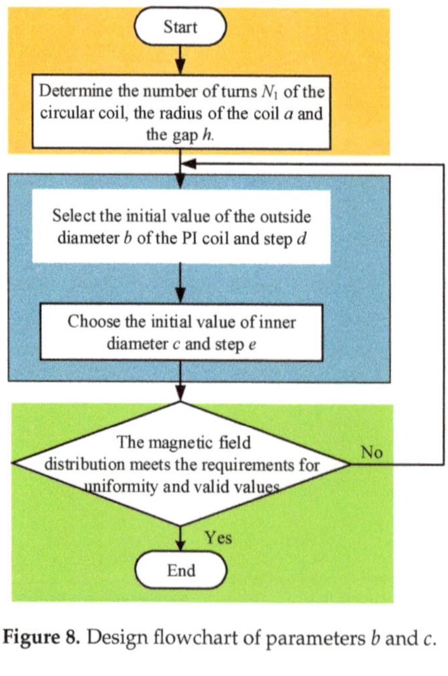

Figure 8. Design flowchart of parameters b and c.

It is finally determined that b = 153 mm, which is the outermost turn of the PI coil when wound tightly to the inside of the Tx coil, and the inner diameter c is selected as 130 mm. Figure 9 shows the magnetic induction intensity distribution on the S-plane when adding a PI coil and using only circular coil.

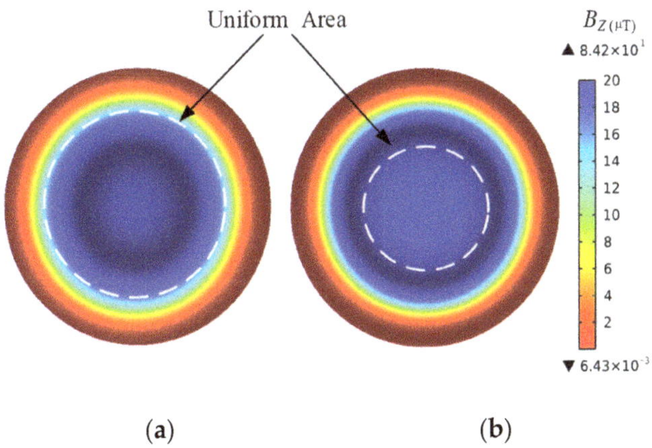

Figure 9. The magnetic induction intensity distribution on the S-plane: (**a**) with PI coil; (**b**) without PI coil.

Comparing Figure 9a,b, it can be clearly seen that the magnetic induction intensity at the center of the corresponding area of the plane has increased, and the distribution of the magnetic induction intensity is more uniform after adding a PI coil.

From the above analysis, it can be seen that when the gap h between the Tx coil and the Rx coil is 50 mm, the PI coil can improve the anti-offset ability of the system; the parameters of the magnetic coupling mechanism are listed in Table 2. When h is 70 mm, the system maintains good anti-offset ability, and there is no need to start a PI coil.

Table 2. The parameters of magnetic coupling mechanism.

Parameters	Definitions	Value
N	Tx coil turn number	12
a	Radius of Tx coil	200 mm
N_2	Rx coil turn number	8
a_1	Radius of Rx coil	100 mm
b	Outside radius of PI coil	153 mm
c	Inner radius of PI coil	130 mm
R_S	Radius of single-turn wire	1.95 mm

3. Circuit Design and Analysis

3.1. System Circuit Structure Design with a PI Coil

The proposed WPT system circuit structure with a PI coil is shown in Figure 10. The DC voltage U_{in} is converted to AC voltage U_1 through a full-bridge inverter consisting of four MOSFETS (S_1–S_4). The power is transmitted from the primary side to the secondary side through the resonant network and electromagnetic coupling mechanism. C_1 and C_3 are the compensation capacitors. The electromagnetic coupling mechanism consists of Tx coil L_1, Rx coil L_3, and PI coil L_2. M_{12}, M_{13}, and M_{23} represent the coupling between each coil, respectively. Switch S_5 controls the switching of the PI coil. The DC voltage on load R_L is obtained by the secondary side compensation network output voltage U_3 through a rectifier consisting of four DIODES (D_1–D_4). Moreover, the S-S compensation topology has the characteristics of simple structure and high efficiency, which is suitable for underwater WPT applications with limited space.

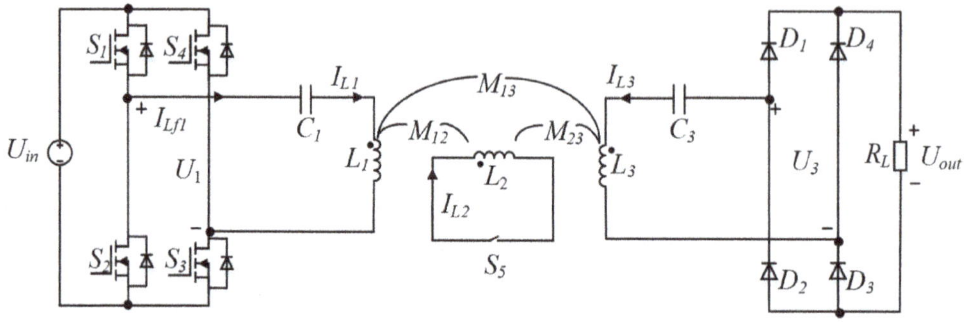

Figure 10. Proposed circuit of the WPT system.

3.2. Circuit Modelling Analysis

To simplify the analysis, the fundamental harmonic approximation is used and all higher-order harmonics are ignored. The simplified model of the proposed WPT system is shown in Figure 11.

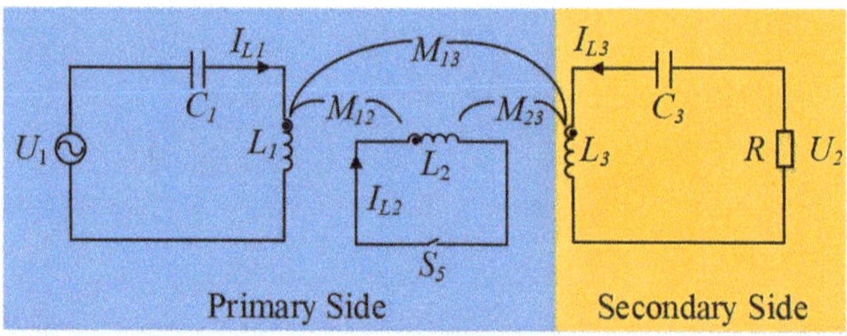

Figure 11. Equivalent circuit of the WPT system.

The root mean square (RMS) value of the fundamental voltage component can be expressed as

$$U_1 = \frac{2\sqrt{2}U_{in}}{\pi} \tag{7}$$

The equivalent load resistance R_L can be calculated as [23]

$$R = \frac{8}{\pi^2} R_L \tag{8}$$

When switch S_5 is turned on, according to Kirchhoff's voltage law, the following equation can be obtained.

$$\begin{bmatrix} U_1 \\ 0 \\ 0 \end{bmatrix} = \begin{bmatrix} j\omega L_1 + \frac{1}{j\omega C_1} & j\omega M_{12} & j\omega M_{13} \\ j\omega M_{12} & j\omega L_2 & j\omega M_{23} \\ j\omega M_{13} & j\omega M_{23} & j\omega L_3 + \frac{1}{j\omega C_3} + R \end{bmatrix} \cdot \begin{bmatrix} I_1 \\ I_2 \\ I_3 \end{bmatrix} \tag{9}$$

where ω represents the angular frequency of the WPT system. When the system is resonant, the relationship between the circuit components is expressed as

$$\begin{cases} \omega L_1 = \frac{1}{\omega C_1} \\ \omega L_3 = \frac{1}{\omega C_3} \end{cases} \tag{10}$$

By substituting (10) into (9), the current can be obtained as

$$\begin{cases} I_1 = \frac{\omega M_{23}^2 + L_2 R j}{\omega M_{12}^2 R + A j} U_1 \\ I_2 = \frac{M_{12} R - \omega M_{13} M_{23} j}{\omega M_{12}^2 R j - A} U_1 \\ I_3 = \frac{L_2 M_{13} - M_{12} M_{23}}{M_{12}^2 R + A j} U_1 \end{cases} \quad (11)$$

where $A = \omega(L_2 M_{13}^2 - 2M_{12}M_{13}M_{23})$. When switch S_5 is turned off, the PI coil no longer functions, and the circuit structure becomes a typical S-S compensation.

3.3. Analysis of the ZPA Characteristics of the Proposed WPT System

Maintaining ZPA operating characteristics to avoid reactive power loss is critical to improving the overall efficiency of the WPT system [24]. The S-S compensation structure has ZPA operation characteristics [25], but when a PI coil is added, the input impedance of the WPT system becomes

$$\begin{cases} Z_{in} = \frac{U_1}{I_{Lf1}} = R_{in} + jX_{in} = \frac{\omega^2 M_{12}^2 M_{23}^2 R - A L_2 R}{\omega^2 M_{23}^4 + L_2^2 R^2} + j \frac{A \omega M_{23}^2 - \omega M_{12}^2 R^2 L_2}{\omega^2 M_{23}^4 + L_2^2 R^2} \\ \alpha_{(\text{rad})} = \arctan \frac{X_{in}}{Z_{in}} \end{cases} \quad (12)$$

where α represents the input impedance angle. The addition of a PI coil brings an imaginary part to the input impedance, and ZPA can no longer be realized. As can be seen from (12), the input impedance angle α is related to the angular frequency ω, equivalent load R, the inductance of PI coil L_2, and coupling between coils. M_{12}, M_{23}, and M_{13} are 1.9 μH, 0.1 μH, and 7.51 μH, respectively. With the calculation of MATLAB, the relationship among the input impedance angle α, load R and system frequency f is shown in Figure 12.

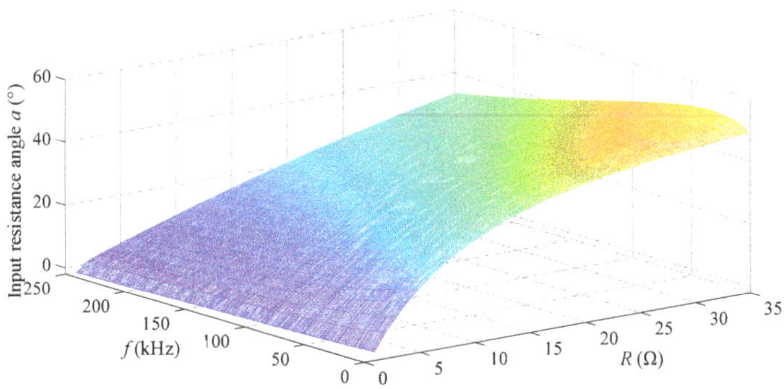

Figure 12. The relationship between input impedance angle α and system frequency f, load R.

As can be seen from Figure 12, the input impedance angle α tends to decrease as the load increases, but the decrease in α can be mitigated by increasing the system frequency f. Considering that when the system frequency is higher than 300 kHz, the propagation of high-frequency electromagnetic fields in seawater will produce large eddy current losses [26], the system operating frequency is selected as 244 kHz.

Equation (12) is simplified to obtain the real and imaginary parts of the input impedance.

$$\begin{cases} real(Z_{in}) = \frac{\omega^2 M_{12}^2 M_{23}^2 R - A L_2 R}{\omega^2 M_{23}^4 + L_2^2 R^2} \\ imag(Z_{in}) = \frac{A \omega M_{23}^2 - \omega M_{12}^2 R^2 L_2}{\omega^2 M_{23}^4 + L_2^2 R^2} \end{cases} \quad (13)$$

Since the value of M_{23} is very small, it can be omitted in the analysis. The imaginary part of the input impedance can be expressed as

$$imag(Z_{in}) = -\frac{\omega M_{12}^2}{L_2} \tag{14}$$

It means that the introduction of a PI coil brings an imaginary part to the input impedance, which is related to M_{12}. By adjusting the value of compensation capacitor C_1, the ZPA characteristics can be maintained in the WPT system and the relationship between C_1 and mutual inductance M_{12} is expressed as

$$j\omega L_1 + \frac{1}{j\omega C_1} = \frac{j\omega M_{12}^2}{L_2} \tag{15}$$

Figure 13 shows the relationship between the input impedance angle and system frequency *f*, load *R* after adjusting C_1. It can be seen that when the operating frequency is set to 244 kHz, the ZPA characteristics of the system will not be affected as the load changes.

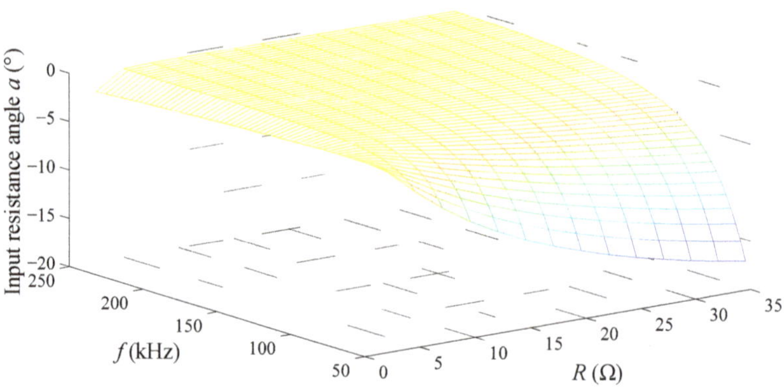

Figure 13. The relationship between input impedance angle α and system frequency *f*, load *R* after adjusting C_1.

3.4. System Efficiency Analysis

This section focuses on the effect of a PI coil on the power transmission characteristics of the WPT system. In the above analysis, the influence of the parasitic resistance of each coil on the system is not taken into account because its value is very small. However, when analyzing system efficiency, it cannot be ignored. In addition, in the marine environment, the impact of eddy current losses caused by the propagation of electromagnetic fields in seawater on WPT system efficiency also needs to be considered. When a PI coil is operating, the power transfer efficiency η and system output power P_{out} can be expressed as

$$\begin{cases} \eta = \frac{|I_3|^2 R}{P_{eddy}+|I_1|^2 R_1+|I_2|^2 R_2+|I_3|^2 (R_3+R)} \\ P_{out} = |I_3|^2 R \end{cases} \tag{16}$$

where R_1, R_2, and R_3 are the parasitic resistance of L_1, L_2, and L_3, respectively. The effect of load changes on power transfer efficiency is shown in Figure 14. It can be seen that when the load changes within 0~50 Ω, the power transfer efficiency of the WPT system after adding a PI coil and without a PI coil are consistent, both increase sharply at first, and then decrease slowly. Meanwhile, the introduction of a PI coil reduces the optimal efficiency of the system by about 3%.

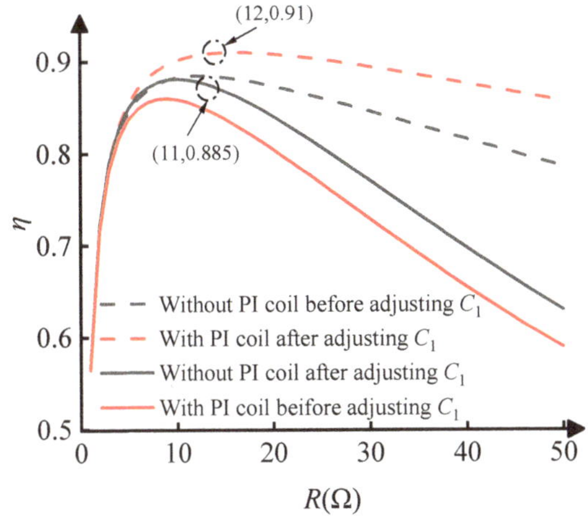

Figure 14. Power transfer efficiency varying with load.

4. Experimental Verification

An experimental platform is built to verify the proposed structure, as shown in Figure 15. To reduce the skin and proximity effects, the Litz wires are employed for the construction of Tx, Rx, and PI coils. They are wound by AWG 38 litz wires with 400 strands. The structure of the Tx coil, Rx coil, and a PI coil are also shown in Figure 15. The water tank is built to realistically simulate the marine environment, and the gap between coils is filled by saltwater with a salinity of 4‰. The electronic load is used to simulate the load change in the WPT system. The parameters of the proposed WPT system are listed in Table 3.

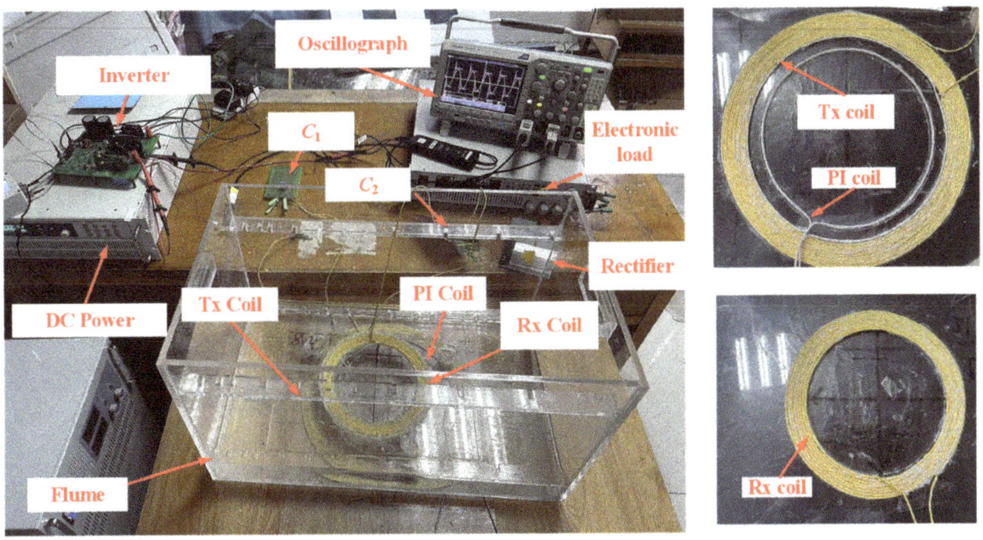

Figure 15. Experimental platform.

Table 3. The parameters of the proposed WPT system.

Parameters	Definitions	Value
L_1	Inductance of Tx coil	94.38 μH
R_1	Resistance of Tx coil	1105.1 mΩ
L_2	Inductance of PI coil	1.68 μH
R_2	Resistance of PI coil	124 mΩ
L_3	Inductance of Rx coil	21.85 μH
R_3	Resistance of Rx coil	303.99 mΩ
L_{f1}	Inductance of compensating inductor	8.93 μH
R_{Lf1}	Resistance of compensating inductor	155.49 mΩ
f	System frequency	244 kHz
U_1	Input voltage	100 V

4.1. Verification of ZPA Characteristics of the Proposed WPT System

The ZPA characteristics of the system are verified by measuring the phase angle of the input voltage and current with an Oscilloscope. Considering that the internal resistance of the batteries does not change much during the charging process, the resistance of the electronic load gradually increases from 10 to 50 Ω. The experimental waveforms of input voltage and current are shown in Figure 16.

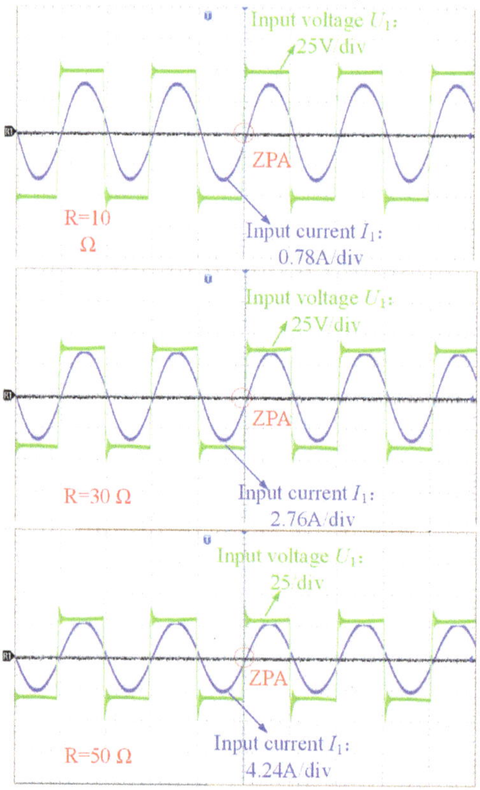

Figure 16. The input voltage and current waveform of the WPT system after adding a PI coil under different loads.

As can be seen in Figure 16, the phase angle of input voltage and current does not change with the variation in load resistance. Through the above analysis and experiment, it can be verified that the WPT system with the addition of a PI coil is still able to maintain the ZPA characteristics after adjusting the compensation capacitor C_1.

4.2. Verification of System Output Characteristics

The introduction of the PI coil changes the output characteristic of the WPT system, and in order to compare the fluctuation of output power, the output power of the system is normalized, as shown in Figure 17. In addition, the comparison test of the proposed WPT system in air and saltwater is added to verify the function of the PI coil.

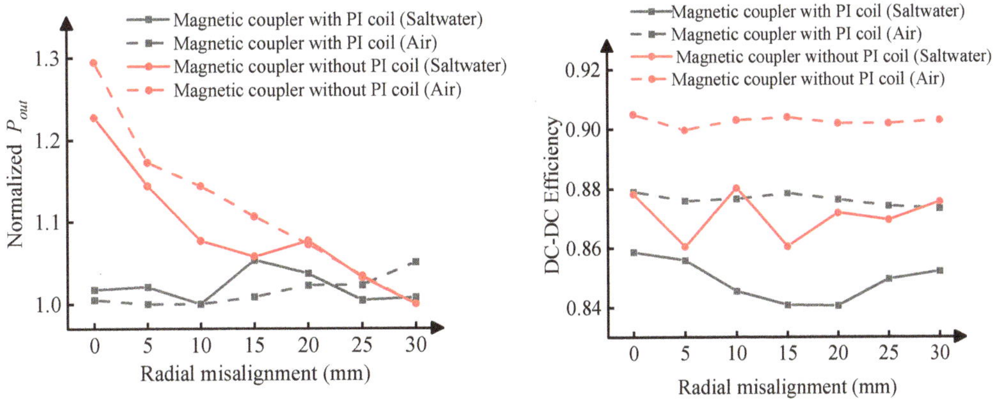

Figure 17. The output power and system efficiency varying with radial misalignments when h = 50 mm: (**a**) normalized output power; (**b**) system efficiency.

Compared with the output power of the WPT system without a PI coil, the fluctuation of output power decreases by 17% with the introduction of a PI coil, as can be seen in Figure 17. The stability of the output power of the system is improved significantly with the addition of a PI coil. A PI coil brings negative effects to the transmission efficiency, which cause approximately a 3% decrease in power transfer efficiency compared with the WPT system without a PI coil. Figure 18 shows the output power and efficiency of the WPT system varying with radial misalignments when only a circular coil is used at the Tx coil when h = 70 mm.

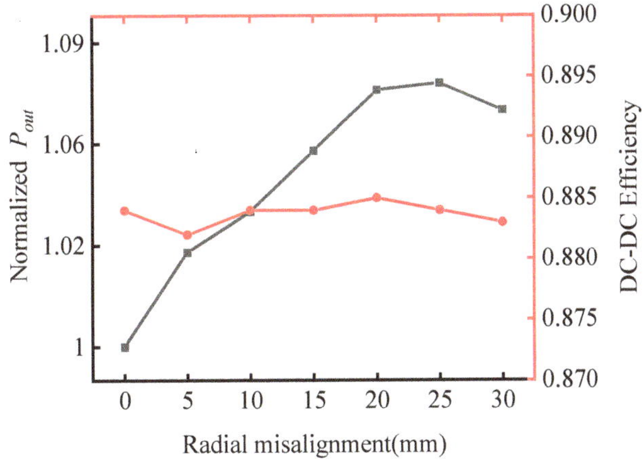

Figure 18. The output power and system efficiency varying with radial misalignments when h = 70 mm.

As can be seen from Figure 18, when h = 70 mm, the WPT system using only a circular coil at the Tx coil can meet the fluctuation of output power within 0.08, while maintaining the power transfer efficiency at 88.5%. Figure 19 shows the output power and efficiency of the WPT system varying with rotational misalignments.

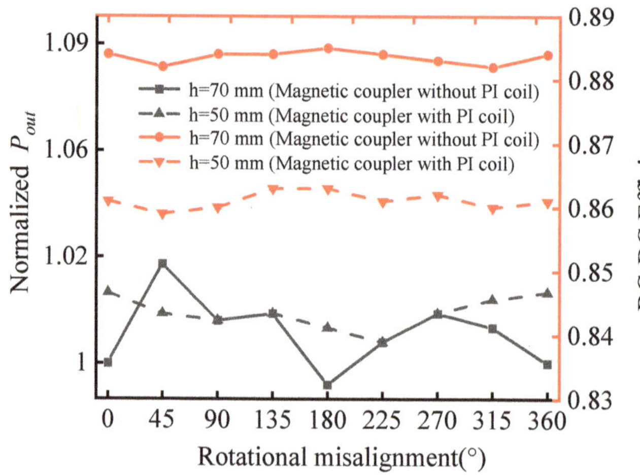

Figure 19. The output power and efficiency varying with rotational misalignments.

As can be seen from Figure 19, the output power and efficiency of the WPT system vary very little as the rotational degree changes. The introduction of a PI coil does not affect the good resistance to rotational misalignments of the circular coil.

5. Discussion

A method to improve the anti-offset ability of the WPT system by using PI coil is proposed in this paper, and the influence of PI coil on the magnetic field distribution of circular coil is studied from the perspective of deviation analysis. However, the principle and function of PI coil have not been deeply analyzed. In future work, the UMF construction method with PI coil will be studied by accurate theoretical analysis and experimental verification. Furthermore, mutual inductance will be taken into consideration in the parameter optimization of PI coil to deal with intricate position misalignments in the WPT system.

Although the stability of the output power of the system under different misalignments has been improved with the introduction of a PI coil, it has a certain negative impact on the power transfer efficiency. In the subsequent work, a design tradeoff among power transfer efficiency, output power stability, and coils' geometric layout will be investigated.

6. Conclusions

A WPT system with circular coils and a PI coil for an autonomous underwater vehicle is proposed in this paper. A PI coil is introduced to enhance or weaken the magnetic field in a specific area and build a UMF in a WPT system. The theoretical analysis shows that the proposed magnetic coupler can effectively eliminate the deviation of magnetic field distribution in comparison with traditional circular spiral coils. Experimental results show that the addition of a PI coil significantly improves the stability of the system's output power and power transfer efficiency. The WPT system's output power changes within 5.5% and the power transfer efficiency maintains at approximately 84.5% in arbitrary radial positions with an axial distance of 50 mm in saltwater. Meanwhile, compared to the WPT system without a PI coil, the fluctuation of output power is reduced by 17%.

Author Contributions: Conceptualization, H.W. and J.L.; methodology, H.W., J.L. and K.Z.; software, J.L. and Z.Y.; validation, H.W., J.L. and Z.Y.; formal analysis, K.Z., J.Y. and J.L.; investigation, H.W. and K.Z.; resources, B.S. and X.T.; data curation, J.L., J.Y. and Z.Y.; writing—original draft preparation, H.W. and J.L.; writing—review and editing, H.W., J.L. and K.Z.; visualization, J.L.; supervision, B.S. and X.T.; project administration, K.Z. and H.W.; funding acquisition, K.Z. and H.W. All authors have read and agreed to the published version of the manuscript.

Funding: This work was supported in part by the National Natural Science Foundation of China under Grant 52171338, in part by the China Postdoctoral Science Foundation under Grant 2021M702638, in part by the Natural Science Basic Research Plan in Shaanxi Province of China under Grant 2023-JC-QN-0475, and in part by the Starting Research Fund for Postdoctoral Researcher from Xi'an University of Technology under Grant 119-451121005.

Institutional Review Board Statement: Not applicable.

Informed Consent Statement: Not applicable.

Data Availability Statement: Not applicable.

Conflicts of Interest: The authors declare no conflict of interest.

References

1. Feng, H.; Tavakoli, R.; Onar, O.C.; Pantic, Z. Advances in High-Power Wireless Charging Systems: Overview and Design Considerations. *IEEE Trans. Transp. Electrif.* **2020**, *6*, 886–919. [CrossRef]
2. Harakare, A.; Barhate, N.; Randad, N.; Varghese, A.G.; Gupta, A.; Dave, P.; Modi, S.; Shrivastava, A.; Khare, L.; Raj, S. Design of Battery Management System for an Autonomous Underwater Vehicle. In Proceedings of the OCEANS 2022—Chennai, Chennai, India, 21–24 February 2022.
3. Painter, H.; Flynn, J. *Current and Future Wet-Mate Connector Technology Developments for Scientific Seabed Observatory Applications*; Oceans: Boston, MA, USA, 2006; pp. 1–6.
4. Mohsan, S.A.H.; Khan, M.A.; Mazinani, A.; Alsharif, M.H.; Cho, H.-S. Enabling Underwater Wireless Power Transfer towards Sixth Generation (6G) Wireless Networks: Opportunities, Recent Advances, and Technical Challenges. *J. Mar. Sci. Eng.* **2022**, *10*, 1282. [CrossRef]
5. Yan, Z.; Zhang, Y.; Song, B.; Zhang, K.; Kan, T.; Mi, C. An LCC-P Compensated Wireless Power Transfer System with a Constant Current Output and Reduced Receiver Size. *Energies* **2019**, *12*, 172. [CrossRef]
6. Kan, T.; Mai, R.; Mercier, P.P.; Mi, C.C. Design and Analysis of a Three-Phase Wireless Charging System for Lightweight Autonomous Underwater Vehicles. *IEEE Trans. Power Electron.* **2018**, *33*, 6622–6632. [CrossRef]
7. Wu, S.; Cai, C.; Chai, W.; Li, J.; Cui, Q.; Yang, S. Uniform Power IPT System With Quadruple-Coil Transmitter and Crossed Dipole Receiver for Autonomous Underwater Vehicles. *IEEE Trans. Ind. Appl.* **2022**, *58*, 1289–1297. [CrossRef]
8. Zeng, Y.; Rong, C.; Lu, C.; Tao, X.; Liu, X.; Liu, R.; Liu, M. Misalignment Insensitive Wireless Power Transfer System Using a Hybrid Transmitter for Autonomous Underwater Vehicles. *IEEE Trans. Ind. Appl.* **2022**, *58*, 1298–1306. [CrossRef]
9. Yan, Z.; Zhang, Y.; Kan, T.; Lu, F.; Zhang, K.; Song, B.; Mi, C.C. Frequency Optimization of a Loosely Coupled Underwater Wireless Power Transfer System Considering Eddy Current Loss. *IEEE Trans. Ind. Electron.* **2019**, *66*, 3468–3476. [CrossRef]
10. Liu, P.; Gao, T.; Zhao, R.; Mao, Z. A Novel Conformal Coil Structure Design of Wireless Power Transfer System for Autonomous Underwater Vehicles. *J. Mar. Sci. Eng.* **2022**, *10*, 875. [CrossRef]
11. Wang, Y.; Song, B.; Mao, Z. Application of Shielding Coils in Underwater Wireless Power Transfer Systems. *J. Mar. Sci. Eng.* **2019**, *7*, 267. [CrossRef]
12. Yan, Z.; Wu, M.; Zhao, C.; Hu, Q.; Zhu, L.; Qiao, L.; Wang, L. Free-Rotation Wireless Power Transfer System Based on Composite Anti-Misalignment Method for AUVs. *IEEE Trans. Power Electron.* **2023**, *38*, 4262–4266. [CrossRef]
13. Yan, Z.; Zhao, C.; Hu, Q.; Wu, M.; Qiao, L.; Zhang, K.; Hu, Y. An Underwater Inductive Power Transfer System with a Compact Receiver and Reduced Eddy Current Loss. *J. Mar. Sci. Eng.* **2022**, *10*, 1900. [CrossRef]
14. Wang, T.; Zhao, Q.; Yang, C. Visual navigation and docking for a planar type AUV docking and charging system. *Ocean. Eng.* **2021**, *224*, 108744. [CrossRef]
15. Wang, S.; Hu, Z.; Rong, C.; Lu, C.; Chen, J.; Liu, M. Planar Multiple-Antiparallel Square Transmitter for Position-Insensitive Wireless Power Transfer. *IEEE Antennas Wirel. Propag. Lett.* **2018**, *17*, 188–192. [CrossRef]
16. Zhang, Y.; Wang, L.; Guo, Y.; Zhang, Y. Optimisation of planar rectangular coil achieving uniform magnetic field distribution for EV wireless charging based on genetic algorithm. *IET Power Electron.* **2019**, *12*, 2706–2712. [CrossRef]
17. Zhang, C.; Wang, W.; Xu, C.; Yang, J. Research on Uniform Magnetic Field Compensation Structure of Array Circular Coils for Wireless Power Transfer. *IEEE Trans. Magn.* **2021**, *57*, 1–5. [CrossRef]
18. Xun, L.; Hui, S.Y. Optimal Design of a Hybrid Winding Structure for Planar Contactless Battery Charging Platform. *IEEE Trans. Power Electron.* **2008**, *23*, 455–463. [CrossRef]

19. Bharadwaj, A.; Srivastava, V.K.; Sharma, A.; Reddy, C.C. A Switchable Multicoil Antenna With Booster Coil to Improve Coverage in WPT Systems. *IEEE Trans. Antennas Propag.* **2022**, *70*, 2490–2498. [CrossRef]
20. Mirbozorgi, S.A.; Maghsoudloo, E.; Bahrami, H.; Sawan, M.; Gosselin, B. Multi-resonator arrays for smart wireless power distribution: Comparison with experimental assessment. *IET Power Electron.* **2020**, *13*, 4183–4193. [CrossRef]
21. Pahlavan, S.; Shooshtari, M.; Jafarabadi Ashtiani, S. Star-Shaped Coils in the Transmitter Array for Receiver Rotation Tolerance in Free-Moving Wireless Power Transfer Applications. *Energies* **2022**, *15*, 8643. [CrossRef]
22. Kim, J.; Kim, J.; Kong, S.; Kim, H.; Suh, I.S.; Suh, N.P.; Cho, D.H.; Kim, J.; Ahn, S. Coil Design and Shielding Methods for a Magnetic Resonant Wireless Power Transfer System. *Proc. IEEE* **2013**, *101*, 1332–1342. [CrossRef]
23. Zhang, W.; Wong, S.; Tse, C.; Chen, Q. Analysis and Comparison of Secondary Series- and Parallel-Compensated Inductive Power Transfer Systems Operating for Optimal Efficiency and Load-Independent Voltage-Transfer Ratio. *IEEE Trans. Power Electron.* **2014**, *29*, 2979–2990. [CrossRef]
24. Yang, L.; Ren, L.; Shi, Y.; Wang, M.; Geng, Z. Analysis and Design of a S/S/P-Compensated Three-coil Structure WPT System With Constant Current and Constant Voltage Output. *IEEE J. Emerg. Sel. Top. Power Electron.* **2022**. *early access*. [CrossRef]
25. Chen, Y.; Zhang, H.; Park, S.-J.; Kim, D.-H. A Switching Hybrid LCC-S Compensation Topology for Constant Current/Voltage EV Wireless Charging. *IEEE Access* **2019**, *7*, 133924–133935. [CrossRef]
26. Bana, V.; Kerber, M.; Anderson, G.; Rockway, J.D.; Phipps, A. In Underwater wireless power transfer for maritime applications. In Proceedings of the 2015 IEEE Wireless Power Transfer Conference (WPTC), Boulder, CO, USA, 13–15 May 2015.

Journal of
*Marine Science
and Engineering*

MDPI

Article

Application of Multi-Cylinder Synchronous Control for Telescopic Mechanism of Marine Steel Pile Cleaning Equipment

Chao Li [1], Nan Pang [2,*], Kai Xu [3,*], Qingling Geng [1], Xiangyu Wang [4], Feihong Yun [5] and Lei Gao [6]

1 Guangdong Mechanical and Electronical College of Technology, Guangzhou 510515, China
2 College of Mechanical and Electrical Engineering, Guilin University of Aerospace Technology, Guilin 541004, China
3 Wuhan Institute of Shipbuilding Technology, Wuhan 430212, China
4 Yantai Research Institute of Harbin Engineering University, Yantai 264006, China
5 College of Mechanical and Electrical Engineering, Harbin Engineering University, Harbin 150001, China
6 School of Economics and Management, Yanshan University, Qinhuangdao 066004, China
* Correspondence: 435818875@hrbeu.edu.cn (N.P.); xukai0705@163.com (K.X.)

Abstract: In order to clean up marine fouling attached to marine steel pile, this paper proposed an innovative configuration scheme of the marine steel pile cleaning equipment by the scraping method and its telescopic mechanism by applying a multi-cylinder synchronous control strategy to the cleaning equipment, and produced a test prototype of the cleaning equipment that could solve the problem of cleaning equipment eccentricity and tilt in the field of ocean engineering. Based on the MATLAB Simulink module, a simulation model of the operation process of the telescopic mechanism of the marine steel pile cleaning equipment was established to complete the evaluation of the multi-cylinder synchronous control performance under multiple working conditions. Through the test, the synchronous working performance of the telescopic mechanism of the cleaning equipment under the no-load condition was preliminarily verified. The test results showed that under the no-load condition, the relative errors between the three cylinders and the target displacement were 0.8%, 0.4%, and 0.2%, respectively, and the cleaning equipment could reach the specified working position at the given working speed. The displacement synchronization error between each cylinder was 0.7 mm, 0.7 mm, and 0.6 mm, respectively, and the displacement synchronization error was controlled within 1 mm. The telescopic mechanism had good synchronization, which can ensure the stability and prevent the eccentricity and tilt during the cleaning equipment operation as well as provide a valuable reference for the manufacturing of cleaning equipment.

Keywords: marine steel pile; cleaning equipment; telescopic mechanism; synchronous control; Simulink simulation; test

Citation: Li, C.; Pang, N.; Xu, K.; Geng, Q.; Wang, X.; Yun, F.; Gao, L. Application of Multi-Cylinder Synchronous Control for Telescopic Mechanism of Marine Steel Pile Cleaning Equipment. *J. Mar. Sci. Eng.* **2023**, *11*, 1010. https://doi.org/10.3390/jmse11051010

Academic Editor: Dong-Sheng Jeng

Received: 9 March 2023
Revised: 21 April 2023
Accepted: 7 May 2023
Published: 8 May 2023

1. Introduction

With the development and utilization of marine resources, a large number of marine infrastructures have been built. Among them, a large number of metal materials have been used for marine steel piles [1]. Most of the marine steel piles standing in the ocean have been corroded by attached fouling organisms including thousands of species [2]. The corrosion pattern of marine steel piles in different marine environmental zones is different, and the wave splash zone is more severely affected by fouling organism attachment and corrosion [3,4]. Because of its special location, these steel structures are affected by many factors, for example, alternating cycles of drying and wetting of the salt film, salt enrichment, sunlight, spray impact, fouling organisms attachment, etc. [5], of which the attachment corrosion of fouling organisms is particularly serious [6]. Over time, a solid, rough, thick crust will form, which can increase the surface area of the steel pile and the resistance it experiences in seawater [7], thus increasing its surface corrosion and reducing its service life [8,9]. The corrosion rate in the wave splash zone is about 3 to 10 times higher than in other corrosion zones. Corrosion prevention techniques are now well-established in this zone [10], but the cleaning equipment for fouling organisms have yet to be improved. Research on underwater cleaning equipment has been carried out in some countries, but some still require the assistance of a diver to operate it [11,12]. Currently, the traditional manual cleaning method is still used in China, which has some drawbacks such as low cleaning efficiency, short operation window, weather influence, and difficult operation. Therefore, the problem of using mechanical equipment to automatically remove fouling organisms has not been resolved.

Most of the current research has focused on ship surface cleaning equipment such as the underwater cleaning equipment developed by the UMC Company in the UK [13]. The equipment generates negative pressure suction through a reverse rotating brush, thereby adsorbing and maintaining it on the surface of the ship; with the assistance of divers, the device can be equipped with four different types of rotary brushes such as silicone bristles (soft), polypropylene bristles (hard), nylon cleaning brushes, and steel wire brushes to clean media with different adhesion forces. Procurement customers can also selectively install metal scraper blades at the front to remove hard and upright fouling organisms, but the equipment still requires diver assistance. Nassiraei et al. [14] from the Kyushu University of Technology in Japan developed a type of wheeled hull surface cleaning equipment. This equipment was based on thrust adsorption technology and equipped with six suction thrusters and two rotating cleaning mechanisms; it can also be controlled remotely, and operators can observe the cleaned surface in real-time through their front and rear cameras as well as record and store images before and after cleaning. The cleaning equipment tested completed basic motion control tests and ship cleaning tests, and the reliability of the cleaning equipment was verified through tests. However, there was a deficiency as it requires divers to perform underwater auxiliary operations. Souto et al. [15,16], from the University of La Coruna in Spain, designed non-magnetic cleaning equipment for complex ship hull surfaces. This equipment adopted a hybrid technology of thrust and negative pressure adsorption and equipped with a rotary brush cleaning method. Two cleaning units can cross complex obstacles on the same surface or crawl from one surface to another surface that was perpendicular to it through a similar "walking" alternating adsorption method during the crawling process. Through the kinematics research, simulation analysis, and underwater hull surface cleaning test, the equipment achieved good test results, but the equipment moved and cleaned slowly.

Most of the cleaning equipment above-mentioned targets the surface of the ship, but there has been relatively little research on cleaning equipment used for offshore steel piles. For instance, the Australian Iasgroup Company [17,18] has developed two cleaning robots, Splash Genius and Splash Genius II, for the cleaning of marine infrastructure platforms. These two cleaning robots can automatically clean the marine infrastructure platforms. Splash Genius adopts the surrounding clamping and climbing technology, and is equipped with a high-pressure water jet to remove the fouling organisms attached to the marine steel piles. Splash Genius II is an upgraded product of Splash Genius, which realized various autonomous control methods based on Splash Genius, thus there are significant savings not only in terms of the labor costs and force, but could also improve the cleaning efficiency. However, there are situations where the high-pressure water jet is not thoroughly cleaned. Woolfrey's [19–21] team at the University of Technology Sydney, Australia developed the Submersible Pile Inspection Robot (SPIR), which is an underwater cleaning robot. SPIR adopts clamping arm and climbing technology equipped with a high-pressure water jet, and can clean cylindrical or square steel piles. However, the SPIR is larger in size and has higher manufacturing costs. From the perspective of the operation process, the cleaning effect was not significant. The domestic research carried out in the field of cleaning marine steel piles started relatively late. At press, some universities have carried out relevant research, for example, CNOOC Energy Development Co. Ltd. And Yang Canjun's team [22–24] at

Zhejiang University developed a cleaning robot called "HOME". HOME is composed of a four-wheel permanent magnet adsorption module and passive roll joint, which ensures the self-adaptive adjustment function. HOME adopted the cavitation jet cleaning method, which means that it can automatically clean the fouling organisms attached to the steel piles within 100 m below the water surface. A panoramic visual navigation system was applied to HOME, which can sense the real-time surrounding environment and cleaning status. However, there were some problems such as incomplete cleaning from the actual cleaning effect, which did not meet our requirements for the follow-up multiple layers of Petrolatum Tape Cover anticorrosion technology (PTC) [25] on steel piles. Wang Liquan's team [26,27] at Harbin Engineering University designed a cleaning robot for marine steel piles that adopted an upper and lower ring structure including upper and lower clamps. Three telescopic hydraulic cylinders were connected between the two ring structures as telescopic mechanisms, which could realize the telescopic crawling of the robot. A gear rotation structure was designed in the middle part to connect the scraping tools to ensure that the scraping tools could rotate at a certain angle in order to continue to clean vertically along the axis of the steel pile. However, the overall structure of the robot was relatively complex, which easily affected the cleaning speed and stability during the robot operation. At Shandong University, Yu Fujie et al. [28] designed an underwater cleaning robot with a manipulator for an offshore jacket called ROVMS. The ROVMS was used to generate the adsorption force through the electromagnetic adsorption unit, and the ROVMS could maintain its distance from the surface to be cleaned. Four rollers were designed for the ROVMS to realize its stable support and prevent its body from colliding with the steel pile in the contact direction of the cleaning surface. The ROVMS adopts the cavitation jet method and has the adaptability of resisting the interference of ocean current, but there is still the possibility of incomplete cleaning.

Currently, most marine steel pile cleaning equipment is equipped with a water jet cleaning method. In this paper, a configuration scheme of marine steel pile cleaning equipment by the scraping method and its telescopic mechanism is proposed by applying a multi-cylinder synchronous control strategy to cleaning equipment, which can effectively avoid the problem of cleaning equipment eccentricity and tilt. If the synchronization accuracy is too high, the control method will lead to an increase in manufacturing costs and unnecessary resource waste; if the synchronization accuracy is too low, it will cause asynchronous expansion and contraction during the operation of the telescopic mechanism, leading to eccentricity and tilt of the cleaning equipment, thereby affecting the cleaning effect of the scraping tools. Therefore, conducting research on the three-cylinder synchronous control method with a telescopic mechanism to ensure the synchronization of the cleaning equipment's telescopic mechanism operation and achieve the effective removal of attached marine fouling is of great significance for the development of marine steel pile cleaning equipment.

2. Cleaning Equipment Structure Scheme and Working Principle

Marine steel pile cleaning equipment is composed of the frame, clamps, hinges and locks, and scraping tools. The overall structure is shown in Figure 1. The frame includes upper and lower rings and telescopic mechanisms. The upper and lower rings are composed of two half rings that can be opened to encircle the steel piles. The cleaning equipment can be held on the steel pile by the clamps. The telescopic mechanism is composed of three telescopic hydraulic cylinders that are connected between the upper and lower ring platforms with bolts. It not only supports the upper and lower ring platforms, but also controls the distance between them. Through the telescopic mechanism of three hydraulic cylinders and the cooperation of two groups of clamps, the cleaning equipment can move along with the steel pile. The scraping tools are composed of four groups of blade, each group including a driven hydraulic cylinder and a scraping blade. The driven hydraulic cylinder drives the scraping blade to clean the attached fouling organisms along the marine steel pile at a certain angle through the telescopic mechanism. The opening and closing of the equipment's upper and lower half rings are controlled by the hinges and locks through the hydraulic cylinder.

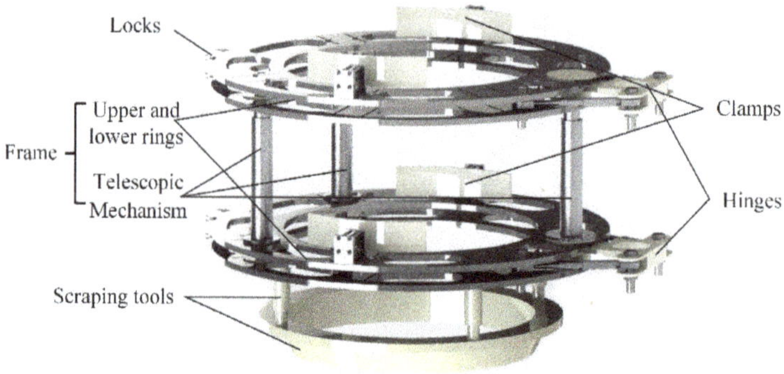

Figure 1. 3D model plan of the marine steel pile cleaning equipment.

The working area of the designed cleaning equipment is the spray splashing area or the tidal range area. Therefore, the equipment should be able to work on and under the water at the same time, and its working range is ±20 m above sea level. The fouling organisms attached to the marine steel piles in this area are relatively serious. The steel pile surface shall be treated as PTC after cleaning. The cleaning equipment designed at present can clean the marine steel piles with a diameter of 1 m. The creeping speed of the cleaning equipment along the axis of the marine steel pile is about 1 m/20 min to 1 m/30 min. The scraping tools are made of an aluminum alloy. In order to reduce the overall weight of the equipment, carbon fiber and aluminum alloy were used as the main frame materials.

With the assistance of the lifting boat, the staff installed the cleaning equipment on the marine steel piles with the upper and lower rings open. After installation, the equipment could automatically clean the marine steel piles. In the initial state, the cleaning equipment is clamped by the upper and lower clamps to hold the equipment onto the steel pile. In the climbing stage, when the equipment is ready to move downward, the lower clamps are released, and the hydraulic cylinder of the telescopic mechanism is extended simultaneously. When the hydraulic cylinder is extended to its limit, the lower clamps are clamped. At the end of climbing, the upper clamps are released, the telescopic mechanism is retracted to its limit, and the upper clamping hydraulic cylinder is clamped. At this time, the cleaning equipment completes its downward movement and moves to a new position on the steel pile. In the cleaning stage, two groups of clamps are clamped, and the scraping tools reciprocates vertically along the steel pile. By repeating the above steps, the steel piles in the entire splash zone can be cleaned. The schematic diagram of cleaning operation is shown in Figure 2.

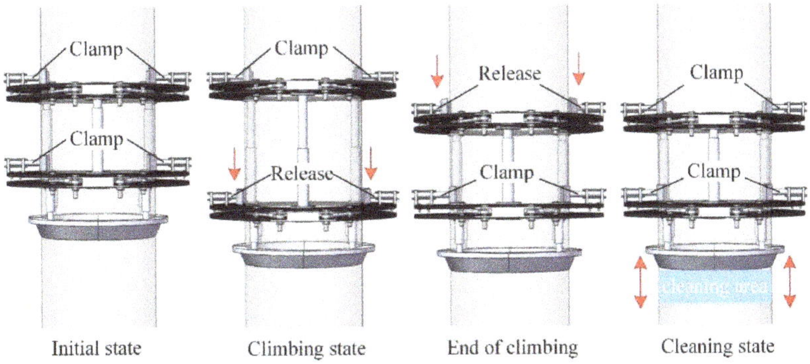

Figure 2. Operation process of the cleaning equipment (The red arrow indicates the direction of movement).

3. Influence of Telescopic Mechanism Synchronization on Scraping Tools

During the operation of the cleaning equipment in the ideal state, when the clamps clamp onto the marine steel pile, the upper and lower rings of the cleaning equipment should be kept horizontal, and the centers of the two rings and the steel pile should be concentric, as shown in Figure 3a, which is a top view of the cleaning equipment. When the center line of the upper and lower rings of the cleaning equipment (referred to as the cleaning equipment center line) coincides with the center line of the steel pile, the cleaning equipment is not inclined, as shown in Figure 3b, which is a cross-sectional view of the cleaning equipment. In this state, the angle of the blade is normal, the included angle between the blade and the steel pile is α_0, and the rake angle of the blade is $\gamma_0 = (90° - \alpha_0)$. Under the condition of no tilt, it is assumed that the cleaning equipment is ready to clean the annual barnacles that have attached to the circumference of the marine steel pile. According to the calculation, the four scraping tools of the cleaning equipment can be subjected to a total reaction force of about 6520 N when cleaning the attached barnacles [26].

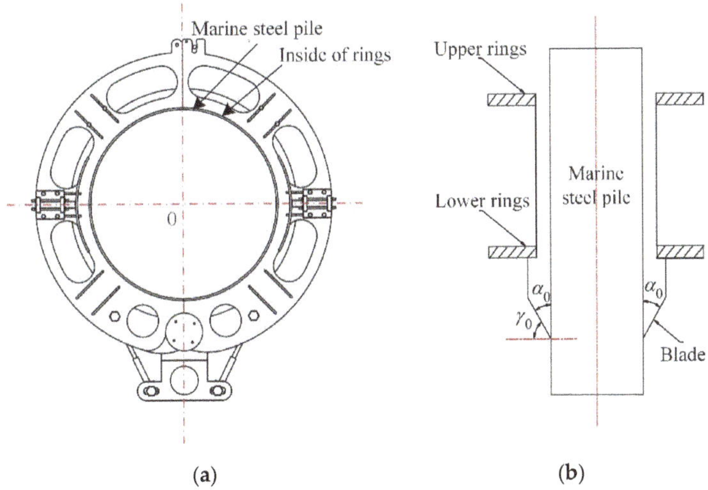

Figure 3. The working position of the cleaning equipment in an ideal state: (**a**) concentric; (**b**) no tilt.

During the operation of the cleaning equipment, due to the synchronization error of the telescopic mechanism, the eccentricity in or tilt of the cleaning equipment may occur as shown in Figure 4. Compared to Figure 3, when eccentric, as shown in Figure 4a, the included angle between the blade and the steel pile surface α_0 changed, so that the angle between the blade on one side and the steel pile was $\alpha_1 < \alpha_0$, and the angle between the opposite blade and steel pile was $\alpha_2 > \alpha_0$, which further caused the change in the rake angle of the blade and affected the cleaning effect of the cleaning equipment. When tilted, as shown in Figure 4b, the included angle of the blade close to the side of the steel pile was $\alpha_3 < \alpha_0$, and the angle between the opposite blade and steel pile was $\alpha_4 > \alpha_0$, which also affected the cleaning effect and the stability of the cleaning equipment. Next, the multi-cylinder synchronous control method was applied to the telescopic mechanism of the cleaning equipment for modeling and analysis.

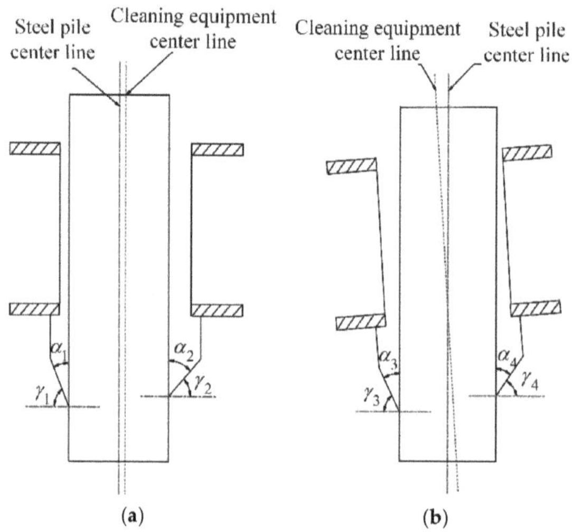

Figure 4. Operation process of cleaning equipment: (**a**) eccentricity of the cleaning equipment; (**b**) tilt of the cleaning equipment.

4. Synchronous Modeling and Simulation of the Telescopic Mechanism

4.1. Modeling of Synchronous Control System of the Telescopic Mechanism

The mathematical model of the synchronous extending process of the telescopic mechanism was established, as shown in Figure 5. The three hydraulic cylinders were connected by flanges. Each hydraulic cylinder could be rotated around the x-axis with a rotation angle of θ_{xi}, and around the y-axis with a rotation angle of θ_{yi}, where $I = 1, 2, 3$. In order to simplify the form of load motion, the following assumptions were made:

(1) Assuming that the piston of the hydraulic cylinder does not rotate in the cavity, there is no degree of freedom to twist around the z-axis for the telescopic load;
(2) Assuming that the synchronization error of the three-cylinder is much smaller than the distance between the installation positions of each hydraulic cylinder, the telescopic load mainly moves in the vertical direction, that is, the z-axis direction, and can be ignored along the x-axis and y-axis directions;
(3) Assuming that the hydraulic cylinder piston is not affected by the lateral force, regardless of the gravity of the piston and cylinder, the output force F_i ($I = 1, 2, 3$) of the piston rod of the hydraulic cylinder is along the direction of the piston rod, and the angle with the z-axis is approximately zero;
(4) Assume that the upper flange of the hydraulic cylinder and the upper ring are connected by a point centered on the flange.

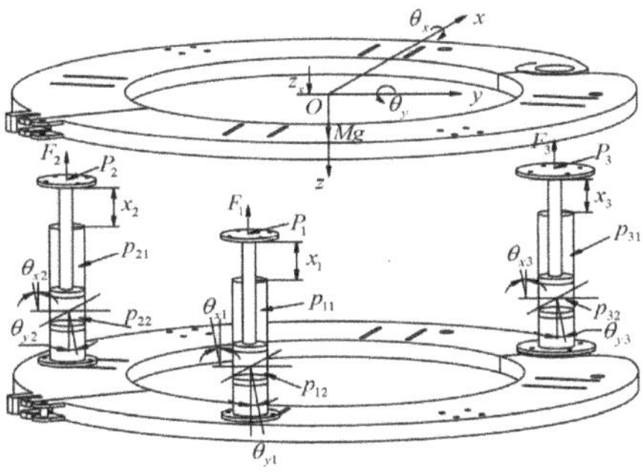

Figure 5. Operation process of the cleaning equipment.

On the basis of the above assumptions, the coordinate system of the system was established as follows:

In the figure, *O-xyz* is the model coordinate system, the center of mass of the above ring is the coordinate origin *O*, and the *x*, *y*, and *z* axes are as shown in Figure 5. In this way, the load movement such as the telescopic upper ring during the crawling process of the cleaning equipment can be simplified into three degrees of freedom: the degree of freedom of vertical motion in the *z*-axis direction, the degree of freedom of rotation around the *x*-axis, and the degree of freedom of rotation around the *y*-axis. That is to say, the movement of the telescopic load during the crawling process of the cleaning equipment can be described by the displacement z_c of the center of mass in the *z*-axis direction, the rotation angle θ_{xi} of the load around the *x*-axis, and the rotation angle θ_{yi} of the load around the *y*-axis.

4.1.1. Equation of Motion of Telescopic Mechanism

When the lower clamps of the cleaning equipment are clamped, the upper clamps are released, and the hydraulic cylinder of the telescopic mechanism is mainly extended. The above ring is the object of load research, which realizes the telescopic motion of the upper ring, the positive direction of each vector, and the force analysis results, as shown in Figure 5. On the three degrees of freedom of the hydraulic cylinder extending, using Newton's second law and the law of rigid body rotation, we obtain [29–31]:

$$\begin{cases} -\sum\limits_{i=1}^{3} F_i + Mg = M\ddot{z}_x \\ -\sum\limits_{i=1}^{3} F_i P_{iy} = J_x\ddot{\theta}_x \\ -\sum\limits_{i=1}^{3} F_i P_{ix} = J_y\ddot{\theta}_y \end{cases} \tag{1}$$

where F_i is the output force of the *i-th* hydraulic cylinder piston, (N); M is the total mass of load, (kg); z_x is the displacement of the center of mass of the load in the *z*-axis direction, (m); P_{ix} is the coordinate component of the P_i point on the *x*-axis in the coordinate system, (m); P_{iy} is the coordinate component of the *y*-axis of the P_i point in the coordinate system, (m); J_x is the moment of inertia of the telescopic load to the *x*-axis, (kg·m^2); J_y is the moment of inertia of the telescopic load to the *y*-axis, (kg·m^2); θ_x is the angle of rotation of the telescopic load about the *x*-axis, (rad), which is positive in the counterclockwise direction along the *Ox*-axis; θ_y is the rotation angle of the telescopic load around the *y*-axis, (rad), which is positive in the counterclockwise direction along the *Oy*-axis.

4.1.2. Telescopic Load and Point Contact Position Equation of Hydraulic Cylinder

Consider that when the telescopic load makes a small rotation in the *O-xyz* coordinate system, the coordinate of any point P_i on the *z*-axis can be approximated by $(z_x, \theta_x, \theta_y)$ as:

$$P_{iz} = z_x + P_{iz0} + P_{iy}\theta_x + P_{ix}\theta_y \tag{2}$$

where P_{iz} is the coordinate component of the *z*-axis of the P_i point in the coordinate system *O-xyz*, (m); P_{iz0} is the coordinate component of the *z*-axis of the initial P_i point in the coordinate system *O-xyz*, (m); z_x is the displacement of the piston rod of the hydraulic cylinder, (m).

Therefore, the approximate expressions of the displacements of the three hydraulic cylinder pistons from Equation (2) can be obtained as:

$$x_i = P_{iz} - P_{iz0} = z_x + P_{iy}\theta_x + P_{ix}\theta_y \tag{3}$$

Select P_1, P_2, and P_3 to form a set of three points that are not on a straight line on the plane. Let P_1, P_2, and P_3 constitute the minimum point set of the telescopic load attitude. Define $x_q = \begin{pmatrix} x_1 & x_2 & x_3 \end{pmatrix}^T$ as the displacement vector of the piston rod of each hydraulic cylinder, then x_q and $\begin{pmatrix} z_x & \theta_x & \theta_y \end{pmatrix}^T$ have the following relationship:

$$x_q = \begin{pmatrix} 1 & P_{1y} & P_{1x} \\ 1 & P_{2y} & P_{2x} \\ 1 & P_{3y} & P_{3x} \end{pmatrix} \begin{pmatrix} z_x \\ \theta_x \\ \theta_y \end{pmatrix} = L_q \begin{pmatrix} z_x \\ \theta_x \\ \theta_y \end{pmatrix} \tag{4}$$

4.1.3. The Equation of Motion of the Piston Rod of a Hydraulic Cylinder

Taking the piston rod of each hydraulic cylinder as the research object, its various vector directions and force results are shown in Figure 5. Considering that the hydraulic cylinder moves approximately vertically, Newton's second law is applied to the piston rod of the hydraulic cylinder in the vertical direction:

$$\begin{cases} p_{iL}A_{i1} + F_i + m_i g - B_{ip}\dot{x}_i = m_i\ddot{x}_i \\ p_{iL} = p_{i1} - p_{i2}A_{i2}/A_{i1} \end{cases} \tag{5}$$

where p_{iL} is the load pressure of the *i-th* hydraulic cylinder, (Pa); p_{iL} is the rodless chamber pressure of the *i-th* hydraulic cylinder, (Pa); p_{i2} is the rod chamber pressure of the *i-th* hydraulic cylinder, (Pa); A_{i1} is the rodless cavity area of the *i-th* hydraulic cylinder, (m); A_{i2} is the rod cavity area of the *i-th* hydraulic cylinder, (m); m_i is the rod cavity area of the *i-th* hydraulic cylinder, (m); B_{ip} is the viscous damping coefficient of the *i-th* hydraulic cylinder, (N·s/m).

4.1.4. Load Pressure Dynamic Characteristic Equation of Asymmetric Hydraulic Cylinder

Since the three-cylinder drive system in the model adopts the form of one-to-one valve-controlled cylinders, the oil flow into each hydraulic cylinder is controlled by its own proportional valve. The dynamic characteristics of the load pressure in the *i-th* hydraulic cylinder can be expressed by the following equation:

$$\begin{cases} \dot{p}_{iL} = \dfrac{\beta_e}{V_{ie}(x_i)}\left[Q_{iL}(x_{iv}, p_{iL}) - C_{it}p_{iL} - A_{i1}\dot{x}_i\right] \\ V_{ie}(x_i) = \dfrac{A_{i1}(S_i + 3x_i)}{7.5} \\ C_{it} = \dfrac{1.5C_{ip} + C_{ep}}{1.25} \end{cases} \tag{6}$$

where p_{iL} is the load pressure of the *i-th* hydraulic cylinder, (Pa); V_{ie} is the equivalent volume of the *i-th* hydraulic cylinder, (m^3); β_e is the oil modulus of elasticity, Pa; Q_{iL} is the load flow of the *i-th* valve, (m^3/s); x_{iv} is the displacement control signal of the *i-th* valve, (m); A_{i1} is the rodless cavity area of the *i-th* hydraulic cylinder, (m); S_i is the total stroke of the *i-th* hydraulic cylinder, (m); C_{it} is the total leakage coefficient of the *i-th* hydraulic cylinder, ($m^5/N{\cdot}s$); C_{ip} is the internal leakage coefficient of the *i-th* hydraulic cylinder, ($m^5/N{\cdot}s$); C_{ep} is the external leakage coefficient of the *i-th* hydraulic cylinder, ($m^5/N{\cdot}s$);

4.1.5. Mathematical Model of Three-Cylinder System

The three-cylinder system can be described according to Equations (4)–(6). Defining $F = (F_1 \;\; F_2 \;\; F_3)^T$ as the force vector of the load acting on the P_i contact point, Equations (1) and (5) can be arranged as:

$$\begin{cases} -L_f F + Mg = M_L \ddot{x}_q \\ R - B_p \dot{x} + F = m\ddot{x} \end{cases} \tag{7}$$

In the formula, $L_f = \begin{pmatrix} 1 & 1 & 1 \\ P_{1y} & P_{2y} & P_{3y} \\ P_{1x} & P_{2x} & P_{3x} \end{pmatrix}$ is the arm matrix of the force F; $M_g = \begin{pmatrix} M_g \\ 0 \\ 0 \end{pmatrix}$ is the gravity matrix of the telescopic load; $M_L = diag(M \quad J_x \quad J_y)L_q^{-1}$ is the inertia matrix of the telescopic load; $R = \begin{pmatrix} p_{1L}A_{11} + m_1 g \\ p_{2L}A_{21} + m_2 g \\ p_{3L}A_{31} + m_3 g \end{pmatrix}$ is the hydraulic cylinder load resultant external force matrix; $B_p = diag(B_{1p} \quad B_{2p} \quad B_{3p})$ is the matrix of viscous damping coefficient of hydraulic cylinder piston; $m = diag(m_1 \quad m_2 \quad m_3)$ is the mass matrix of the hydraulic cylinder piston.

Equation (7), after eliminating F, obtains the motion equation of the telescopic load, namely:

$$(M_L + L_f m)\ddot{x}_q + L_f B_p \dot{x}_q = L_f \tau \tag{8}$$

In the formula, $\tau = L_f^+ M_g + R$, L_f^+ is the generalized inverse of L_f.

Let $M_{xq} = M_L + L_f m$, $B_{xp} = L_f B_p$, the mathematical model of the three-cylinder system can be obtained by combining Equations (6) and (8), namely:

$$\begin{cases} M_{xq}\ddot{x} + B_{xp}\dot{x}_q = L_f \tau \\ \dot{p}_{iL} = \beta_e K_{iq} \psi_i(x_i, p_{iL}) x_{iv} + \pi(x_i) p_{iL} + h_i(x_i)\dot{x}_i \end{cases} \tag{9}$$

In the formula, $K_{iq} = C_d \omega_1$, $h(x_i) = \frac{\beta_e A_{i1}}{V_{ie}(x_i)}$.

Among them, C_d is the valve port flow coefficient; ω_1 is the area gradient of valve port, (m^2/m).

In the formula, $\psi(x_i) = \frac{1}{V_{ie}(x_i)}\sqrt{\frac{4\alpha_{XV} p_s - \frac{x_{iv}}{|x_{iv}|} p_{iL}}{3\rho}}$, $\pi(x_i) = -\frac{C_{it}\beta_e}{V_{ie}(x_i)}$.

Among them, α_{XV} is the commutation factor of the valve, when $x_v > 0$, $\alpha_{XV} = 1$, when $x_v < 0$, $\alpha_{XV} = \eta$, $\eta = A_{i2}/A_{i1}$; ρ is the density of oil, (kg/m^3), for Equation (9), when $L_f = \begin{pmatrix} 1 & 1 & 1 \\ P_{1y} & P_{2y} & P_{3y} \\ P_{1x} & P_{2x} & P_{3x} \end{pmatrix}$, $L_q = \begin{pmatrix} 1 & P_{1y} & P_{1x} \\ 1 & P_{2y} & P_{2x} \\ 1 & P_{3y} & P_{3x} \end{pmatrix}$ is taken, which is the mathematical model of the three-cylinder system at this time. At this time, the expressions of M_{xq}, B_{xp}, and τ are:

$$M_{xq} = M_L + L_f m =$$

$$\begin{pmatrix} m_1 + \frac{M(P_{2x}P_{3y} - P_{3x}P_{2y})}{P_{1x}P_{2y} - P_{2x}P_{1y} - P_{1x}P_{3y} + P_{3x}P_{1y} + P_{2x}P_{3y} - P_{3x}P_{2y}} & \frac{-M(P_{1x}P_{3y} - P_{3x}P_{1y})}{P_{1x}P_{2y} - P_{2x}P_{1y} - P_{1x}P_{3y} + P_{3x}P_{1y} + P_{2x}P_{3y} - P_{3x}P_{2y}} & \frac{M(P_{1x}P_{2y} - P_{2x}P_{1y})}{P_{1x}P_{2y} - P_{2x}P_{1y} - P_{1x}P_{3y} + P_{3x}P_{1y} + P_{2x}P_{3y} - P_{3x}P_{2y}} \\ \frac{-J_x(P_{2x} - P_{3x})}{P_{1x}P_{2y} - P_{2x}P_{1y} - P_{1x}P_{3y} + P_{3x}P_{1y} + P_{2x}P_{3y} - P_{3x}P_{2y}} & m_2 P_{2y} + \frac{J_x(P_{1x} - P_{3x})}{P_{1x}P_{2y} - P_{2x}P_{1y} - P_{1x}P_{3y} + P_{3x}P_{1y} + P_{2x}P_{3y} - P_{3x}P_{2y}} & \frac{-J_x(P_{1x} - P_{2x})}{P_{1x}P_{2y} - P_{2x}P_{1y} - P_{1x}P_{3y} + P_{3x}P_{1y} + P_{2x}P_{3y} - P_{3x}P_{2y}} \\ \frac{J_y(P_{2y} - P_{3y})}{P_{1x}P_{2y} - P_{2x}P_{1y} - P_{1x}P_{3y} + P_{3x}P_{1y} + P_{2x}P_{3y} - P_{3x}P_{2y}} & \frac{-J_y(P_{1y} - P_{3y})}{P_{1x}P_{2y} - P_{2x}P_{1y} - P_{1x}P_{3y} + P_{3x}P_{1y} + P_{2x}P_{3y} - P_{3x}P_{2y}} & m_3 P_{3x} + \frac{J_y(P_{1y} - P_{2y})}{P_{1x}P_{2y} - P_{2x}P_{1y} - P_{1x}P_{3y} + P_{3x}P_{1y} + P_{2x}P_{3y} - P_{3x}P_{2y}} \end{pmatrix} \tag{10}$$

$$B_{xp} = L_f \cdot diag(B_{1p} \quad B_{2p} \quad B_{3p}) \tag{11}$$

$$\tau = \begin{pmatrix} A_{11}p_{1L} + m_1 g + \frac{M_g(P_{2x}P_{3y} - P_{3x}P_{2y})}{P_{1x}P_{2y} - P_{2x}P_{1y} - P_{1x}P_{3y} + P_{3x}P_{1y} + P_{2x}P_{3y} - P_{3x}P_{2y}} \\ A_{21}p_{2L} + m_2 g - \frac{M_g(P_{1x}P_{3y} - P_{3x}P_{1y})}{P_{1x}P_{2y} - P_{2x}P_{1y} - P_{1x}P_{3y} + P_{3x}P_{1y} + P_{2x}P_{3y} - P_{3x}P_{2y}} \\ A_{31}p_{3L} + m_3 g + \frac{M_g(P_{1x}P_{2y} - P_{2x}P_{1y})}{P_{1x}P_{2y} - P_{2x}P_{1y} - P_{1x}P_{3y} + P_{3x}P_{1y} + P_{2x}P_{3y} - P_{3x}P_{2y}} \end{pmatrix} \tag{12}$$

4.2. Synchronous Control Simulation of Telescopic Mechanism

Based on the above theoretical analysis, the operation simulation model of the telescopic mechanism driven by three cylinders was established including three-channel load coupling modules and three single-channel electro-hydraulic servo drive modules. The single channel electro-hydraulic servo drive module includes a PID controller, proportional valve module, valve control cylinder load flow module, flow continuity module, and hydraulic cylinder piston dynamic characteristics module. Using the basic modules provided in MATLAB Simulink in the subsystem mode, the classic PID simulation model of the three-cylinder telescopic drive system was established. The established simulation model of the three-cylinder telescopic drive system is shown in Figure 6.

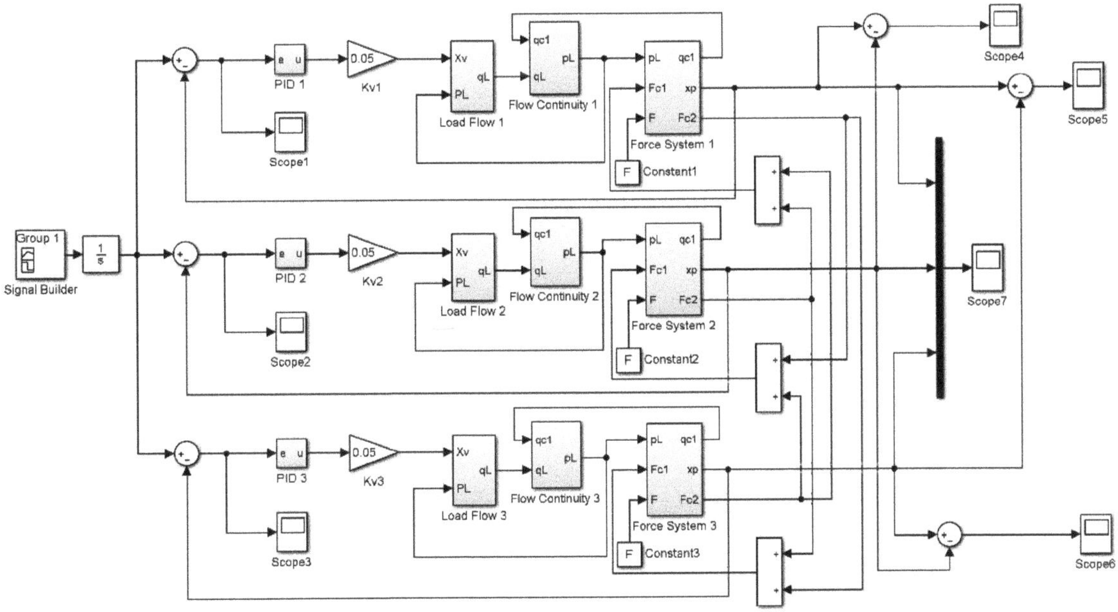

Figure 6. Synchronous telescopic system of three hydraulic cylinders.

According to the Simulink simulation model of the three-cylinder system obtained from the above analysis, the specific simulation experiments were carried out as follows: during the simulation, the hydraulic cylinder accelerated for 0.5 s to reach the working speed of 100 mm/s, and then moved at the working speed, with the maximum working stroke of 500 mm, and the simulation running time of 10 s.

Based on the analysis of the working principle of the marine steel pile cleaning equipment, it can be seen that the three-cylinder drive system had three typical working conditions: the no-load operation of the telescopic mechanism, the operation of the telescopic mechanism under the scraping reaction load, and the operation of a single hydraulic cylinder under the scraping reaction load. The displacement synchronization error curve of the three-cylinder was obtained, as shown in Figure 7.

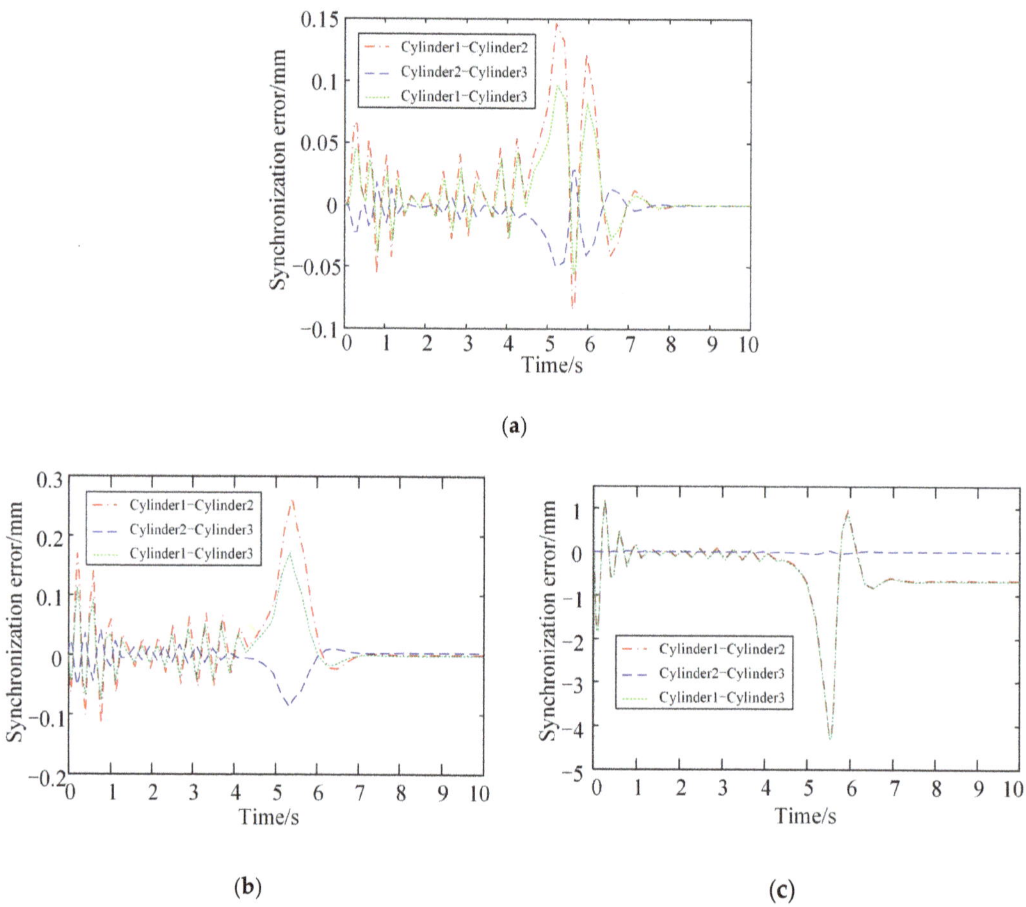

Figure 7. Three-cylinder synchronization error tracking simulation curve: (**a**) no-load operation of the telescopic mechanism; (**b**) the operation of the telescopic mechanism under the scraping reaction load; (**c**) the operation of a single hydraulic cylinder under the scraping reaction load.

Under the no-load condition of the telescopic mechanism, it can be seen from the displacement synchronization error curve of the three cylinders in Figure 7a that when the 0.5 s hydraulic cylinder started, the fluctuation of the synchronization error tracking curve between each cylinder started to increase, and the maximum error of the two cylinders was about 0.05 mm. During the hydraulic cylinder extending process, the synchronization error between the two cylinders was relatively small, which was kept at about 0.05 mm. When the hydraulic cylinder stopped extending from 4 s to 8 s, and was about to reach the limit displacement of 500 mm, the piston entered the deceleration stage, the displacement synchronization error curve between each cylinder fluctuated most obviously, and the maximum synchronization error of cylinder 1 and cylinder 2 was about 0.15 mm, the maximum synchronization error of cylinder 1 and cylinder 3 was about 0.1 mm, the maximum synchronization error of cylinder 2 and cylinder 3 was about 0.025 mm; the synchronous error tracking curve between each cylinder was basically stable during the operation stage. When the extending limit position was reached, the displacement error between each cylinder was finally maintained at 0 mm.

Under the loaded condition of the telescopic mechanism by the scraping reaction force, it can be seen from the displacement synchronization error curve of the three cylinders in Figure 7b that when the 0.5 s hydraulic cylinder started, the fluctuation of the synchronization error tracking curve between each cylinder started to increase, and the maximum error of the two cylinders was about 0.15 mm. During the hydraulic cylinder extending process, the synchronization error between the two cylinders was relatively small, and the synchronization error was kept at about 0.05 mm. When the hydraulic cylinder stopped extending from 4 s to 8 s, and was about to reach the limit displacement of the 500 mm waiting position, the piston entered the deceleration stage, the displacement synchronization error curve between each cylinder fluctuated most obviously, and the maximum synchronization error of cylinder 1 and cylinder 2 was about 0.28 mm, the maximum synchronization error of cylinder 1 and cylinder 3 was about 0.18 mm, the maximum synchronization error of cylinder 2 and cylinder 3 was about 0.1 mm; the synchronous error tracking curve between each cylinder was basically stable during the operation stage. When the extending limit position was reached, the displacement error between each cylinder was finally maintained at 0 mm.

Under the condition that a single hydraulic cylinder (hydraulic cylinder 1) is loaded with the scraping reaction force, it can be seen from the displacement synchronization error curve of three cylinders in Figure 7c that the synchronization error between cylinder 2, cylinder 3 and cylinder 1 was relatively large and the synchronization error between cylinder 2 and cylinder 3 was almost zero. When the hydraulic cylinder started, the synchronous error tracking curve fluctuation between cylinder 2 and cylinder 3 started to increase, and the maximum error of the two cylinders was about 1.8 mm. During the hydraulic cylinder extending process, the synchronization error between the two cylinders was relatively small, and the synchronization error was maintained at about 0.05 mm; When the hydraulic cylinder stopped extending from 4 s to 8 s, and was about to reach the limit displacement of 500 mm, the piston entered the deceleration stage, the displacement synchronization error curve between each cylinder fluctuated most obviously, and the maximum synchronization error between cylinder 2, cylinder 3, and cylinder 1 was about 4 mm, the synchronization error of cylinder 2 and cylinder 3 was about 0 mm; the synchronization error tracking curve between each cylinder was basically stable in the operation stage. When the extending limit position was reached, there was a 1 mm synchronization error between cylinder 2, cylinder 3, and cylinder 1, and the displacement error between cylinder 2 and cylinder 3 was finally maintained at 0 mm.

It can be seen from the above analysis that In the three cases, the synchronous error of the multi-cylinder displacement was small during the operation of the telescopic mechanism, and good synchronization could be achieved, which not only effectively guarantees the stability of the cleaning operation of the cleaning equipment, but also avoids the eccentricity and tilt of the cleaning equipment during the climbing and cleaning process.

5. Synchronous Test of the Telescopic Mechanism

5.1. Development and Test of Cleaning Equipment

Based on the design scheme and simulation results of the marine steel pile cleaning equipment proposed in this paper, a multi-cylinder driving telescopic mechanism test prototype was developed. During the manufacturing and processing of the test prototype, the working principle and system composition of the pneumatic transmission and hydraulic transmission were similar, and both could achieve the purpose of cleaning the climbing, so a cylinder with a relatively low economic cost was selected to replace the hydraulic cylinder in this design scheme to achieve the same effect. Although the air was compressible, the action speed of the cylinder was easily affected by the load, and the stability was not as good as that of the hydraulic transmission [32]. If the cylinder can achieve the test effect stably, it would be even better when the hydraulic cylinder with better stability and more accurate control is used.

After assembling the clamps, telescopic mechanism, and scraping tools, the final developed cleaning equipment test prototype was obtained, as shown in Figure 8. In order to make its placement stable, the whole equipment was inverted, the scraping tools were located at the top, and the whole equipment was placed on the ground.

Figure 8. Cleaning equipment test prototype.

In order to verify the synchronization effect of simulation when the telescopic mechanism was under no-load, the telescopic mechanism was equipped with their respective displacement sensors, as shown in Figure 9a.

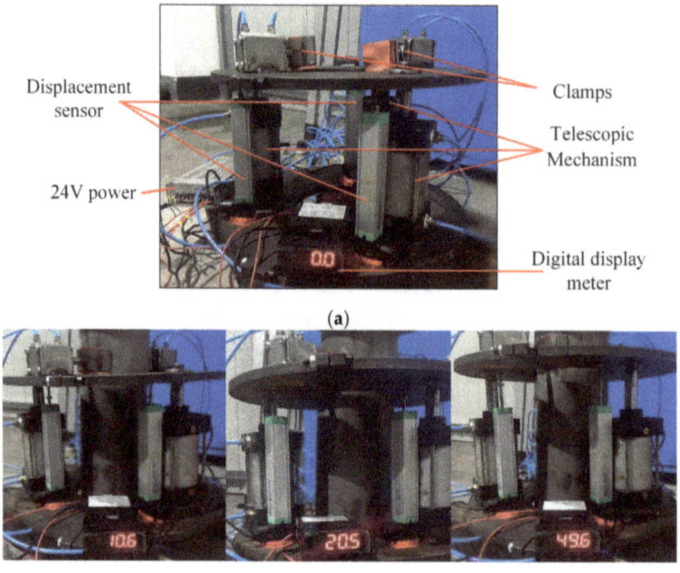

Figure 9. Three-cylinder synchronous test device diagram: (**a**) three-cylinder synchronous test device; (**b**) three-cylinder synchronous test.

The displacement sensor was fixed on the upper and lower support rings of the cleaning equipment to monitor the telescopic displacement of each cylinder in real-time. The three displacement sensors were connected to their respective digital display meters, and the telescopic displacement of each sensor can be fed back in real-time through the readings of their digital displays.

During the test, the clamps and telescopic mechanism switch of the cleaning equipment were operated through the console. The extending process of the three cylinders was monitored in real-time through the displacement sensor to carry out the telescopic synchronization test on the cleaning equipment.

In this test, the cleaning equipment prototype was produced with the telescopic mechanism gradually extending to its extreme position under the clamping state of the lower clamps, and the cleaning equipment was in an upward creeping state for testing. The test steps were as follows:

First, we controlled the lower clamps of the cleaning equipment to clamp and the upper clamps to release, so that the cleaning equipment was in the clamping state and stable on the steel pile; second, we ensured that the telescopic mechanism gradually extended, and the cleaning equipment started to climb up along the axis of the steel pile. During the climbing process, we monitored the extending displacement of the three cylinders in real-time, as shown in Figure 9b, took one cylinder as an example, and recorded the extending displacement data of the other two cylinders in the same way. Finally, when the telescopic mechanism reached its limit, the upper clamps clamped, and the upper and lower clamps were in the clamping state. This means that the cleaning equipment has completed an upward climbing movement along the axial direction of the steel pile, so the cleaning equipment can move up or down along the axial direction of the steel pile.

5.2. Test Results and Discussion

According to the synchronous displacement data of three cylinders, the extending displacement curve of three cylinders can be drawn under the same coordinate, as shown in Figure 10.

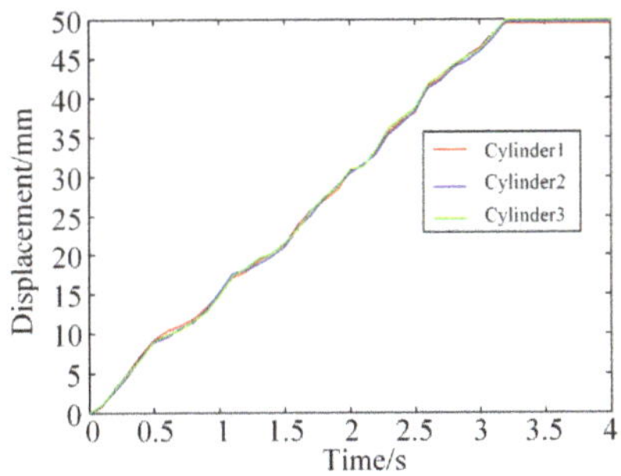

Figure 10. Synchronous displacement curve of the three-cylinder test.

It can be seen from Figure 10 that the displacement curve fluctuated slightly during the extending of the three cylinders, but the synchronization was good. The three cylinders almost reached the 50 mm position at the same time. After reaching this position, although the displacement of the three cylinders remained stable with the increase in time, the three hydraulic cylinders moved 49.6 mm, 49.8 mm, and 49.9 mm, with relative errors of 0.8%, 0.4%, and 0.2%, respectively. According to the data recorded by the displacement sensor, we depicted the error curve of the relative target displacement between each cylinder, as shown in Figure 11.

It can be seen from Figure 11 that the maximum displacement synchronization error of the cylinder 1 and cylinder 2 test reached 0.7 mm, the cylinder 2 and cylinder 3 test reached 0.7 mm, and the cylinder 1 and cylinder 3 test reached 0.6 mm. According to the test, the maximum displacement error between two different cylinders was controlled at 0.7 mm, the minimum displacement error was 0.1 mm, and the synchronous displacement error between the three cylinders was well controlled.

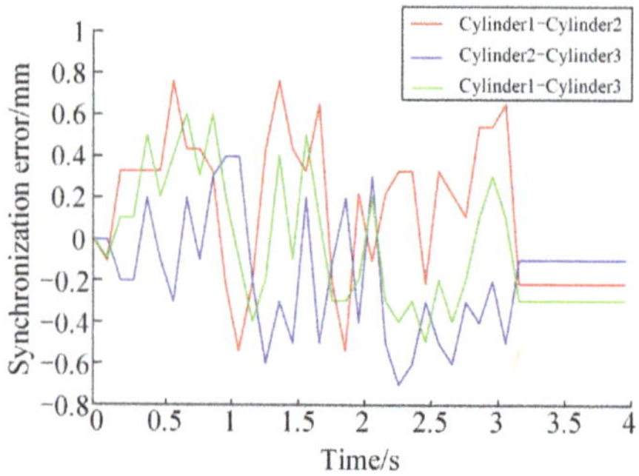

Figure 11. The three-cylinder test displacement synchronization error.

To sum up, when the extension of the telescopic mechanism reached the limit position of 50 mm, the relative error between the three cylinders and the target displacement was 0.8%, 0.4%, and 0.2%, respectively, to ensure that the equipment reached the specified working position at the given working speed. At the same time, the maximum error between the two cylinders was 0.7 mm, which effectively controlled the synchronization error between the two cylinders, and the synchronization of the telescopic mechanism was preliminarily verified.

6. Conclusions

In this paper, a configuration scheme of marine steel pile cleaning equipment and its telescopic mechanism with a multi-cylinder synchronous control strategy was proposed. Based on the MATLAB Simulink module, a simulation model of the operation process of the marine steel pile cleaning equipment was established, and its multi-cylinder synchronous control performance evaluation under multiple working conditions completed. The synchronous working performance of the telescopic mechanism of the cleaning equipment under no-load conditions was preliminarily verified through the test, and the following conclusions were obtained:

(1) In order to clean up the fouling organisms attached to the offshore infrastructure, this paper designed a new configuration of marine steel pile cleaning equipment using the scraping method and its telescopic mechanism with a multi-cylinder synchronous control strategy and process and produced a test prototype.
(2) The simulation model of the operation process of the marine steel pile cleaning equipment was established, and the simulation of its multi-cylinder synchronous control under multiple working conditions was completed. Through the simulation solution, the maximum displacement synchronization error between each cylinder of the telescopic mechanism under the no-load condition was 0.15 mm. When the three hydraulic cylinders were loaded by the scraping reaction force, the maximum synchronization error between each cylinder was about 0.28 mm; when a single hydraulic cylinder was loaded with the scraping reaction force, the maximum synchronization error between each cylinder was about 4 mm. In all three cases, the telescopic mechanism achieved good synchronization in the simulation process.
(3) Through the no-load test of the telescopic mechanism, the synchronization of the telescopic mechanism of the prototype of the cleaning equipment was preliminarily verified. The test results showed that the relative errors between the three cylinders and the target displacement were 0.8%, 0.4%, and 0.2%, respectively, to ensure that the equipment reached the designated working position at the given working speed. The displacement synchronization error of each cylinder was controlled within 1 mm, and the telescopic mechanism had good telescopic synchronization, which can prevent the eccentricity and tilt of the cleaning equipment during its operation while ensuring its stability.

Author Contributions: Formal analysis, C.L. and N.P.; Methodology, C.L. and X.W.; Writing—original draft preparation, C.L.; Writing—review and editing, K.X. and F.Y., Validation, Q.G. and L.G.; Supervision Q.G. and L.G.; Funding acquisition, C.L., N.P., Q.G., X.W. and F.Y. All authors have read and agreed to the published version of the manuscript.

Funding: This research was funded by the High Level Talents Project of Guangdong Mechanical and Electrical Polytechnic (Gccrcxm-202209); the National Natural Science Foundation of China (62241304); The Basic Ability Enhancement Program for Young and Middle-Aged Teachers of Guangxi (2023KY0821); the Key Scientific Research Projects of Colleges and Universities in Guangdong Province (2022KQNCX178); the Shandong Province Key Research and Development Program (2021CXGC010706); the Heilongjiang Provincial Natural Science Foundation of China (LH2021E046); the China Postdoctoral Science Foundation, grant number (2020M670889); the Key R&D Program of Shandong Province (2021JMR0302).

Institutional Review Board Statement: Not applicable.

Informed Consent Statement: Not applicable.

Data Availability Statement: Not applicable.

Conflicts of Interest: The authors declare no conflict of interest.

References

1. Hopkins, G.; Davidson, I.; Georgiades, E.; Floerl, O.; Morrisey, D.; Cahill, P. Managing biofouling on submerged static artificial structures in the marine environment–assessment of current and emerging approaches. *Front. Mar. Sci.* **2021**, *8*, 759194. [CrossRef]
2. Azis, P.K.A.; Al-Tisan, I.; Al-Daili, M.; Green, T.N.; Ba-Mardouf, K.; Al-Qahtani, S.A.; Al-Sabai, K. Marine macrofouling: A review of control technology in the context of an on-line experiment in the turbine condenser water box of Al-Jubail Phase-I power MSF plants. *Desalination* **2003**, *154*, 277–290. [CrossRef]
3. Magin, C.M.; Cooper, S.P.; Brennan, A.B. Non-toxic antifouling strategies. *Mater. Today* **2010**, *13*, 36–44. [CrossRef]
4. Yebra, D.M.; Kiil, S.; Dam-Johansen, K. Antifouling technology—Past, present and future steps towards efficient and environmentally friendly antifouling coatings. *Prog. Org. Coat.* **2004**, *50*, 75–104. [CrossRef]
5. Ren, W.; Chen, Y.; Xv, H.; Chen, Y.; Xie, Z.; Chen, Y. Study of Corrosion Mechanism and Corrosion Status of Offshore Wind Power Equipment. *Ship Eng.* **2021**, *43*, 1–5.
6. Telegdi, J.; Trif, L.; Románszki, L. Smart anti-biofouling composite coatings for naval Applications. In *Smart Composite Coatings and Membranes*; Woodhead Publishing: Cambridge, UK; Elsevier: Amsterdam, The Netherlands, 2016; pp. 123–155.
7. Zan, Y.; Guo, R.; Yuan, L.; Ma, Q.; Zhou, A.; Wu, Z. Experimental study of a suspended subsea module at different positions in the splash zone. *Mar. Struct.* **2021**, *77*, 102935. [CrossRef]
8. Heaf, N.J. The Effect of Marine Growth On The Performance of Fixed Offshore Platforms in the North Sea. In Proceedings of the 11th Annual Offshore Technology Conference, Houston, TX, USA, 30 April–3 May 1979; pp. 255–267.
9. Campos, R.M.; Islam, H.; Ferreira, T.R.S.; Soares, C.G. Impact of heavy biofouling on a nearshore heave-pitch-roll wave buoy performance. *Appl. Ocean Res.* **2020**, *107*, 102500. [CrossRef]
10. Wei, S.C.; Xu, B.S.; Liang, X.B.; Wang, Y.J.; Liu, Y. Research on Corrosion-Resistance of High Velocity Arc Spray Coatings on Surface of Steel Structure in Splash Zone Environment. *Mater. Sci. Forum* **2011**, *675–677*, 1291–1294.
11. Albitar, H.; Dandan, K.; Ananiev, A.; Kalaykov, I. Underwater robotics: Surface cleaning technics, adhesion and locomotion systems. *Int. J. Adv. Robot. Syst.* **2016**, *13*, 7. [CrossRef]
12. Song, C.; Cui, W. Review of underwater ship hull cleaning technologies. *J. Mar. Sci. Appl.* **2020**, *19*, 415–429. [CrossRef]
13. Morrisey, D.; Woods, C. *In-Water Cleaning Technologies: Review of Information*; Ministry for Primary Industries, Manatū Ahu Matua: Wellington, New Zealand, 2015.
14. Nassiraei, A.A.F.; Sonoda, T.; Ishii, K. Development of ship hull cleaning underwater robot. In Proceedings of the 2012 Fifth International Conference on Emerging Trends in Engineering and Technology, Himeji, Japan, 5–7 November 2012; pp. 157–162.
15. Souto, D.; Faiña, A.; López-Peña, F.; Duro, R.J. Morphologically intelligent underactuated robot for underwater hull cleaning. In Proceedings of the 2015 IEEE 8th International Conference on Intelligent Data Acquisition and Advanced Computing Systems: Technology and Applications (IDAACS), Warsaw, Poland, 24–26 September 2015; pp. 879–886.
16. Souto, D.; Faiña, A.; López-Peña, F.; Duro, R.J. Lappa: A new type of robot for underwater non-magnetic and complex hull cleaning. In Proceedings of the 2013 IEEE International Conference on Robotics and Automation, Karlsruhe, Germany, 6–10 May 2013; pp. 3409–3414.
17. Iasgroup Corporation. Splash Genius II Cleaning System. Available online: http://www.ias-group.com.au/view/video-articles/splash-genius-ii-remotely-operated-cleaning-tool (accessed on 12 February 2023).
18. Iasgroup Corporation. Splash Genius Cleaning System. Available online: http://www.ias-group.com.au/view/video-articles/splash-genius-cleaning-tool (accessed on 12 February 2023).

19. Woolfrey, J.; Lu, W.; Vidal-Calleja, T.; Liu, D.J.O.E. Clarifying clairvoyance: Analysis of forecasting models for near-sinusoidal periodic motion as applied to AUVs in shallow bathymetry. *Ocean. Eng.* **2019**, *190*, 106385. [CrossRef]
20. Le, K.; To, A.; Leighton, B.; Hassan, M.; Liu, D. The spir: An autonomous underwater robot for bridge pile cleaning and condition assessment. In Proceedings of the 2020 IEEE/RSJ International Conference on Intelligent Robots and Systems (IROS), Las Vegas, NV, USA, 24 October 2020–24 January 2021; pp. 1725–1731.
21. Submersible Pile Inspection Robot. Available online: https://www.youtube.com/watch?v=hFtW2cXaHYk&t=24s (accessed on 12 February 2023).
22. Fan, J.; Yang, C.; Chen, Y.; Wang, H.; Huang, Z.; Shou, Z.; Jiang, P.; Wei, Q. An underwater robot with self-adaption mechanism for cleaning steel pipes with variable diameters. *Ind. Robot: Int. J.* **2018**, *42*, 193–205. [CrossRef]
23. Huang, Z.; Chen, Y.; Yang, C.; Fan, J.; Jiang, P. Teleoperate system of underwater cleaning robot based on HUD. In Proceedings of the 2017 11th Asian Control Conference (ASCC), Gold Coast, Australia, 17–20 December 2017; pp. 2675–2679.
24. Jiang, P.; Wei, Q.; Chen, Y.; Yang, C.; Fan, J.; Shou, Z.; Huang, Z. Real-time panoramic system for underwater cleaning robot. In Proceedings of the 2018 IEEE 9th International Conference on Mechanical and Intelligent Manufacturing Technologies (ICMIMT), Cape Town, South Africa, 10–13 February 2018; pp. 155–159.
25. Jie, P.; Baorong, H.; Jiali, L.; Weiguo, Z. Advance in application of petrolatum tape cover anticorrsion technology in China. *Corros. Prot.* **2015**, *36*, 1170–1173.
26. Chao, L.; Gang, W.; Chen, K.; Peng, J.; Liquan, W.; Xiangyu, W.; Feihong, Y. Analysis of removing barnacles attached on rough substrate with cleaning robot. *J. Mar. Sci. Eng.* **2020**, *8*, 569. [CrossRef]
27. Liquan, W.; Chao, L.; Kaiyun, C.; Gang, W. Mechanical research on removing adhesion barnacles with a cleaning robot for marine steel piles. *J. Harbin Eng. Univ.* **2021**, *42*, 259–265.
28. Fujie, Y.; Qingzhong, L.; Yao, W.; Yuan, C. Optimization of tool orientation for improving the cleaning efficiency of offshore jacket-cleaning systems. *Appl. Ocean Res.* **2021**, *112*, 102687.
29. Mo, W.; Liu, N.; Li, L.; Han, H. Application of PID control in hydraulic synchronous system of cleaning equipment. In *Journal of Physics: Conference Series*; IOP Publishing: Bristol, UK, 2021; pp. 62–64.
30. Kim, G.S.; Lee, D.J. Study on Optimization of Valve Parameter for Multi-Cylinder Synchronous Control System. *Adv. Sci. Lett.* **2015**, *21*, 2985–2988. [CrossRef]
31. Jing, N.; Hongliang, W.; Zhen, M.; Xiangqi, L. DCAT-NDI Control on Synchro-motion System with Multi-cylinder. *J. Hangzhou Dianzi Univ. Nat. Sci.* **2015**, *35*, 14–22.
32. Xiaoyao, Z.; Tao, W.; Bo, W. Synchronization Control of a 3-DOF Pneumatic Translational Parallel Robot. *Chin. Hydraul. Pneum.* **2021**, *45*, 129–134.

Journal of
*Marine Science
and Engineering*

MDPI

Article

Mechanical Behavior and Sealing Performance Study of Subsea Connector Core-Sealing Components under the Combined Action of Internal Pressure, Bending Moment, and Axial Load

Xiaoquan Hao [1], Feihong Yun [1,*], Kefeng Jiao [1], Xi Chen [2], Peng Jia [1], Xiangyu Wang [1] and Liquan Wang [1]

1 College of Mechanical and Electrical Engineering, Harbin Engineering University, Harbin 150001, China; haoxiaoquan@hrbeu.edu.cn (X.H.); jiaokefeng@hrbeu.edu.cn (K.J.); jiapeng@hrbeu.edu.cn (P.J.); wangxiangyu@hrbeu.edu.cn (X.W.); wangliquan@hrbeu.edu.cn (L.W.)
2 College of Mechanical and Electrical Engineering, Heilongjiang Institute of Technology, Harbin 150050, China; chenxi_1113652@hrbeu.edu.cn
* Correspondence: yunfeihong@hrbeu.edu.cn; Tel.: +86-159-4568-3623

Abstract: A complete subsea production system (SPS) is assembled by interconnecting subsea manufacturing facilities through subsea connectors. To ensure the reliability and longevity of the SPS, it is imperative to thoroughly investigate the mechanical behavior and sealing performance of the subsea connector's core-sealing components. In this study, the loading conditions of the subsea clamp connector are examined to analyze the load transfer relationship between its components under different modes. A mathematical model for the load transfer between locking torque and sealing contact pressure is developed for the preloading mode, and the concept of mechanical transfer efficiency is introduced. Another mathematical model for the load transfer between the locking torque and the design pressure is developed for the operation mode. Furthermore, a three-dimensional full-size finite element model of the subsea clamp connector is established to analyze the effects of complex loads on the mechanical behavior and sealing performance of its core-sealing components. The simulation results indicate that internal pressure loading positively affects the sealing of the subsea connector, and that the stress distribution in the core-sealing components under bending moment loading exhibits significant asymmetric characteristics. Additionally, the superposition of axial tensile loads reduces the effect of the bending moment on the strength of the core seal member but further weakens the seal. Finally, an experimental system is designed to validate the simulation results.

Keywords: subsea clamp connector; core-sealing components; load transfer; mechanical behavior; sealing performance

Citation: Hao, X.; Yun, F.; Jiao, K.; Chen, X.; Jia, P.; Wang, X.; Wang, L. Mechanical Behavior and Sealing Performance Study of Subsea Connector Core-Sealing Components under the Combined Action of Internal Pressure, Bending Moment, and Axial Load. *J. Mar. Sci. Eng.* **2023**, *11*, 1691. https://doi.org/10.3390/jmse11091691

Academic Editors: Weicheng Cui, Lian Lian and Dahai Zhang

Received: 7 August 2023
Revised: 24 August 2023
Accepted: 25 August 2023
Published: 27 August 2023

1. Introduction

The subsea production system has emerged as one of the primary methods for developing oil and gas resources in deepwater regions, and is continuously advancing towards ultra-deepwater with higher production equipment demands. The subsea production system consists of essential equipment, including Christmas trees, jumpers, wellheads, pipe manifolds, and others [1]. These components are constructed onshore and installed at designated locations on the seafloor and then connected via subsea connectors to form a complete subsea production system. During operation, the subsea connectors are exposed to high temperature and pressure, as well as internal and external loads from connected equipment and ocean currents [2]. The installation of subsea production systems in deep and ultra-deepwater environments presents complex terrain and operational challenges. Any leakage or failure of subsea connectors under load can lead to substantial repair costs and irreversible damage to the marine environment.

Scholars have extensively researched the sealing performance and design optimization of subsea connectors' core-sealing components, namely, the hub and sealing gasket. For instance, Cascales et al. [3] proposed a mathematical model equivalent to the Murray and Stuart model, which led to an approximate formula for evaluating hub rotation. Dekker et al. [4] developed a design method for pipe clamp connector connections, which had a more reasonable design pressure. Abid et al. [5] showed that bolted flange connectors sealing performance was related to both internal pressure and transient thermal loads. Murali Krishna et al. [6] analyzed the loading and unloading characteristics of a bolted flange joint experimentally and examined the effect of contact stress distribution on the sealing performance using three-dimensional finite element analysis. Takagi et al. [7] predicted the leakage amount by analyzing the spiral wound gasket contact stress distribution and the stress in the hub using the three-dimensional elasto-plastic finite element method. Wang et al. [8] used experimental and finite element methods to study the bending behavior of flange connections under pure bending action, revealing the distribution of contact pressure at the connection end plates. Abid et al. [9] investigated the strength of the hub structure of a non-gasketed bolt flange joint under the combined effect of internal pressure, axial and bending loading, and the sealing capacity of the gasket with different taper angles. Peng et al. [10] optimized and analyzed the mathematical model of the locking mechanism of the subsea clamp connector and established the mathematical model of the force transfer of the subsea clamp connector, which was verified by the finite element method. Nelson et al. [11] proposed an empirical relationship equation for bolt preload that guarantees minimum compressive stress after comparing the sealing performance of single and twin gasket and studying the contact pressure of single and twin gasketed flange joint under bolt preload and internal pressure. Yun et al. [12] established a mathematical model of metal seal contact stress and verified it by finite element analysis and experimental studies. Zhang et al. [13], based on the Hertzian contact theory, derived the analytical equation for the compression of the collector connector and proposed the compression limit equation. Chen et al. [14] analyzed the stress distribution characteristics of the main sealing components of the threaded connector by establishing a three-dimensional elasto-plastic finite element model based on the sealing mechanism of the taper-to-arc seal. Wang et al. [15] established the critical conditions of the sealing performance of the new submarine pipeline mechanical connector and the calculation formula of the sealing contact pressure by analyzing the mechanism of the metal static seal of the connector. Li et al. [16] analyzed the mechanical properties of metal seals of subsea wellhead connectors under preloading and operating mode and determined the theoretical relationship between contact stress and metal seal structural parameters and operating pressure. Liu et al. [17] experimentally investigated the mechanical properties of square steel pipe column joints with bolted flange connections (without sealing role) under combined tensile, bending, and shear loads. Liu et al. [18] developed a new fractal porous media model to analyze the leakage principle of the metal seals of subsea connectors and calculated the permeability of the seal components based on this media model and the length-dependent mechanical model. Duan et al. [19] proposed a stress analysis method (SAM) to solve the theoretical design problem of the main parameters of the hub structure of the subsea connector by establishing a deformation continuum relationship between a thick-walled cylinder and a hollow ring plate. Li et al. [20] theoretically analyzed the load-carrying capacity of subsea wellhead connectors when subjected to external pressure and the bending moment under different operating modes and loading conditions. Yun et al. [21] developed a heat transfer model to investigate the sealing capability of lens-type sealing structures of subsea clamp connectors under the influence of external seawater and internal oil and gas temperature loads. Meanwhile, Zhang et al. [22] proposed an analytical calculation method (ACM) to evaluate the thermal-structural coupling strength and sealing performance of subsea wellhead connectors. Li et al. [23] derived the radial temperature distribution function of a subsea wellhead connection and investigated the effects of internal pressure and temperature on its metal sealing capacity through experimental studies.

The findings from these studies are crucial for improving the reliability and safety of subsea wellhead connectors and advancing the subsea oil and gas industry.

Many previous studies have concentrated on the analysis of the metal seal contact of subsea connectors. These studies have utilized various methods, including theoretical approaches, finite element methods, and experimental methods to investigate the microscopic contact mechanism of the sealing surface, the effect of contact stress, and related parameters such as contact surface width and deformation, as well as the effect of temperature and load on the seal. However, after the subsea connector is connected, it is exposed to a combination of internal oil and gas pressure and external loads (axial and moment loads), which can impact the sealing performance and mechanical behavior of the core-sealing components of the subsea connector. Regarding the investigation into the impact of external loads, the primary focal points encompass diverse connector types, including the gasketless bolted flange connector (primarily serving a connection role), bolted flange connector, subsea collet connector, and subsea wellhead connector. Notably, the subsea clamp connector, as designed within this paper, has yet to be a subject of such research. Amidst the scrutinized connectors, the emphasis has primarily been on exploring the influence of internal pressure or axial tension loads on the connector's sealing function under external loading. However, there has been a notable absence of research regarding the sealing and strength characteristics when the connector is subjected to the combined effect of internal pressure, axial tension, and bending loads.

In this paper, we investigate the sealing and strength performance of the subsea clamp connector under complex loads. Firstly, we analyze the load transfer relationship between each structure under preloading and operation modes, and establish a mathematical model accordingly. Secondly, we develop a simulation model for the core-sealing component of a six-inch subsea connector and conduct finite element analysis to investigate its sealing and strength performance under the combined effects of internal pressure, bending moment load, and axial load. Lastly, we compare the experimental results with the finite element analysis results to validate the accuracy of our mathematical model and finite element simulation.

2. Subsea Connector Structure and Working Principle

The subsea connector structure examined in this study, depicted in Figure 1, is suitable for deployment in shallow water and deep-water environments, and can be installed automatically through ROV-assisted means. The connector is a horizontally operating clamp-type connector, comprising a core structure that primarily consists of a mobile hub, fixed hub, metal sealing gasket with lens-type structure, tightened bolt, and clamp. The fixed hub is welded to the subsea pipe and pre-positioned on the subsea production facility, while the mobile hub is typically welded to both ends of the jumper. The jumper and mobile end are placed together at the designated subsea location and connected to the fixed hub on both sides, forming a complete subsea connection.

In this study, Figure 2a illustrates the state of the subsea connector before locking, while Figure 2b depicts the state after locking. The connector operates by rotating the tensioning bolt using an ROV-operated torque spanner. This action causes the clamp to move radially and hold the mobile and fixed hubs, which creates contact between the tapered hub surface and the spherical surface of the metal lens type sealing gasket, leading to the formation of a seal. The load transfer mechanism in this structure involves multiple components, including the tightened bolt, clamp, hub, and sealing gasket, among others. Due to the complexity and non-uniformity of the load transfer process, it is essential to investigate the load transfer in this structure. Moreover, the subsea connector is exposed to both internal oil and gas pressure from the pipeline and external loads, which can further impact the mechanical properties of the core structure.

Figure 1. Practical application for the subsea connector.

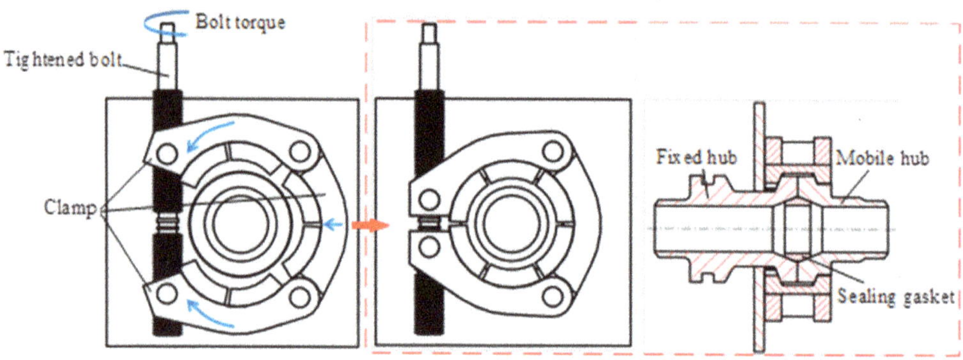

(a) (b) (c)

Figure 2. Subsea connector working principle: (**a**) before locking; (**b**) after locking; and (**c**) cross-section.

3. Study of Subsea Connector Load Transfer under Different Work Modes

This paper investigates two working modes of the subsea connector: preloading mode and operating mode. In preloading mode, the metal seal achieves pipe sealing by being subjected to pressure on both sides of the moving end flange and fixed end flange. In operating mode, the connector is subjected to both internal oil and gas pressure and external loads, resulting in a force state that is different from the preload.

The transfer of loads from the bolt to the clamp remains consistent in both modes, therefore, it is analyzed first. Subsequently, the force transfer from the clamp to the metal sealing gasket is analyzed separately in each mode.

3.1. Analysis of Bolt to Clamp Force Transfer

The model of the subsea connector is simplified, as shown in Figure 3. The clamps are simplified to hinges in the model for calculation, with pin 1 between clamp A and clamp B, and pin 2 between clamp B and C. Clamp A and C are connected with pins 3 and 4, respectively, and pins 3 and 4 are connected closer by the tightened bolt.

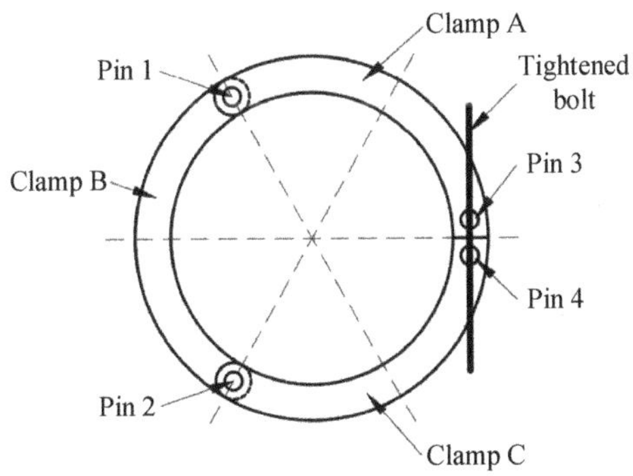

Figure 3. Simplified model of the clamp.

A force analysis of the connection between the tightened bolt and the pin 3 is carried out to obtain the load transfer relationship between the tightening torque T_1 of the approach bolt and the tension F_1 of the pin 3 on the clamp A.

$$T = 2F_1 \cdot \tan(\omega + \rho_v)\frac{d_2}{2} \tag{1}$$

where ρ_v is the equivalent friction angle of the bolt thread, $\rho_v = \tan^{-1}(\mu_1/\cos\lambda)$, λ is the inter-thread friction coefficient. ω is the lead angle, $\omega = \tan^{-1}(p/\pi d_2)$, p is the pitch, and d_2 is the pitch diameter of thread.

3.2. Analysis of Force Transfer from Clamp to Sealing Gasket under Preloading Mode

The hub is acted upon by three clamps uniformly distributed at 120° intervals through the tightened bolt, and Figure 4 illustrates the forces exerted by the tightened bolt on the clamps and by the hub on the clamps. The assumptions made in this analysis are as follows: (1) the linear contact normal force f_{n1}, f_{n2} of the clamp is uniformly distributed around the circumference of the clamp, (2) the loads are in a state of equilibrium after the clamp connection is completed, and (3) the loads are transferred within the annular contact area between the clamp and the flange, and the clamp contact diameter ϕ_{c1} is the average of the flange outer diameter ϕ_0 and the effective inner diameter ϕ_i. Firstly, the analysis of the clamp without considering the frictional action is carried out.

$$\begin{gathered} |F_1| = |F_2| = |F_3| = |F_4| = |F_5| = |F_6| \\ |f_{n1}| = |f_{n2}| \\ \phi_{c1} = (\phi_0 + \phi_i)/2 \end{gathered} \tag{2}$$

where F_2 is the force of pin 3 on clamp A, F_3, F_4 is the force of pin 1 on clamp A and B, respectively, and F_5, F_6 is the force of pin 2 on clamp B and C, respectively. f_{n1} can be decomposed into the line contact radial force f_{r1} and axial force f_{a1} of the fixed hub on the clamp, and similarly, f_{n2} can be decomposed into the line contact radial force f_{r2} and axial force f_{a2} of the mobile hub on the clamp.

The frictional action existing between the clamps and the hub under preloading conditions is analyzed. As the stiffness of the clamps is significant and they do not move completely radially with respect to the flange, it is reasonable to assume that no deformation will occur. After the connector is connected, each clamp tends to move closer to the center, and the frictional force acting on the clamp is opposite to its tendency to move inward. Figure 5 illustrates the movement of clamp A, which tends to rotate 30° clockwise inward along the y-axis. Consequently, the direction of friction in the shaft section is 30° clockwise outward along the y-axis. Similarly, the direction of friction in clamp B is outward along the x-axis, while the direction of friction in clamp C is 30° counterclockwise outward in the negative direction of the y-axis.

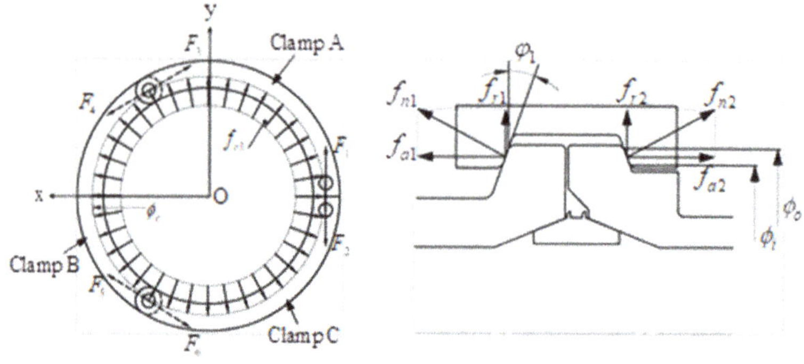

(**a**) (**b**)

Figure 4. Force analysis of clamps: (**a**) radial force analysis of clamps and (**b**) force analysis of clamp shaft sections.

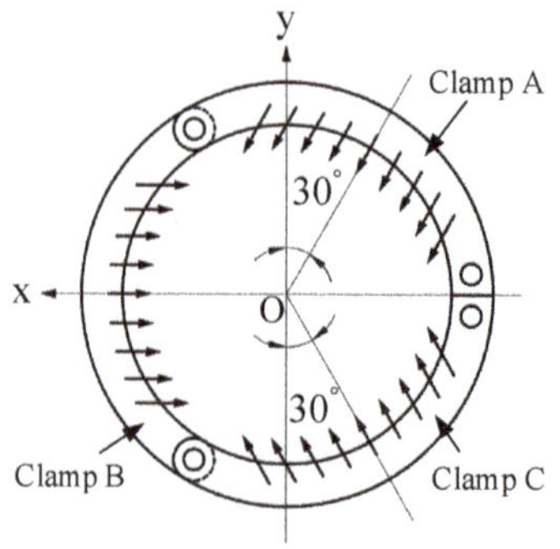

Figure 5. Clamp movement trend.

As the three clamps are equally distributed around the hub with a center angle of 120° and exert the same magnitude of force, differing only in direction, we will focus on the mechanical analysis of clamp B below. Figure 6 shows that the angle β between the contact segment area and the clamp along the x-axis is a function of the circumferential position and is expressed as an angle α. For calculation convenience, the angle range for α is chosen to be 30° to 150°.

$$\tan\beta = \tan\varphi \sin\alpha (30^\circ \leq \alpha \leq 150^\circ) \quad (3)$$

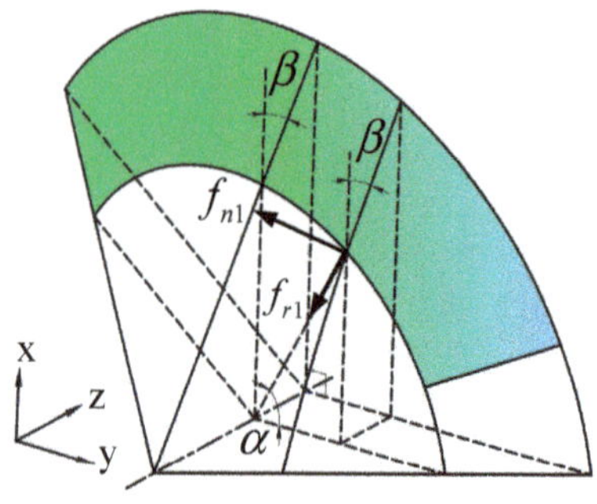

Figure 6. Clamp B contact area with hub.

The equilibrium equation is established in the x-axis direction of clamp B. $f_{r1} \sin\alpha$ is used as the component of f_{n1} along the x-axis, and $\mu_2 f_{n1} \cos\beta$ is used as the friction force along the x-axis, where μ_2 is the friction coefficient.

$$\int_{\pi/6}^{5\pi/6} (f_{r1} \sin\alpha + \mu_2 f_{n1} \cos\beta) \frac{\phi_{c1}}{2} d\alpha = F_6 \cos\frac{\pi}{6} \tag{4}$$

Combining Equation (2) with (4) and using $f_{a1} = f_{n1} \cos\varphi_1 = f_{r1}/\tan\varphi_1$, we obtain the following:

$$f_{a1} = \frac{9\sqrt{3}F_1\sqrt{4+\tan^2\varphi_1}\sqrt{4+3\tan^2\varphi_1}}{9\sqrt{3}\phi_{c1}\sin\varphi_1\sqrt{4+\tan^2\varphi_1}\sqrt{4+3\tan^2\varphi_1} + \pi\mu_2\phi_{c1}(4\sqrt{4+3\tan^2\varphi_1} + 2\sqrt{4+\tan^2\varphi_1})} \tag{5}$$

Denoting $\sqrt{4+\tan^2\varphi_1}$ as ψ_1 and $\sqrt{4+3\tan^2\varphi_1}$ as ψ_2, the above equation can be simplified as follows:

$$f_{a1} = \frac{9\sqrt{3}F_1\psi_1\psi_2}{9\sqrt{3}\phi_{c1}\sin\varphi_1\psi_1\psi_2 + \pi\mu_2\phi_{c1}(4\psi_2 + 2\psi_1)} \tag{6}$$

The total axial frictional force component ΔF_{a1} resulting from friction is analyzed. Integrating the axial friction component $\mu_2 f_{n1} \sin\beta$ in the range of angle α 30° to 90° creates the effect of 1/6 of the frictional force due to friction in the axial direction, and the total axial force due to friction is six times the value of this integral.

$$\Delta F_{a1} = 6\int_{\pi/6}^{\pi/2} \mu_2 f_{n1} \sin\beta \frac{\phi_{c1}}{2} d\alpha = \frac{27\sqrt{3}\mu_2 F_1\psi_1\psi_2[\pi - 2\arctan(\frac{\sqrt{5+3\cos 2\varphi_1}}{\sqrt{6}\sin 2\varphi_1})]}{9\sqrt{3}\sin 2\varphi_1\psi_1\psi_2 + 2\pi\mu_2\cos\varphi_1(4\psi_2 + 2\psi_1)} \tag{7}$$

Calculation and analysis of Equation (7) under preloading mode: The axial friction generated by the friction has a negative effect on the clamp under preloading mode. Therefore, the total axial force F_{A1} exerted by the hub on the clamp under preloading mode is given by the following:

$$F_{A1} = \pi\phi_{c1} f_{a1} - \Delta F_{a1} = \frac{18\sqrt{3}\pi F_1\cos\varphi_1\psi_1\psi_2 - 27\sqrt{3}\mu_2 F_1\psi_1\psi_2[\pi - 2\arctan(\frac{\sqrt{5+3\cos 2\varphi_1}}{\sqrt{6}\sin 2\varphi_1})]}{9\sqrt{3}\sin 2\varphi_1\psi_1\psi_2 + 2\pi\mu_2\cos\varphi_1(4\psi_2 + 2\psi_1)} \tag{8}$$

The force analysis of the mobile hub and the fixed hub is shown in Figure 7, where ϕ_{c2} is the contact diameter of the hub in contact with the metal sealing gasket. The reaction force f_{T3}, f_{T4} of the clamp B acts on its contact area with the hub, and the reaction force of the metal sealing gasket acts on its contact area with the hub with a horizontal reaction force of f_{A5}, f_{A6}, $f_{A5} = -f_{A6}$. The total axial force F_{A3} of the clamp on the hub and the total axial force F_{A5}, F_{A6} of the metal sealing gasket on the hub are equal in magnitude. Assume that the forces between the hub and the sealing gasket are uniformly distributed.

$$F_{A1} = F_{A3} = \pi\phi_{c2} f_{A5} \tag{9}$$

A force analysis is conducted on the metal sealing gasket, as depicted in Figure 8. The seal contact area experiences hub compression and friction. Under preloading conditions, the hub has a tendency to move outward with respect to the metal sealing gasket axis. The metal sealing gasket experiences tangential frictional forces from the fixed and mobile hubs, denoted as f_{f1} and f_{f2}, respectively, acting outward relative to the metal sealing gasket axis. The contact pressures are f_{N1}, f_{N2}, and $|f_{N1}| = |f_{N2}|$.

$$f_{A5} = f_{A7} = f_{T7}\cos(\varphi_2 - \gamma) = \frac{f_{n1}}{\cos\gamma}\cos(\varphi_2 - \gamma) \tag{10}$$

where f_{T7} is the combined force of the contact pressure f_{N1} and friction force f_{f1} between the fixed hub and the sealing gasket, γ is the friction angle of the contact surface of the metal sealing gasket, φ_2 is the inclination angle of the contact surface of the hub and the metal sealing gasket, f_{A7} is the horizontal extrusion force, namely, locking force or preload, and the horizontal component of resultant force f_{T7}, $|f_{A7}| = |f_{A8}|$.

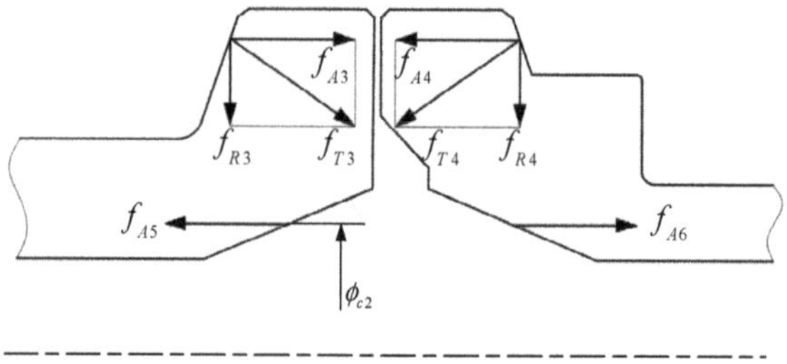

Figure 7. Flange force analysis.

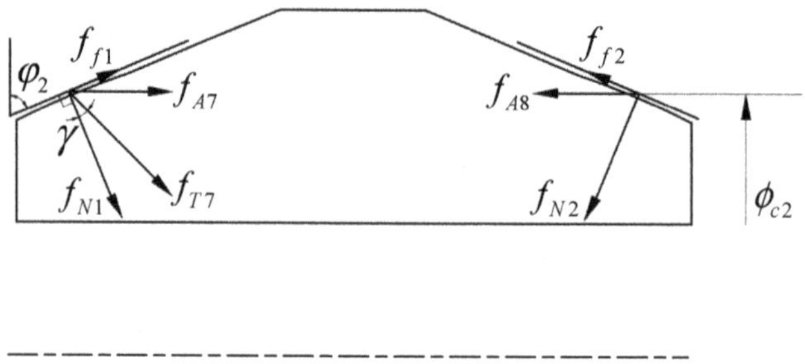

Figure 8. Force analysis of metal sealing gasket.

The relationship between the contact pressure f_{N1} of the metal sealing gasket and the torque T of the tightened bolt can be obtained by combining (1), (8), (9), and (10):

$$f_{N1} = \frac{18\sqrt{3}\pi\cos\varphi_1\cos\gamma\psi_1\psi_2 - 27\sqrt{3}\mu_2\cos\gamma\psi_1\psi_2[\pi - 2\arctan(\frac{\sqrt{5+3\cos 2\varphi_1}}{\sqrt{6}\sin 2\varphi_1})]}{2\pi d_2\phi_{c2}\tan(\omega+\rho_v)\cos\varphi_1\cos(\varphi_2-\gamma)[9\sqrt{3}\sin\varphi_1\psi_1\psi_2 + \pi\mu_2(4\psi_2+2\psi_1)]}T \quad (11)$$

Equation (11) shows the relationship between the tightened torque of the subsea connector and the average contact pressure generated in the contact area of the sealing gasket under preloading conditions. The magnitude of the contact pressure f_{N1} is related to the structural parameters and deformation of the metal sealing gasket, which is one of the factors determining the sealing performance, and another determining factor is the contact width of the sealing area [13,24].

$$\eta = \frac{T}{f_{N1}} = \frac{2\pi d_2\phi_{c2}\tan(\omega+\rho_v)\cos\varphi_1\cos(\varphi_2-\gamma)[9\sqrt{3}\sin\varphi_1\psi_1\psi_2 + \pi\mu_2(4\psi_2+2\psi_1)]}{18\sqrt{3}\pi\cos\varphi_1\cos\gamma\psi_1\psi_2 - 27\sqrt{3}\mu_2\cos\gamma\psi_1\psi_2[\pi - 2\arctan(\frac{\sqrt{5+3\cos 2\varphi_1}}{\sqrt{6}\sin 2\varphi_1})]} \quad (12)$$

η can be defined as the connector mechanical transfer efficiency, which reflects the connector locking mechanism force transfer performance.

3.3. Analysis of Force Transfer from Clamp to Sealing Gasket under Operation Mode

The analysis investigates the frictional forces between the clamp and the hub during operation mode, as illustrated in Figure 9. When subjected to internal oil and gas pressure, the clamp A tends to rotate outward by 30° clockwise along the y-axis, resulting in a friction direction of clamp A that is 30° clockwise inward along the y-axis in the shaft section. Similarly, clamp B tends to rotate outward along the x-axis, which creates a friction direction of clamp B that is inward along the x-axis in the shaft section. The movement of clamp C tends to rotate 30° counterclockwise outward along the negative direction of the y-axis, so the direction of friction in the shaft section is 30° counterclockwise in the negative direction of the y-axis. Likewise, the tendency of clamp C is to rotate 30° outward counterclockwise along the negative direction of y-axis, resulting in a friction direction that is 30° inward counterclockwise in the negative direction of the y-axis in the shaft section.

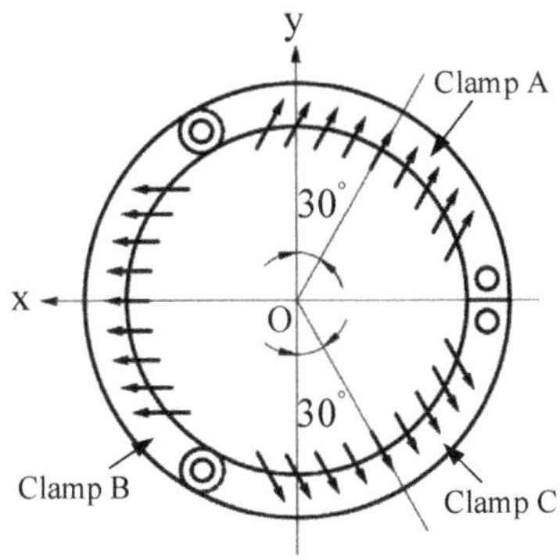

Figure 9. Trend in clamp movement.

As the forces of the three clamps have the same magnitude and distribution, only the directional differences require separate analysis. For simplicity, the force analysis in this study is conducted only for clamp B. To differentiate the forces under preloading and operation modes, all forces under operation mode are denoted with a prime symbol "′".

Establish the equilibrium equation in the x-axis direction of the clamp B, $f_{r1}{}' \sin\alpha$ as the component of $f_{n1}{}'$ along the x-axis and $\mu_2 f_{n1}{}' \cos\beta$ as the frictional force along the x-axis, where μ_2 is the friction coefficient.

$$\int_{\pi/6}^{5\pi/6} \left(f_{r1}{}' \sin\alpha - \mu_2 f_{n1}{}' \cos\beta\right) \frac{\phi_{c1}}{2} d\alpha = F_6 \cos\frac{\pi}{6} \tag{13}$$

Combining Equation (2) with (13) and using $f_{a1}{}' = f_{n1}{}' \cos\varphi_1 = f_{r1}{}'/\tan\varphi_1$, we obtain the following:

$$f_{a1}{}' = \frac{9\sqrt{3}F_1\psi_1\psi_2}{9\sqrt{3}\phi_{c1}\sin\varphi_1\psi_1\psi_2 - \pi\mu_2\phi_{c1}(4\psi_2 + 2\psi_1)} \tag{14}$$

The total axial frictional component $\Delta F_{a1}{}'$ due to friction is analyzed. By integrating the axial friction component $\mu f_{n1}{}' \sin\beta$ over the range of angle α 30° to 90°, the effect of 1/6 of the frictional force in the axial direction due to friction is obtained.

$$\Delta F_{a1}{}' = 6\int_{\pi/6}^{\pi/2} \mu_2 f_{n1}{}' \sin\beta \frac{\phi_{c1}}{2} d\alpha = \frac{27\sqrt{3}\mu_2 F_1\psi_1\psi_2[\pi - 2\arctan(\frac{\sqrt{5+3\cos 2\varphi_1}}{\sqrt{6}\sin 2\varphi_1})]}{9\sqrt{3}\sin 2\varphi_1\psi_1\psi_2 + 2\pi\mu_2\cos\varphi_1(4\psi_2 + 2\psi_1)} \tag{15}$$

In the operation mode, the hub tends to compress and separate the clamp due to the pressure of oil and gas inside the pipe. At this time, the axial component of the friction force exerts a positive effect on the clamp. Thus, the total axial force $F_{A1}{}'$ generated on the clamp under the operation mode is given by the following:

$$F_{A1}{}' = \pi\phi_{c1} f_{a1}{}' + \Delta F_{a1}{}' \tag{16}$$

Mechanical analysis of the mobile hub and fixed hub, the two are subjected to the same force state, as shown in Figure 10, where $f_{A1}' = -f_{A3}' = f_{A4}'$; f_{ph5}, f_{ph6} is the equivalent horizontal force of the internal pressure p_{in} acting on the hub, and $f_{ph5} = -f_{ph6}$; the reaction force of the clamp B f_{T3}', f_{T4}' acting on its contact area with the hub; the reaction force of the metal sealing gasket acts on its contact area with the hub, and the horizontal reaction force is f_{A5}', f_{A6}', and $f_{A5}' = -f_{A6}'$; ϕ_{c2} is the contact diameter of the hub in contact with the metal sealing gasket. Assume that the forces between the hub and the sealing gasket are uniformly distributed.

$$F_{A1}' = F_{A3}' = \pi\phi_{c2}(f_{A5}' + f_{ph5}) \tag{17}$$

$$f_{ph5} = \frac{\pi\phi_{c2}^2}{4}p_{in}\Big/(\pi\phi_{c2}) = \frac{\phi_{c2}}{4}p_{in} \tag{18}$$

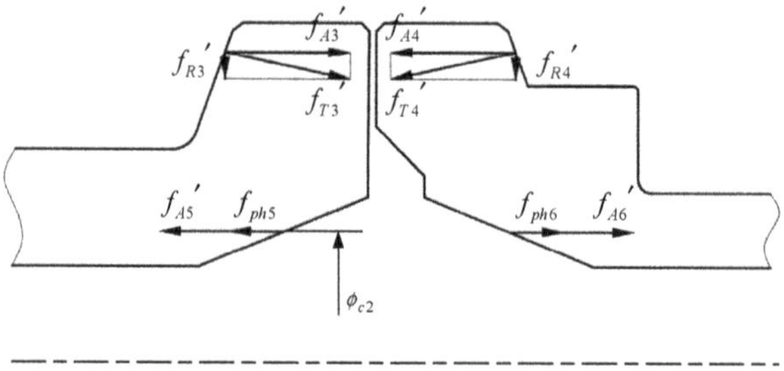

Figure 10. Operation mode hub force analysis.

f_{A5}' can also be referred to as the residual preload force under the operation mode [12], which can be expressed as

$$f_{A5}' = b\overline{q}\tan(\varphi_2 - \gamma) \tag{19}$$

where, p_{in} is the internal design pressure, which is 34.5 MPa in this study, b is the width of the contact area when the sealing gasket is pressed, $b = 16\overline{q}r/\pi E^*$, r is the radius of the sealing spherical surface of the metal sealing gasket, E^* is related to the material of the two components in contact, the material of the hub in this paper is F22, and the material of the metal sealing gasket is Incoloy 825, so it is $E^* = 1.128 \times 10^5$ MPa, $\overline{q}$ is the average contact pressure of the sealing contact area, $\overline{q} = mp_{in} = 224.5$ MPa under the operation mode, which is the minimum preloading specific pressure, and m is the gasket coefficient, which indicates the ratio of the contact pressure applied on the unit effective area to the internal pressure, and in this paper, the metal seal material is Incoloy 825, so $m = 6.5$ [12].

The relationship between the tightening torque and the design pressure of the subsea connector in the operation mode can be obtained by combining (1), (14)–(19) with a factor of three as a safety margin (employed to withstand external loads):

$$T = \frac{3F_1\left(\pi\phi_{c2}b\overline{q}\tan(\varphi_2 - \gamma) + \frac{\pi\phi_{c2}^2}{4}p_{in}\right)\tan(\omega + \rho_v)d_2}{\pi\phi_{c1}f_{a1}' + \Delta F_{a1}'} \tag{20}$$

4. Finite Element Simulation

In order to investigate the sealing performance and mechanical behavior of the core-sealing components of the subsea clamp connector under loads of internal pressure, tension, and bending moment, a full-size three-dimensional finite element model was employed for finite element simulation analysis. The initial loading conditions used in the simulation study were derived from the analysis presented in Section 1, which served as the foundation for the simulation study.

4.1. Three-Dimensional Finite Element Model

Ansys Workbench is used to simulate the six-inch subsea clamp connector using a complete three-dimensional finite element model. The main parameters of the six-inch subsea clamp connector are shown in Table 1. In this paper, the connector model is simplified to a certain extent, only studying the core components of the connector, such as the mobile hub, fixed hub, and metal sealing gasket. The simplified model is shown in Figure 11, and the six clamps are used only to simulate the role of real loading.

Table 1. Six-inch subsea connector geometry dimensions.

Name of Parameter	Parameter Value	Name of Parameter	Parameter Value
Lead angle ω	1.27°	Clamp inclination angle φ_1	20°
Equivalent friction angle ρ_v	8.83°	Sealing gasket inclination angle φ_2	23°
Pitch diameter of thread d_2	57.402 mm	Thread friction coefficient μ_1	0.15
Clamp contact diameter ϕ_{c1}	295 mm	Friction coefficient between components μ_2	0.15
Seal contact diameter ϕ_{c2}	167 mm	Friction angle of sealing gasket γ	8.53°

(a) (b)

Figure 11. Simplified model and meshing: (**a**) simplified model and (**b**) sealing gasket.

For the analysis, Solid 185 elements were utilized to model all parts. The steel-to-steel friction coefficient is set at 0.15, and the model contains a total of 210,636 elements and 257,940 nodes. Hexahedral meshes were employed for all parts, as illustrated in Figure 11.

The material used for the hub and clamp is F22, while the metal sealing gasket is made of Incoloy 825, and the pipe is made of X65, as presented in Table 2. During operation, the metal sealing gasket may undergo plastic deformation at the contact area with the hub, resulting in nonlinear material behavior. Hence, a nonlinear material model, specifically a bilinear isotropic hardening model, was employed for analysis. The material curve for this model is shown in Figure 12.

Table 2. Material properties.

Material	Yield Strength (MPa)	Tensile Strength (MPa)	Poisson Ratio	Density (g/cm^3)	Elastic Modulus (GPa)
X65	450	535	0.3	7.85	207
Incoloy 825	241	586	0.275	8.14	205
F22	552	689	0.286	7.85	211

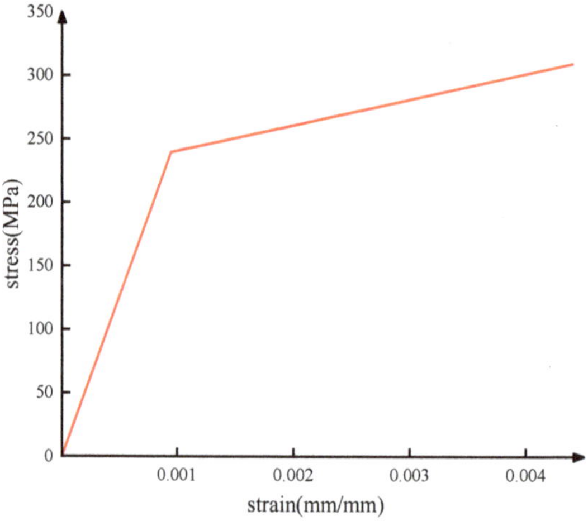

Figure 12. Bilinear isotropic hardening stress–strain curve for Incoloy 825.

To evaluate the performance of the subsea connector, two factors are considered, i.e., connection strength and sealing capacity. The connection strength is evaluated based on the von Mises stress of the hubs and clamps, which must not exceed their yield strength. The von Mises stress of the metal sealing gasket may exceed its yield strength, but must not exceed its tensile strength. Sealing capacity is evaluated based on the average contact pressure and contact width of the metal sealing gasket and hub [25,26].

$$\begin{cases} \overline{q} \geq 6.5p_{in} = 224.5 \text{ Mpa} \\ b \geq 2 \text{ mm} \end{cases} \tag{21}$$

where connector design pressure p_{in}: 34.5 MPa, and b is the seal contact width.

4.2. Boundary Conditions

The contact surface between the fixed hub and external components is fixed in accordance with the actual working conditions, and internal pressure is applied to the pipe, hub, and metal sealing gasket. An axial load is applied to the end of the pipe, while a bending load is applied at a distance of 821.5 mm from the centerline of the fixed hub, as depicted in Figure 13. The application points of the axial tensile load and bending moment load are distant from the sealing location. The Saint-Venant principle [27] indicates that the impact on the stress distribution in the core-sealing components of the connector can be disregarded.

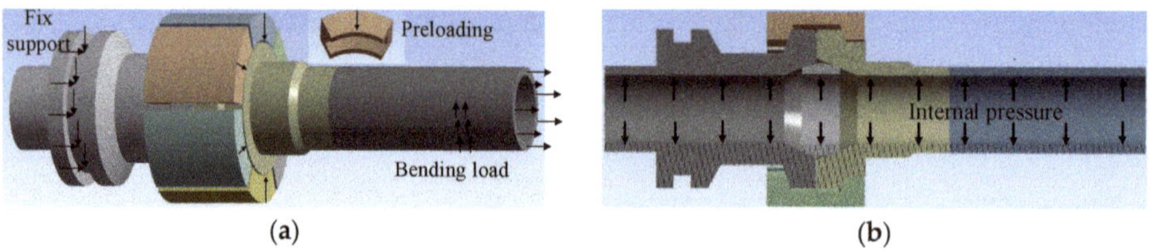

Figure 13. Boundary condition loading: (**a**) axial tensile load and bending moment loading and (**b**) internal pressure loading.

To simulate the loading process of the subsea connector during actual operation, four loading steps have been established. In the first loading step (S_{pre}), by substituting the design pressure of 34.5 MPa into Equation (20), the preload torque is determined to be

1799.39 Nm. After this step is completed, preloading is achieved, and the sealing condition is formed. In the second loading step (S_{in}), 34.5 MPa internal pressure is applied. In the third loading step (S_b), a bending moment of 174 kNm is applied. Finally, in the fourth loading step (S_{ax}), an axial tensile tension of 212 kN is applied. These loading steps are presented in Table 3.

Table 3. Load loading steps.

Step	Load	The Parameter Value
S_{pre}	Preloading	1799.39 Nm
S_{in}	Internal pressure	34.5 MPa
S_b	Bending moment	174 kNm
S_{ax}	Axial tensile load	212 kN
S_o	Unloading axial tension and bending load	/

To investigate the stress distribution in the core components of the seal, namely, the mobile hub, fixed hub, and sealing gasket, eight paths are established with their definitions, and codes presented in Table 4. Figure 14 depicts the schematic diagram of these paths along with their respective starting points.

Table 4. The definitions and code of critical node paths.

Path Code	The Definition of Path	The Starting Point
P−1	Mobile hub circulation direction	P1
P−2	Mobile hub axial side under compression	P2
P−3	Mobile hub axial side under tension	P3
P−4	Fixed hub circulation direction	P4
P−5	Fixed hub axial side under compression	P5
P−6	Fixed hub axial side under tension	P6
P−7	Connector compression side sealing gasket contact surface	P7 (Near the fixed hub)
P−8	Connector tension side sealing gasket contact surface	P8 (Near the fixed hub)

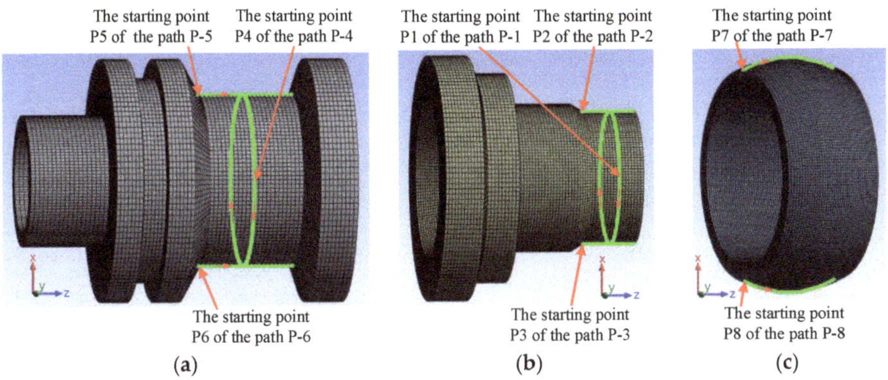

Figure 14. Diagram of the critical path and the corresponding starting point. (**a**) fixed hub; (**b**) mobile hub; and (**c**) sealing gasket.

4.3. Analysis of FE Simulation Results

Figure 15a–d depicts the distribution of von Mises stress for the core-sealing components of the subsea connector under four distinct working conditions.

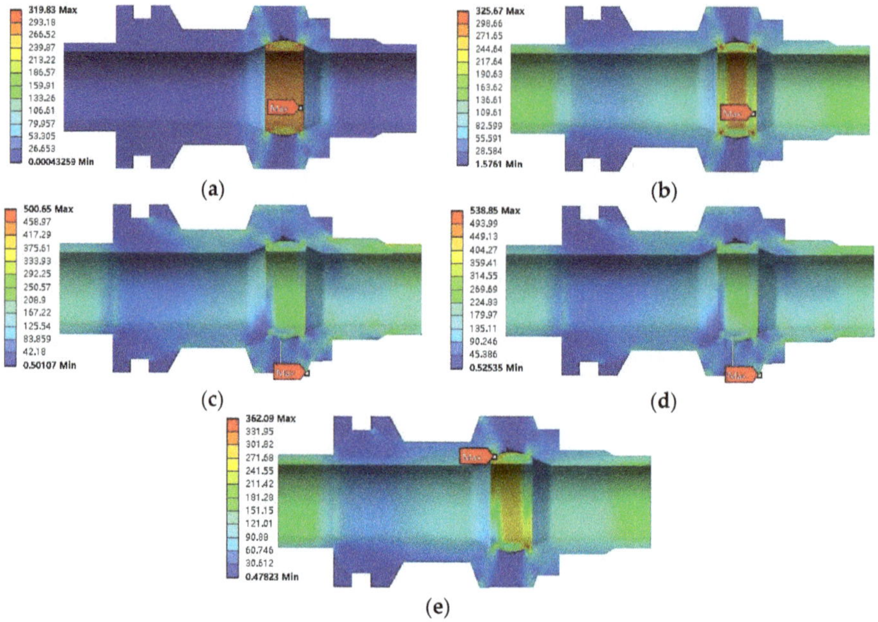

Figure 15. Distribution of von Mises stress under different working conditions: (**a**) S_{pre}; (**b**) S_{in}; (**c**) S_b; (**d**) S_{ax}; and (**e**) S_o.

4.3.1. The Mechanical Properties Analysis on Hub Critical Path

Figure 16a illustrates the stress distribution along the circumferential critical path of the mobile hub. The loading of the S_b condition results in an asymmetric stress distribution in the circumferential position, with the average stress on the compressive side being 61.4 MPa higher than that on the tensile side. This indicates that the superposition of bending moments has a greater effect on the compressive side of the mobile hub circumferential than on its tensile side.

The stress distribution on the axial critical path of the mobile hub is illustrated in Figure 16b. Upon loading the S_b condition, the stresses in the P−2 and P−3 paths gradually decrease, with the maximum and average stresses in the P−2 path being 393.04 MPa and 311.7 MPa, respectively, and the maximum and average stresses in the P−3 path being 317.56 MPa and 229.79 MPa, respectively. This indicates that the superposition of bending moments has a greater effect on the mobile hub near the seal, and the effect on the compressed side is greater than that on the tensile side. Upon loading the S_{ax} condition, the average stress in the P−2 path decreases by 24.36 MPa, while the average stress in the P−3 path increases by 24.47 MPa. This suggests that the superposition of the axial tensile load can attenuate the uneven stress distribution generated by the bending moment for the mobile hub in the axial direction.

The stress distribution on the path of the fixed hub is illustrated in Figure 16c,d. The analysis process is the same as described above, and it leads to the same conclusions. The axial stress of the fixed hub exhibits a distinctly different behavior compared to that of the mobile hub. Specifically, the stress level of the mobile hub exceeds that of the fixed hub following the application of both bending moment and axial tensile load. This suggests that greater emphasis should be placed on the design of the mobile hub during structural engineering endeavors.

Hub maximum rotation when bending moment load is applied is shown in Figure 17. The maximum rotation angle of the hubs appears on the tension side and increases with the increase in bending moment. The maximum rotation angle generated between the hubs is 1.43°. The influence on seals is analyzed in the following section.

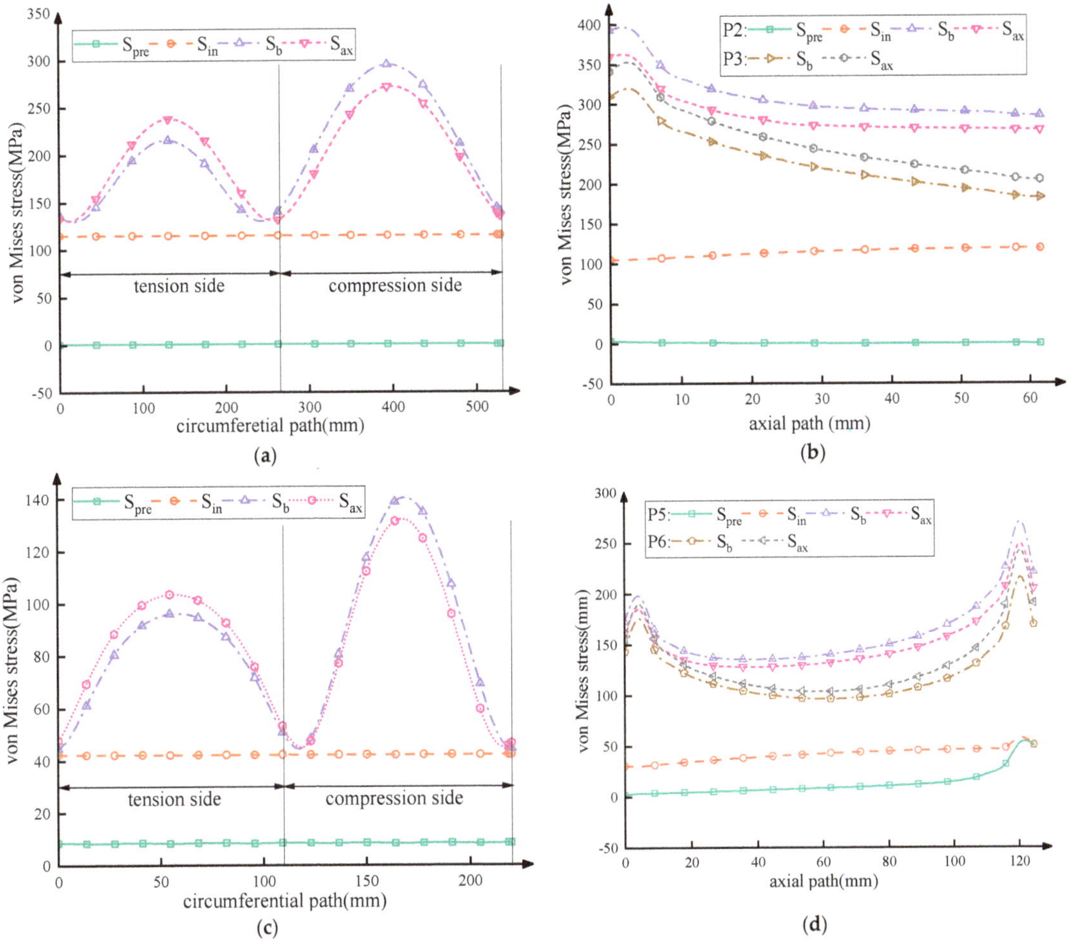

Figure 16. Critical path von Mises stress distribution: (**a**) P−1; (**b**) P−2, P−3; (**c**) P−4; and (**d**) P−5, P−6.

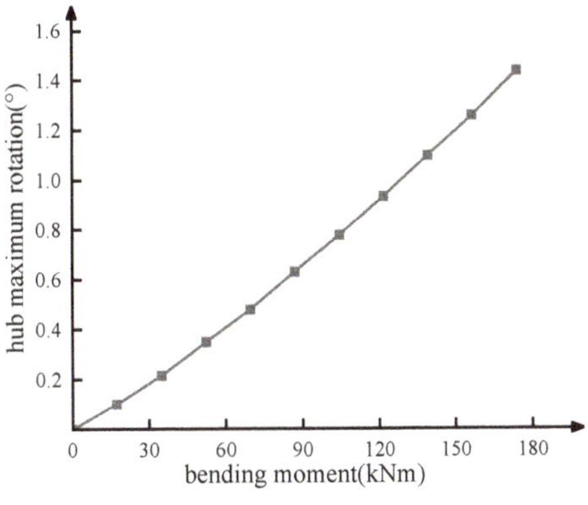

Figure 17. Hub maximum rotation.

4.3.2. Mechanical Properties Analysis of the Metal Seal

The stress distribution on the critical path of the metal seal is depicted in Figure 18a. Under the S_b loading condition, the stress on the compressive side of the metal sealing gasket increases along the P−7 path, reaching a maximum increase of 12.54 MPa, while the stress on the tensile side decreases with a maximum decrease of 30.69 MPa. Similarly, the stress on the tensile side of the P−8 path also experiences a decrease, with a maximum decrease of 52.25 MPa, whereas the stress on the compressive side increases, reaching a maximum increase of 23.58 MPa. Notably, the tensile side of the contact position with the fixed hub bears the greatest impact of the bending moment. Under the S_o loading condition, the stresses on the pressurized side of the P−7 and P−8 paths increase further due to plastic strain generation.

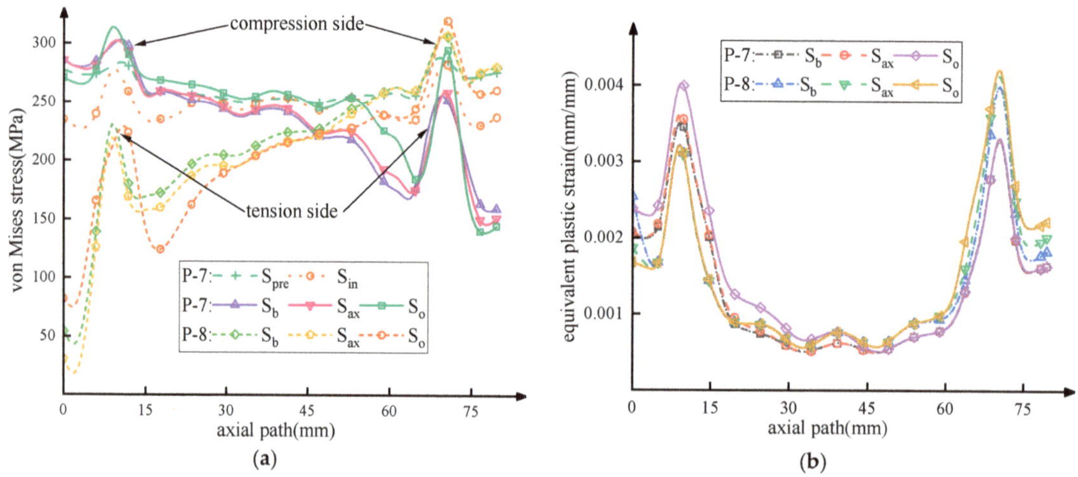

Figure 18. (**a**) Von Mises stress distribution. (**b**) Equivalent plastic strain.

The equivalent plastic strains on the critical paths of the seals are shown in Figure 18b. After loading the S_b condition, there are obvious plastic strains on the pressurized side of the P−7 and P−8 paths, and the maximum equivalent strains are 0.00335 and 0.00397, respectively. After loading the S_{ax} condition, the distributions of plastic strains on the P−7 and P−8 paths are basically the same as that of the Sb condition. After loading the S_o condition, the plastic strain on the pressurized side of the P−7 path and the P−8 path increases further, which is due to the fact that after releasing the external load, the seals are equivalent to carrying out the loading action.

4.3.3. The Contact Properties Analysis of the Metal Seal

The contact pressure of the metal sealing gasket is shown in Figure 19, and the contact pressure in the seal contact area is parabolic in distribution regardless of the working conditions, which is consistent with the sealing mechanism of the metal sealing cone to arc surface [14]. The average contact pressure and seal contact width in the seal contact area are shown in Table 5. Upon loading the S_b condition, the contact pressure and contact width of the metal sealing gasket on the tensile side are reduced compared to the results observed under preload conditions, particularly on the tensile side of the P−7 path, where they decrease to 260.08 MPa and 3.8 mm, respectively. This suggests that the superposition of bending moment has a significant impact on the seal, especially on the tensile side of the sealing gasket. Upon loading the S_{ax} condition, the contact pressure and seal width of the metal sealing gasket on the tensile side further decrease, particularly on the tensile side of the P−8 path, where they decrease to 240.25 MPa and 2.7 mm, respectively, approaching critical sealing conditions. This indicates that the superposition of axial tensile load further reduces the reliability of the subsea connector seal. After loading the S_o condition, which releases the effects of loads unfavorable to the seal, the seal contact pressure and contact zone width increase. Although the external loads are all axisymmetric, the seal contact pressure has become less symmetric due to the appearance of plastic deformation.

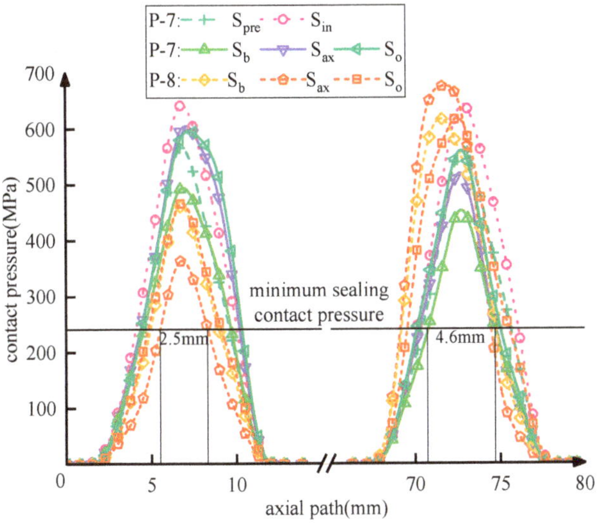

Figure 19. Contract pressure of the metal seal.

Table 5. The seal characteristic of sealing surface.

Work Condition	S_{pre}	S_{in}	S_b (P−7)	S_b (P−8)	S_{ax} (P−7)	S_{ax} (P−8)	S_o (P−7)	S_o (P−8)
average contact pressure (close to the fixed hub, MPa)	340.17	407.70	318.08	262.75	395.16	240.25	401.23	268.35
average contact pressure (close to the mobile hub, MPa)	340.17	407.70	260.08	417.55	297.02	409.53	310.74	421.64
seal band width (close to the fixed hub, mm)	5.1	5.8	5.0	4.0	5.7	2.7	5.7	4.2
seal band width (close to the mobile hub, mm)	5.1	5.8	3.8	5.4	3.5	5.3	3.7	4.9

In summary, the core-sealing components of the six-inch subsea connector form the seal under the action of preload. The superposition of internal pressure, bending moment, and axial force has an adverse effect on its strength, especially for the metal sealing gasket. These loads cause a certain degree of plastic deformation in the contact area but do not lead to destructive failure. Although the seal contact surface after load superposition is close to the minimum sealing requirements, no seal failure occurs.

5. Experimental Study

To investigate the mechanical behavior of the core-sealing components of the subsea connector under external loads and their effect on the sealing performance, an equivalent sealing experimental system was designed. The experimental results were then compared with the finite element results for validation.

5.1. Experimental System

5.1.1. Experimental Equipment

Figure 20 showed the schematic diagram of the experiment that loaded the subsea connector with internal pressure and bending moment load. The experimental equipment consisted of a subsea connector, a bending moment test facility, a hydraulic pump for generating internal fluid pressure, a hydraulic station for applying bending moment, a torque wrench, a sensor with 0.1 MPa accuracy, a static strain system, and strain gauges. The Class 4 torque wrench provided the 1799.39 Nm torque required for the initial seal, which was consistent with the value loaded analytically during the finite element simulation.

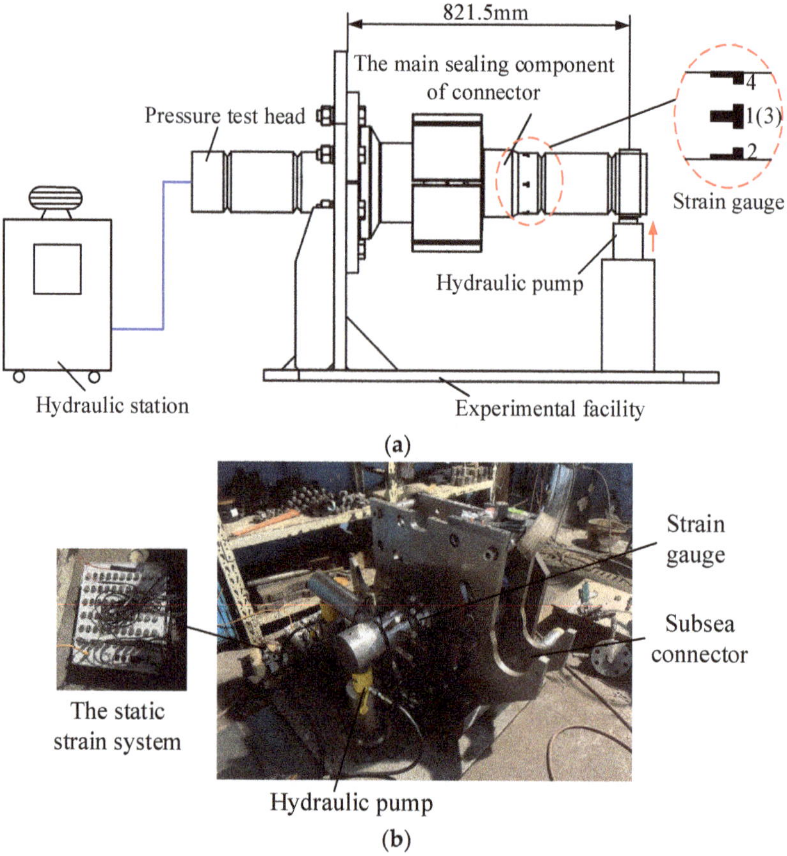

Figure 20. Schematic diagram of external load loading experiment: (**a**) schematic diagram of the experimental program and (**b**) experimental equipment.

Eight strain gauges of the same type were positioned as depicted in Figure 20a. Four sets of strain gauges were arranged at 90° intervals in the circumferential critical position of the movable hub. Each set included one strain gauge in the axial direction and one in the circumferential direction.

5.1.2. Experimental Procedure

The experiment was conducted with loaded internal pressure and moment load, and it was repeated three times to minimize the impact of experimental errors. According to API Spec 6A [28], stable pressure and effective sealing capacity are achieved when the pressure change per hour is not greater than 5% of the experiment or 3.45 MPa/h

(500 psi/h), whichever is smaller, under pressurization, as long as no leakage occurs. The specific experimental procedures are outlined as follows:

1. Filling the hydraulic station with experimental water and starting the pressurization process;
2. Stopping the pressurization process when the pressure sensor reached 100% of the rated working pressure (34.5 MPa), and holding the pressure for 15 min while checking for any leakage in the connector. The pressure data and strain data were recorded after the pressure-holding period;
3. Applying the bending moment up to the theoretical limit (174 kNm) and holding the pressure for 15 min while checking for any leakage in the connector. The pressure data and strain data were recorded after the holding pressure was completed;
4. Unloading the moment and pressure;
5. Repeating the above steps (1)–(4) three times.

5.2. Discussion of Experimental Result

5.2.1. Sealing Capacity Discussion

The pressure change under internal pressure and bending moment loading was shown in Table 6. The experimental results showed that the internal pressure change after internal pressure and bending moment loading was very small, the maximum pressure drop was 0.2 MPa, and the maximum pressure change was 0.57%, which satisfied the sealing requirements, and there was no leakage during the whole experiment, which proved the good sealing performance.

Table 6. Pressure variation condition.

Number of Experiments	Working Condition	Pressure before Pressure-Holding (MPa)	Pressure after Pressure-Holding (MPa)	Decreasing Pressure (MPa)	Pressure Variation (%)
1	Sin	35.0	35.0	0	0
	Sb		34.8	0.2	0.57
2	Sin	35.0	34.8	0.2	0.57
	Sb		34.8	0.2	0.57
3	Sin	34.9	35.0	0	0
	Sb		34.9	0.1	0.29

The pressure variations resulting from internal pressure and bending moment loading were presented in Table 6. The experimental findings demonstrated that the internal pressure changes after the application of internal pressure and bending moment loads were minimal, with a maximum pressure drop of 0.2 MPa and a maximum pressure variation of 0.57%. These changes were within the acceptable range of sealing requirements specified by the API Spec 6A [28]. Furthermore, no leakage was observed during the entire experiment, indicating the excellent sealing performance of the subsea connector.

5.2.2. Strength Performance Discussion

The stress results of three experiments conducted at the location of the strain gauge under preload, internal pressure, and bending moment loading were compared with the simulation results, as presented in Table 7 and Figure 21. The findings are summarized as follows:

(1) Under S_{pre} conditions, the maximum discrepancy between the experimental and simulation results was 0.1 MPa, with a maximum error of 1%, which is negligible.

(2) After loading S_{in} condition, the finite element analysis predicted a stress result of 115.3 MPa, while the experimental results ranged between 117.64–121.71 MPa, both higher than the finite element analysis result. The error ranged between 2.1% to 5.4%, primarily due to the initial internal pressure being larger than the one used in the finite element analysis.

(3) After loading the S_b condition, the experimental results exceeded those of the finite element analysis, and the error range increased to between 4.3 and 6.6% compared to the S_{pre} and S_{in} conditions. The increase in error was mainly due to two factors: the internal pressure of loading exceeding the design pressure of 34.5 MPa and the slight deformation of the experimental device under the bearing bending moment.

The maximum error between the experimental and finite element analysis results under the combined effect of internal pressure and bending moment loading was 6.6%, which confirmed the accuracy of the finite element analysis, irrespective of the source of the error.

Table 7. Strain gauge location stress results.

Number of Experiments	Working Condition	Strain Gauge Position	Experimental Stress (MPa)	Finite Element Analysis Stress (MPa)	Error (%)
1	S_{pre}	1	0.90	0.91	1
		2	0.90		1
		3	0.91		0
		4	0.91		0
	S_{in}	1	118.69	115.30	2.5
		2	119.16		3.3
		3	120.71		4.6
		4	117.94		2.3
	S_b	1	146.57	139.09	5.4
		2	235.38	221.24	6.4
		3	145.01	139.09	4.3
		4	315.33	297.28	6.1
2	S_{pre}	1	0.91	0.91	0
		2	0.91		0
		3	0.91		0
		4	0.92		1
	S_{in}	1	119.24	115.30	3.4
		2	121.53		5.4
		3	120.60		4.6
		4	120.49		4.5
	S_b	1	147.04	139.09	5.7
		2	235.50	221.24	6.4
		3	148.11	139.09	6.4
		4	317.03	297.28	6.6
3	S_{pre}	1	0.90	0.91	1
		2	0.91		0
		3	090		1
		4	0.92		0
	S_{in}	1	117.98	115.30	2.3
		2	118.63		2.6
		3	117.64		2.1
		4	119.28		3.5
	S_b	1	145.38	139.09	4.5
		2	233.56	221.24	5.6
		3	146.46	139.09	5.3
		4	315.89	297.28	6.3

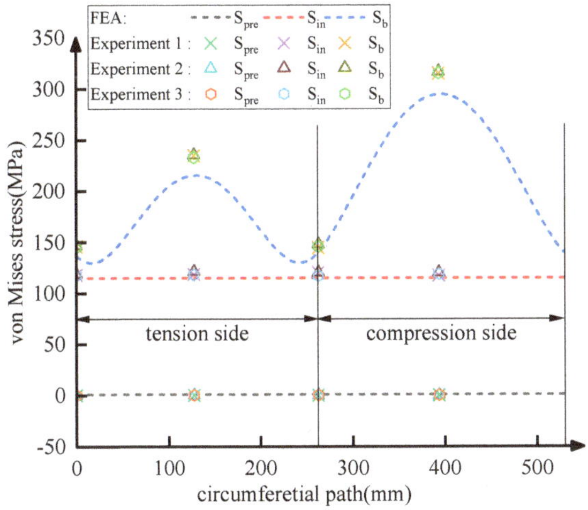

Figure 21. Comparison of experimental stress with simulation.

6. Conclusions

This study presents a theoretical analysis of the intricate load transfer relationship among the subsea clamp connector structures under various working conditions. Finite element simulation is utilized to examine the strength and sealing effectiveness of the key sealing components of six-inch subsea clamp connector subjected to 34.5 MPa internal pressure, 174 kNm bending moment, and axial 212 kN tensile load. Subsequently, an experimental setup was devised to verify the accuracy of the finite element simulation. The primary findings are as follows:

(1) The paper presents a theoretical analysis of the load transfer relationship between the subsea connector components under different working conditions. The analysis considers various factors such as friction between clamp and hub, hub and sealing gasket, and contact angle and diameter. The study establishes a mathematical model between the locking torque and contact pressure under the preloading mode, and proposes the concept of mechanical transfer efficiency. Additionally, a mathematical model of the load transfer between the locking torque and design pressure of the subsea connector is developed based on the sealing criterion under operational mode. The theoretical models provide a basis for the structural design, optimization, and mechanical behavior analysis of subsea connectors.

(2) Simulation results studying the effect of complex load superposition on the mechanical behavior and sealing performance of the core-sealing components of six-inch subsea clamp connector demonstrate the following: (a) The connector will not lead to failure and loss of leak tightness under 34.5 MPa internal pressure, 174 kNm bending moment and 212 kN axial tensile load. (b) After the moment load superposition, the maximum stresses in both the mobile hub and the fixed hub occur on the pressurized side. The contact pressure and width of the contact zone are reduced on both sides of the seal in tension, which greatly impacts the seal. (c) The superposition of axial tensile load will weaken the unevenness of stress distribution caused by the bending moment at the connection of the two hubs, but will further aggravate the unevenness of stress distribution away from the connection of the two hubs. For the metal sealing gasket, this will further reduce the starting contact pressure and contact width, especially for the tensile sides.

(3) An experimental setup was developed to test the strength and sealing performance of the six-inch subsea clamp connector under the influence of 34.5 MPa internal pressure and 174 kNm moment loading and to compare the results with those obtained from finite element analysis. The experimental findings are as follows: (a) The sealing performance was excellent as there was no leakage observed during the experiments, and the maximum pressure drop was only 0.2 MPa in all three trials. (b) The maximum discrepancy between

the critical stress positions of the mobile hub obtained from the experimental and simulation results was 6.4%, indicating good agreement and verifying the accuracy of the simulation.

This paper presents a theoretical mathematical model used as input conditions for finite element simulations and experimental analysis. The experimental results demonstrate the consistency between the theoretical mathematical model and the finite element simulations, which validates the rationality of the theoretical mathematical model. However, direct validation of the theoretical mathematical model and analysis and correction of any resulting errors require further investigation in future work. It is worth noting that when subjected to a large load, plastic deformation occurs near the seal contact surface of the metal sealing gasket, which is confirmed by the simulation results. Thus, the influence of plasticity will be taken into consideration in future research. In future studies, we will consider a more accurate axisymmetric fine mesh model and a more accurate multilinear isotropic hardening model.

Author Contributions: Conceptualization, F.Y. and X.H.; methodology, F.Y., X.H. and L.W.; validation, P.J., X.C. and F.Y.; formal analysis, X.C., K.J. and F.Y.; investigation, F.Y. and X.H.; resources, F.Y. and L.W.; data curation, X.H., L.W. and X.W.; writing—original draft preparation, L.W., X.W. and F.Y.; writing—review and editing, P.J., K.J. and X.H.; funding acquisition, F.Y. All authors have read and agreed to the published version of the manuscript.

Funding: This research was funded by the National Natural Science Foundation Of China, grant number 52001089; Heilongjiang Provincial Natural Science Foundation Of China, grant number LH2021E046; and National Natural Science Foundation of China, grant number 52001116.

Institutional Review Board Statement: Not applicable.

Informed Consent Statement: Not applicable.

Data Availability Statement: Data sharing not applicable.

Conflicts of Interest: The authors declare no conflict of interest.

References

1. Wu, J.-H.; Zhen, X.-W.; Liu, G.; Huang, Y. Optimization design on the riser system of next generation subsea production system with the assistance of DOE and surrogate model techniques. *Appl. Ocean Res.* **2019**, *85*, 34–44. [CrossRef]
2. Vedachalam, N.; Srinivasalu, S.; Ramesh, R.; Aarthi, A.; Ramadass, G.; Atmanand, M. Review and reliability modeling of maturing subsea hydrocarbon boosting systems. *J. Nat. Gas Sci. Eng.* **2015**, *25*, 284–296. [CrossRef]
3. Cascales, D.H.; Militello, C. An accurate simple model to evaluate integral flange rotation. *Int. J. Press. Vessels Pip.* **1987**, *30*, 151–159. [CrossRef]
4. Dekker, C.J.; Stikvoort, W.J. Improved design rules for pipe clamp connectors. *Int. J. Press. Vessels Pip.* **2004**, *81*, 141–157.
5. Abid, M.; Chattha, J.A.; Khan, K.A. Finite element analysis of a gasketed flange joint under combined internal pressure and thermal transient loading. In Proceedings of the ASME Pressure Vessels and Piping Conference, San Antonio, TX, USA, 22–26 July 2007; pp. 261–267.
6. Krishna, M.M.; Shunmugam, M.; Prasad, N.S. A study on the sealing performance of bolted flange joints with gaskets using finite element analysis. *Int. J. Press. Vessels Pip.* **2007**, *84*, 349–357. [CrossRef]
7. Takagi, Y.; Torii, H.; Omiya, Y.; Kobayash, T.; Sawa, T. FEM Stress Analysis and the Sealing Performance Prediction of Pipe Flange Connections under External Bending Moments and Internal Pressure. *J. Solid Mech. Mater. Eng.* **2013**, *7*, 486–495. [CrossRef]
8. Wang, Y.; Zong, L.; Shi, Y. Bending behavior and design model of bolted flange-plate connection. *J. Constr. Steel Res.* **2013**, *84*, 1–16. [CrossRef]
9. Abid, M.; Awan, A.W.; Nash, D.H. Determination of load capacity of a non-gasketed flange joint under combined internal pressure, axial and bending loading for safe strength and sealing. *J. Braz. Soc. Mech. Sci. Eng.* **2014**, *36*, 477–490. [CrossRef]
10. Peng, F.; Duan, M.; Wang, J.; Zhu, Y.; Wang, X. Optimisation method for mathematical model of deepwater collet connector locking mechanism. *Ships Offshore Struct.* **2015**, *11*, 575–590. [CrossRef]
11. Nelson, N.R.; Prasad, N.S. Sealing behavior of twin gasketed flange joints. *Int. J. Press. Vessels Pip.* **2016**, *138*, 45–50.
12. Yun, F.; Wang, L.; Yao, S.; Liu, J.; Liu, T.; Wang, R. Analytical and experimental study on sealing contact characteristics of subsea collet connectors. *Adv. Mech. Eng.* **2017**, *9*, 1–14. [CrossRef]
13. Zhang, K.; Huang, H.; Duan, M.; Hong, Y.; Estefen, S.F. Theoretical investigation of the compression limits of sealing structures in complex load transferring between subsea connector components. *J. Nat. Gas Sci. Eng.* **2017**, *44*, 202–213. [CrossRef]
14. Chen, W.; Di, Q.; Zhang, H.; Chen, F.; Wang, W. The sealing mechanism of tubing and casing premium threaded connections under complex loads. *J. Pet. Sci. Eng.* **2018**, *171*, 724–730. [CrossRef]

15. Wang, L.-Q.; Wei, Z.-L.; Yao, S.-M.; Guan, Y.; Li, S.-K. Sealing Performance and Optimization of a Subsea Pipeline Mechanical Connector. *Chin. J. Mech. Eng.* **2018**, *31*, 18. [CrossRef]
16. Li, Y.; Zhao, H.; Wang, D.; Xu, Y. Metal sealing mechanism and experimental study of the subsea wellhead connector. *J. Braz. Soc. Mech. Sci. Eng.* **2019**, *42*, 26.
17. Liu, X.-C.; Cui, F.-Y.; Jiang, Z.-Q.; Wang, X.-Q.; Xu, L.; Shang, Z.-X.; Cui, X.-X. Tension–bend–shear capacity of bolted-flange connection for square steel tube column. *Eng. Struct.* **2019**, *201*, 109798.
18. Liu, M.; Zhang, L.; Wang, L.; Liu, H.; Sun, Y.; Wang, Y. The leakage analysis of submarine pipeline connecter based on a new fractal porous media model. *Desalin. Water Treat.* **2020**, *188*, 390–399. [CrossRef]
19. Duan, M.; Zhang, K.; Soares, C.G.; Paik, J.K. Theoretical investigation on hub structure design of subsea connectors. *Thin-Walled Struct.* **2020**, *159*, 107036. [CrossRef]
20. Li, Y.; Su, H.; Wang, Y.; Mou, L.; Wang, Q.; Ren, Y. Research on bearing capacity and sealing contact characteristics of the subsea wellhead connector. *Proc. Inst. Mech. Eng. Part M J. Eng. Marit. Environ.* **2022**, *237*, 153–165. [CrossRef]
21. Yun, F.; Liu, D.; Xu, X.; Jiao, K.; Hao, X.; Wang, L.; Yan, Z.; Jia, P.; Wang, X.; Liang, B. Thermal–Structural Coupling Analysis of Subsea Connector Sealing Contact. *Appl. Sci.* **2022**, *12*, 3194. [CrossRef]
22. Zhang, K.; Cheng, H.; Liu, J.; Wang, H. Analytical calculation method for predicting contact loads and structural strength of metallic gasket of subsea connectors under thermal loads. *Proc. Inst. Mech. Eng. Part M J. Eng. Marit. Environ.* **2023**. [CrossRef]
23. Li, Y.; Su, H.; Jiang, W.; Cai, Z.; Chen, J. Sealing performance of subsea wellhead connector under thermal-structural coupling. *Ocean Eng.* **2023**, *270*, 113504. [CrossRef]
24. Zhao, H.; Chen, R.; Luo, X.; Duan, M.; Lu, Y.; Fu, G.; Tian, H.; Ye, D. Metal sealing performance of subsea X-tree wellhead connector sealer. *Chin. J. Mech. Eng.* **2015**, *28*, 649–656. [CrossRef]
25. Yun, F.; Wang, G.; Yan, Z.; Jia, P.; Xu, X.; Wang, L.; Sun, H.; Liu, W. Analysis of Sealing and Leakage Performance of the Subsea Collet Connector with Lens-Type Sealing Structure. *J. Mar. Sci. Eng.* **2020**, *8*, 444. [CrossRef]
26. Murtagian, G.R.; Fanelli, V.; Villasante, J.A.; Johnson, D.H.; Ernst, H.A. Sealability of Stationary Metal-to-Metal Seals. *J. Tribol.* **2004**, *126*, 591–596. [CrossRef]
27. Horgan, C.O. Recent Developments Concerning Saint-Venant's Principle: An Update. *Appl. Mech. Rev.* **1989**, *42*, 295–303. [CrossRef]
28. *A.S. 6A*; Specification for Wellhead and Tree Equipment. American Petroleum Institute: Washington, DC, USA, 2018.

MDPI AG
Grosspeteranlage 5
4052 Basel
Switzerland
Tel.: +41 61 683 77 34

Journal of Marine Science and Engineering Editorial Office
E-mail: jmse@mdpi.com
www.mdpi.com/journal/jmse

www.ingramcontent.com/pod-product-compliance
Lightning Source LLC
Chambersburg PA
CBHW040822250926
51850CB00007BA/1234
* 9 7 8 3 7 2 5 8 1 6 0 3 3 *